Handmade Electronic Music

DISCLAIMER

Despite stringent efforts by all concerned in the publishing process, some errors or omissions in content may occur. Readers are encouraged to bring these items to our attention where they represent errors of substance. The publisher and author disclaim any liability for damages, in whole or in part, arising from information contained in this publication.

This book contains references to electrical safety that must be observed. *Do not use AC power for any projects discussed herein.*

The publisher and the author disclaim any liability for injury that may result from the use, proper or improper, of the information contained in this book. We do not guarantee that the information contained herein is complete, safe, or accurate, nor should it be considered a substitute for your good judgment and common sense.

Handmade Electronic Music

The Art of Hardware Hacking

SECOND EDITION

Nicolas Collins

Illustrated by Simon Lonergan

NEW YORK AND LONDON

First edition published 2006
by ABC-CLIO, Inc.

This edition published 2009
by Routledge
711 Third Avenue, New York, NY 10017

Simultaneously published in the UK
by Routledge
2 Park Square, Milton Park, Abingdon, Oxon OX14 4RN

Routledge is an imprint of the Taylor & Francis Group, an informa business

Typeset in Bembo and Helvetica Neue
by Florence Production Ltd, Stoodleigh, Devon

Library of Congress Cataloging in Publication Data
Collins, Nicolas.
Handmade electronic music: the art of hardware hacking/
Nicolas Collins; illustrated by Simon Lonergan.—2nd ed.
p. cm.
Includes bibliographical references and index.
1. Electronic apparatus and appliances—Design and construction—
Amateurs' manuals. 2. Electronic musical instruments—Construction.
I. Title.
TK9965.C59 2009
786.7—dc22 2008040865

ISBN10: 0–415–99609–0 (hbk)
ISBN10: 0–415–99873–5 (pbk)
ISBN10: 0–203–87962–7 (ebk)

ISBN13: 978–0–415–99609–9 (hbk)
ISBN13: 978–0–415–99873–4 (pbk)
ISBN13: 978–0–203–87962–7 (ebk)

Printed and bound in the United States of America by Publishers Graphics, LLC on sustainably sourced paper.

Dedication

In memory of Michel Waisvisz (1949–2008), who showed us how to touch electronics.

Praise for the Second Edition

"With wit, wisdom and enviable clarity, Nicolas Collins guides the would-be hardware hacker through the possibilities and pitfalls of playing with electricity. Those who follow his guidance assiduously will not only be able to make noise that is both personal and instilled with the virtue of self-discovery; they will also gain an education and most important of all, stay alive."

David Toop

"Nic Collins has provided an informative and gently structured doorway through which anyone can enter the limitless world of possibilities to be discovered in a raw, hands-on approach to sculpting original, electronic arts hardware. Even starting with little experience, a motivated reader can emerge with invaluable circuit building, hacking and bending skills, while also gaining an enhanced understanding of what goes on inside the boxes and behind the panels of artist-invented, electronic music devices."

David Rosenboom, Composer-Performer,
Richard Seaver Distinguished Chair in Music and Dean,
The Herb Alpert School of Music,
California Institute of the Arts

"A friendly portal into the seemingly arcane art form of circuit bending and building, rich with insights into the history and spirit of experimental electronic music. Chock full of projects, ideas, and inspirations . . . enough to keep your neighborhood circuit bender out of trouble for years to come."

Mark Trayle

Praise for the First Edition

"Here we have, at last, an electronics book that caters to people who have ideas first, and electronics second. Collins offers a splendidly integrative look into the history of 'sound art,' basic electronics, and junk revisioning."

Meara O'Reilly, *MAKE* Magazine and makezine.com

"There are times in the history of any art form when its true visionaries set down in words, the blueprint behind an entire generation of genius. Collins has done just that with *Handmade Electronic Music*, an essential manifesto of know-how, trade secrets, and aesthetic accomplishment leaping off from Cage and Tudor and landing in today's classroom."

Thom Holmes, author of *Electronic and Experimental Music*

Contents

Foreword to First Edition

DAVID BEHRMAN

The appearance of Nic Collins' *Handmade Electronic Music* has made me feel nostalgia for the sixties, when I was young and first heated up a soldering iron.

There was a mixture of exhilaration and wonder that my generation felt, those of us who worked on a grass-roots level with new technology in music in the sixties and seventies, as we taught ourselves about the fresh marvels then made available for the first time ever: the transistor, a little later the integrated circuit, then the microcomputer.

Most musicians in the sixties and seventies didn't make their own circuitry. They had other things to do. The few of us who did were aligning ourselves into the tinkerer-inventor tradition handed down from earlier artists who had built things, questioned the establishment, and found new sounds or tuning systems: artists like the Futurists, like Henry Cowell, Conlon Nancarrow, and Harry Partch.

Nic, the talented author of this manual, is roughly a generation away from me—he started building circuits as a way to make music in 1972. When I started around 1965—learning mostly from two artists who were friends and mentors—David Tudor and Gordon Mumma—there were no music synths for sale; when Nic started, synths existed but were out of reach unless you had a fat budget from a university or a record company.

In the sixties, I learned from Tudor and Mumma that you didn't have to have an engineering degree to build transistorized music circuits. David Tudor's amazing music was based partly on circuits he didn't even understand. He liked the sounds they made, and that was enough.

In the old days there wasn't any distinction between high tech and low tech. The early analog synths were made by creative individuals like Bob Moog and Don Buchla; even the early microcomputers were mostly made by garage start-ups and there wasn't so much difference between these and the craft shops that had made lutes, guitars, or violins for centuries. There had always been a good relationship between performing musicians and the craftspeople who made instruments—whether those were mbiras, clarinets, or gamelans. That relationship was comfortable—it was on a human scale and almost personal.

Only in recent decades have music instruments and software become corporate, mostly mass-produced and mass-marketed, and only recently are the computers used for music generally the same ones found in tens of millions of business establishments.

It isn't surprising that there had to be a reaction among artists to this corporate stain, if one could put it that way, that has spread into the fabric of music.

It's been interesting for me to learn that some independent-minded young artists won't even go near a computer when they think about doing their music. Their instincts tell them to rebel against this "obedient" mode in which artists—like everyone else—are pushed into continually buying, from ever-growing corporations, the latest computer and the latest software packages and then spending a vast number of hours learning how to use them.

There's an inescapable love–hate ambivalence about working as an artist with high-technology tools. On the one hand, computers and digital music-making devices have never been as miraculously powerful and reliable as they are today. They've gotten much cheaper than they used to be. Some software packages like Max/MSP are not really corporate products in the bad sense and they are infinitely personable and endlessly fascinating. I'm amazed when I compare audio recording or post-production work today with the way it was in the sixties when I worked on the night shift at Columbia Records. It used to take three burly professional engineers an hour to accomplish with bulky fifty-thousand-dollar machines what one can do today alone with a laptop in ten seconds.

But on the other hand, if you think about the "laptop music" style of performance, which is currently in vogue, you might notice that there could be a problem, even if the music sounds good, with watching a person sitting in front of a computer and operating the mouse and keyboard. It is just too depressingly similar to what hundreds of millions of workers have to do from nine to five at the office. When evening comes and we go to the concert, we might like to experience something different, something visceral, something that is a direct result of muscular energy. We might like the relief of something zany and crazy. As Antonin Artaud said, there are plenty of people in the real world with two arms and two legs; in the theater we would like to see creatures with three.

Nic Collins' book helps us create creatures with three arms and three legs. It carries the maverick, inventor-tinkerer tradition of Harry Partch, Henry Cowell, and David Tudor into the twenty-first century. And it does so in a light, deft way—its charmingly simple, casual instructions hide the fact that its author is a sophisticated fellow who has done a lot of thinking, conversing, and music-making in the course of his travels and explorations.

Now that we're all stuck in the twenty-first century whether we want to be or not, we have amazing new high-tech devices to work with, but we have to accept our ambivalent relationship with these products of our corporate world. From the past we have the universe of acoustic instruments as well as the tinkerer's arsenal which is explored in this manual. The reassuring smell of heated solder remains. The vise-grip is still with us. So is the alligator clip. The good old soldering iron, the resistor and capacitor, the voltmeter, the color-coded wires—these remain. The fingers that Nic tells us how to use to coax the hidden treasures out of unknown circuit boards—they're still with us. The finger isn't obsolete. The ear isn't obsolete.

Acknowledgments

This book began as a series of handouts for a course I offered in the summer of 2002 at The School of the Art Institute of Chicago. Over the next few years, after four sessions of the course at SAIC and a dozen workshops elsewhere in the United States and Europe, those handouts gradually evolved into the first edition of this book, published in the spring of 2006. Two years and 20 workshops later I began work on the second edition. I owe a great debt to all those students who unwittingly served as guinea pigs while I struggled to clarify my prose and develop projects that would appeal to a diverse group, and be within the reach of would-be-hackers from a wide range of technical backgrounds.

There would have been no course and no book without the support of my colleagues at SAIC, especially Anna Yu, facilities manager of the Department of Art and Technology Studies, who generously made lab space, tools, and parts available to a class that was not at its best when it came to tidy-up time; and Robb Drinkwater, the Sound facilities manager, who repeatedly hunted down oddball supplies and offered constant support to puzzled students. Professor Frank DeBose of the Department of Visual Communication kindly gave access to his department's computer lab for work on the illustrations. The School's Deans of Faculty—Carol Becker and, later, Lisa Wainwright—patiently tolerated my absences when I traveled for the workshops that helped hone this text.

I greatly appreciate the enthusiasm and support of those who risked their budgets (and possibly their reputations) producing my workshops in Hardware Hacking: Phil Hallet of Sonic Arts UK initiated and helped organize the first of these in January 2004 in cooperation with Gill Haworth and Dani Landau at The Watershed in Bristol, Simon Waters of the Music Department of the University of East Anglia in Norwich, and Pedro Rebelo of the Sonic Arts Research Center at Queens University in Belfast. Daniel Schorno, René Wassenburg, Nico Bes, and Michel Waisvisz brought my workshop to STEIM (the Vatican of Hacking) in Amsterdam, in a truly coals-to-Newcastle gesture.

Other intrepid producers were Annemie Maes, Ferdinand Du Bois, and Guy van Belle at x-med-k in Brussels, Belgium; Chris Brown and John Bischoff at Mills College in Oakland, CA; Mark Trayle and Lorin Parker at the California Institute of the Arts, in Valencia, CA; Roland Roos and Monya Pletsch at Hochschule für Gestaltung und Kunst and Dorkbot in Zurich, Switzerland; Hans Tammen, Carol Parkinson, Martin Baumgartner, and Ricardo Arias at Harvestworks in New York City; Ulrich Müller and Christof Hoefig at T-U-B-E in Munich, Germany; Kenneth Fields at the Beijing Central Conservatory; Mike Rosenthal at the Bent Festival in New York City; Liz Mason at

Quimby's Bookstore in Chicago, IL; Valerian Maly at Frei Akademie der Hochschule für Musik und Theater in Bern, Switzerland; Dominic Landwehr at the *Home Made* Festival in Zurich, Switzerland; Larry Crane at TapeOpCon in Tucson, AZ; Ricardo Miranda at The College of New Jersey, NJ; Andrea Szigetvári at the *Making New Waves* Festival in Budapest, Hungary; Bob Bielecki at the Milton Avery School of the Arts, Bard College; Seth Cluett at Princeton University; Roberto Garcia Piedrahita at Universidad Nacional in Bogotá, Colombia; France Cadet, Laurent Costes and Peter Sinclair at École Supérieure d'Art in Aix-en-Provence, France; Steffano Bassanese at the Conservatorio di Musica in Cuneo, Italy; Veniero Rizzardi at the Centro d'Arte in Padova, Italy; Akihiro Kubota at the Tama Art School in Tokyo, Japan; Gustavo Matamoros at ISAW in Miami, Florida; and Enrico Merlin at the Centro Didattico Musica Teatro Danza in Roverto, Italy. The feedback I received from these workshops was invaluable, and I thank the producers and students alike for the opportunities.

Most of the technology and aesthetics in this book, as well as many of the circuit designs and artists highlighted, date from the 1970s—the early days of homemade electronic music. I was fortunate to have started in the field at that time, and this text chronicles my evolution as a self-serving luthier during the heyday of "chipetry." There are many composers, artists, engineers, teachers, and colleagues who helped me on my way, by providing information, schematics, bench space, parts, and encouragement. Wherever possible I have tried to credit individuals in the text, but for their vital, ongoing support over the years I would like to single out David Behrman, Bob Bielecki, Ralph Burhans, Sukandar Kartadinata, Ron Kuivila, Alvin Lucier, Paul De Marinis, Bob Sakayama, David Tudor, and Bob White. I am greatly in debt to all the artists who contributed information, images, video clips and sound files to the book and DVD, and who took the time to answer my questions.

Richard Carlin at Routledge not only took on the first edition enthusiastically, but made suggestions for changes and additions that make the book much more useful and enjoyable. The manuscript got into his hands through the spirited support of Thom Holmes. David Behrman, Chris Brown, Jonathan Impett, and Robert Poss gave me useful criticism on the text and the projects it covered. Constance Ditzel, who inherited the book when she became Music Acquisitions editor at Routledge, graciously encouraged me to revise and expand the text for a new edition, and bought into my hare-brained scheme of including a DVD.

The early drafts of this book were filled with clumsy scribbles that elicited head scratching and laughter from my students and my children alike. Fortunately one student, Simon Lonergan, came to my rescue. Simon spent countless hours patiently translating my doodles and suggestions into the illustrations and photos in this book. He then returned to Chicago to assist me on the new photos and figures for the second edition, with the birth of his daughter imminent. Another overqualified student, James Murray, took on the task of shooting, editing, and mastering the DVD included in the second edition. I am full of awe and gratitude towards both.

My wife, Susan Tallman, and my children Ted and Charlotte, put up with too many physical absences as I conducted workshops around the world, and tolerated my frequent mental absences while I wrote and revised this text—they are far more accommodating and supportive than I deserve. Finally, if I can write clearly enough to be understood it is entirely the result of the years of patient advice and editing by Susan. For this gift I owe her my greatest thanks, and a fresh red pen.

Introduction

This book teaches you how to tickle electronics. It is a guide to the creative transformation of consumer electronic technology for alternative use. We live in a cut-and-paste world: CMD-X and CMD-V give us the freedom to rearrange words, pictures, video, and sound to transform any old thing into our new thing with tremendous ease. But, by and large, this is also an "off-line" world, whose digital tools, as powerful as they might be, are more suitable to preparing texts, photo albums, movies, and CDs in private, rather than on stage. These days "live electronic music" seems to be hibernating, its tranquil countenance only disturbed from time to time by the occasional, discrete click of a mouse.

My generation of composers came of age before the personal computer, at a time when electronic instruments were far too expensive for anyone but rock stars or universities, but whose building blocks (integrated circuits) were pretty cheap and *almost* understandable. A small, merry (if masochistic) band, we presumed to Do-It-Ourselves. We delved into the arcane argot of engineering magazines, scratched our heads, swapped schematics, drank another beer, and cobbled together homemade circuits—most of them eccentric and sloppy enough to give a "real" engineer dyspepsia. These folk electronic instruments became the calling cards of a loose coalition of composers that emerged in the mid-1970s, after John Cage, David Tudor, and David Behrman, and before Oval, Moby, and Matmos. By the end of the 1970s the microcomputers that would eventually evolve into Apples and PCs had emerged from the primordial ooze of Silicon Valley, and most of us hung up our soldering irons and started coding, but the odd circuit popped up from time to time, adding analog spice to the increasingly digital musical meal.

Computers are wonderful, don't get me wrong, but the usual interface—an ASCII keyboard and a mouse—is awkward, and makes the act of *performing* a pretty indirect activity, like trying to hug a baby in an incubator. "Alternative controllers" (such as those made by Donald Buchla and artists working at STEIM) are a step in the right direction, but sometimes it's nice to reach out and *touch* a sound. This book lifts the baby out of the basinet and drops her, naked and gurgling, into your waiting arms, begging to be tickled.

The focus is on sound—making performable instruments, aids to recording, and unusual noisemakers—though some projects have a strong visual component as well. No previous electronic experience is assumed, and the aim is to get you making sounds as soon as possible.

After learning basic soldering skills, you will make a variety of listening devices: acoustic microphones, contact mikes, coils for picking up stray electromagnetic fields, tape heads. Then you will lay hands upon, and modify, cheap electronic toys and other found circuitry—the heart and soul of hacking. You'll build some circuits from scratch: simple, robust oscillators that can be controlled through a variety of means (light, touch, knobs, switches), and combined to create rich electronic textures at minimum cost and difficulty. With the confidence instilled by such a delicious din you will proceed with circuits to amplify, distort, chop, and otherwise mangle any sounds, be they electronic in origin or not: electric guitars, amplified voice, CDs, radio, environmental ambience, etc. You will then move on to designs for linking sound with visual material, and some convenient "glue" circuits, useful for putting disparate parts together for performance, recording, or interfacing to computers. There are several appendices that direct you to sources of supplies and further resources for information. Eleven "Art & Music" sidebars, numerous illustrations, and a DVD place the technology in historical and aesthetic context: over 100 hackers, benders, musicians, artists, and inventors from around the globe are represented in the text, illustrations, audio tracks, video clips.

In selecting the specific projects to include in this book I was guided by a handful of fundamental assumptions and goals:

1. To keep you alive. All the projects in this book are battery powered; none plug into the potentially lethal voltage running through your walls. This makes the early stages of unsupervised electronic play activity considerably safer, and less daunting for the beginner.
2. To keep things simple. We work with a small number of very simple "axiomatic" circuits and concepts that can be combined with great permutational richness as you proceed and gain experience, but are easy to understand and quick to get running at the beginning. The point is to make cool sounds as quickly as possible.
3. To keep things cheap. By limiting ourselves to a few core designs we minimize the quantity and cost of supplies needed to complete this book. You don't need a full electronics lab, just a soldering iron, a few hand tools, and about $50 worth of parts that you can easily obtain online. By focusing on toys and other simple consumer electronics we also minimize the threat of "catastrophic loss" in the early, unpredictable days of freestyle hacking: a Microjammer sets you back considerably less than a vintage Bass Balls.
4. To keep it stupid. You will find here an absolute minimum of theory. We learn to design by ear, not by eye, gazing at sophisticated test instruments or engineering texts. Ignorance is bliss, so enjoy it.
5. To forgive and forget. There's no "right way" to hack. I will try to steer you away from meltdowns, but have included designs that are robust, forgiving of wiring errors, and accept a wide range of component substitutions if you don't have the preferred part. Most of these circuits are starting points from which you can design many variations with no further help from me—if you love a hack, let it run free.

As a result of these guidelines, this is a distinctly non-standard introduction to electronic engineering. Many of the typical subjects of a basic electronics course, such as the worrisomely vague transistor and the admittedly useful little thing called an op-amp, are left unmentioned. After turning over the last page, you will emerge smarter, if weirder, than when you first opened the book. You will have acquired some rare skills, and ones that are exceedingly useful in the pursuit of unusual sounds. You will have significant gaps in your knowledge, but these gaps can be filled by a less structured stroll through resources easily available in books and online (as described in the Appendices). And everything electronic you choose to do after this book will be easy, I promise. Why? Because you will be fearless. You will have the confidence to survey those presumptuous "No user serviceable parts inside!" labels and laugh. You will be a hacker.

NOTES ON THE SECOND EDITION

This revised and expanded second edition of *Handmade Electronic Music* includes new projects, a DVD, and a great many more examples of artists' work. The revision process has benefited from the intervening years. Subsequent to submitting the first manuscript in 2005 I gave some two-dozen workshops and courses using the book as the primary text, and the feedback I received from students and presenters was invaluable. Moreover, even in three short years technology has changed sufficiently to render some portions of the original book quaint, if not actually obsolete. And, perhaps most importantly, encounters with numerous practitioners in the expanding field of handmade electronic music have contributed greatly to the artistic content and context of the text.

The most significant changes to the book are:

- Corrections of errors that slipped into the first edition, as well as revisions where components have become obsolete, changed price or source, etc. Many of the urls and other references in the Notes and Reference section and Appendices have been updated.
- Clarifications of sections of the original text. The projects in this book grew primarily out of teaching, and some benefited from commentary I added in the classroom to compensate for lapses in the first edition. Similarly, drawings and photos have been added to illustrate points that drove me to the whiteboard when words proved inadequate.
- Revisions of the figures. We have cropped many of the original photos in order to increase the resolution of the critical details of a circuit. The more complex breadboard layouts have been freshly rendered as line drawings, as have the designs that are new to this edition.
- A greater presence of artists' work. I've included more photographs and descriptions of physical realizations of the book's various projects, by as many hackers, benders, and ex-students as I could persuade to contribute. The sidebars, which add a real-world perspective to the occasionally abstract technological discussions in the main text, have been expanded, and a new chapter has been added providing an overview of recent developments in the fields of handmade electronic music and art. The accompanying DVD includes an extensive gallery of artists' work (see below).
- New projects and designs. In response to requests from students, and to boredom or curiosity on my part, I have added a number of new projects to the book. Subsumed within the original chapter structure, they may not be immediately apparent to those of you who have read the first volume. The main additions are:

- phantom power and stereo wiring options for electret microphones;
- a mellow triangle-wave output from our basic CMOS oscillators, to complement the usual, more abrasive square waves;
- sweepable filters;
- an envelope follower with applications in noise gates, duckers, expanders, compressors, and tacky Disco drum synthesizers;
- a pitch-tracking circuit;
- a versatile sequencer;
- alternative power supplies, including solar cells, potatoes, and simple generators;
- more variations on many of the basic circuits, described in detail.

I have chosen, however, to adhere to the design philosophy of the original book, limiting myself to those circuits that can run off a single battery, are forgiving of errors, embrace extensive variation, and combine, Lego-like, with one another. I continue to rely heavily on the quirky misapplication of CMOS digital circuitry, and have thus far avoided introducing op-amps, transistors, and several other common components. I prefer to cover designs not frequently found elsewhere in books or online.

- A DVD. The first edition of this book included an audio CD with 20 tracks contributed by musicians and sound artists from around the world. These tracks were not mere "sound examples" of projects from the book, but real pieces of music using the same or similar devices as we create from the text. Impressed by the proliferation of hacked and bent instruments on YouTube, I decided to include a DVD in the new edition. The disk is divided into three parts:

  - Video demonstrations by me of 13 projects from the book. An excellent way to get started if you're not comfortable with book-learnin' and I don't happen to be giving a workshop in your hometown.
  - A gallery section consisting of 87 video clips by a plethora of artists, illustrating their instruments and performances. These were selected from responses to an internationally circulated call for submissions. Each clip was limited to 60-seconds' maximum duration; the resulting hour and a half gives a pretty good overview of what is happening now (the summer of 2008) in the burgeoning field of handmade electronic music.
  - The audio files from the original CD.

Six years after succumbing to pressure from students to present a course in what I then regarded as an idiosyncratic and somewhat anachronistic pastime of dubious general appeal, I find that the book continues to be popular with a surprisingly diverse group of readers. It has been embraced by readers of *Wired*, *Computer Music Journal*, and *Tape Op* alike; by attendees of Bent Festivals, Make Faires, and NIME conferences. It has been adopted as a textbook in colleges around the world, from the USA and Britain, to Japan, New Zealand, and Australia—some universities have even designed whole new courses based on the book. I hope that the corrections and updates, the added projects and illustrations, and the DVD combine to expand the appeal and increase the usefulness of *Handmade Electronic Music*. Complaints or compliments are welcome.

PART I

Starting

CHAPTER 1

Getting Started: Tools and Materials Needed

You will need certain tools and materials to undertake the projects in this book. I have kept the supplies to an absolute minimum—none of the fancy test equipment and drawers full of teeny parts found in a typical electronics lab; a few basic hand tools, and a modest collection of easily obtainable components will see you through.

LISTENING

Whereas electronic engineering is typically taught with *visual* reinforcement—staring at an oscilloscope, meters, or a computer screen—we will work by *ear*, as befits the development of sonic circuitry. A *monitor amplifier* thus becomes your primary tool. Whereas it would be nice to listen to our boops and beeps through a 250-watt Bryston amplifier and a pair of Altec 604E loudspeakers, I advocate the use of a small, battery-powered amplifier (see Figure 1.1). It is cheaper, but more importantly it is *safer*: many of our experiments entail touching electronic circuitry with damp fingers, and those fingers should be kept far, far away from the 120 (or 240) volts streaming into any device with a power cord.

We need a fair amount of gain at the input to our amplifier—especially at the beginning of this book, where we start out making a variety of microphones with pretty low output levels. A typical pair of battery-power speakers intended for amplifying a computer, MP3 player, portable CD player or other line level device will not suffice for the projects in the "Listening" section of the book, although they will come in very handy when we move on to circuit bending (in "Touching") and building our own circuits ("Building" and later sections). Instead, consider acquiring one of those wee bitty guitar amps made by Fender, Marshall, Dan Electro, etc.—they look like little lunchboxes, or the guitarist's equivalent of a shrunken head. The cheapest one I've found comes from Radio Shack (#277–1008, $18.19). It also has a very useful jack for an external speaker, which comes in handy in Chapter 8. The more expensive ones pitched at guitarists, on the other hand, have the advantages of a bigger speaker, a tone control, overdrive/distortion, and a more robust and useful 1/4 inch input jack (the Radio Shack amp uses 1/8 inch inputs).

If you are feeling slightly adventurous, the most economical and flexible solution is to buy a low power (< 1 watt) amplifier kit from any of a number of online retailers.

Figure 1.1 Some battery-powered mini-amplifiers.

These kits include all components, a tidy little printed circuit board, and instructions on where to place which part (see Figure 1.2). This is an excellent way to bootstrap your soldering skills, while saving some money. See Appendices A and B for suggestions on where to find kits, then jump ahead to Chapter 6 for advice on how to solder. Besides the financial and pedagogic advantages of building your own tool, you can connect to these amplifiers using clip leads instead of patch cords, so it's faster and cheaper to test out your projects. The Altoid tin (which will re-appear throughout this book with comet-like regularity) makes a very practical housing for a small circuit board and a 9-volt battery. Or you can pack the circuit board and a speaker into some kind of mini faux guitar amplifier and begin your hacking career by dazzling your friends with your design aesthetic (see Figure 27.5 in Chapter 27).

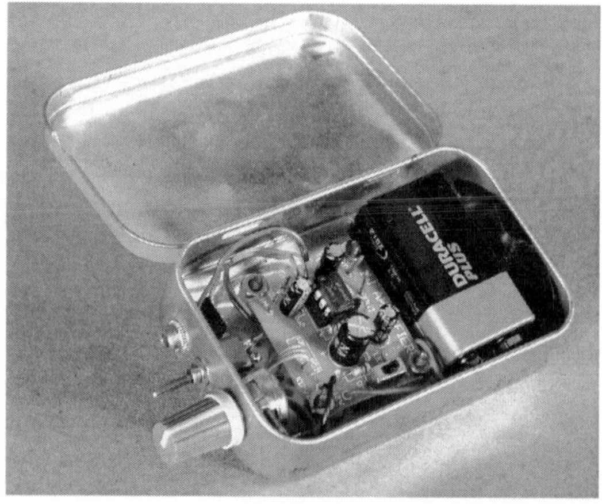

Figure 1.2 A low-power amplifier kit, assembled circuit board (left) and mounted in an Altoid tin (right).

Many of the things we will build produce a very wide range of frequencies. Some of these frequencies are more delicious than others. A cheap graphic equalizer footpedal, such as sold for guitarists, is an excellent tool for separating the yolks from the whites, sonically speaking (you can glimpse one on the right side of Figure 26.1, at the beginning of Chapter 26).

TOOLS

You'll need some basic tools (see Figure 1.3). Many will already be in your collection if you've ever had to change a washer, wire up a lamp, or serve in the Swiss Army. None are expensive—the only place you might want to splurge a little is on a better-than-terrible soldering iron.

- A soldering iron, with a very fine point, 25–60 watts. Not a soldering gun—that's for VoTech classes. Don't get a cheap iron—it makes it very frustrating to learn soldering. Weller makes good ones that are reasonably priced and have replaceable, interchangeable tips.
- Solder—fine, rosin core—not "acid-core" solder, that's for plumbers.
- Diagonal cutters, small, for cutting wire and component leads down by the circuit board.
- Wire strippers (unless you have the perfect gap between your front teeth)—simple, adjustable manual kind for light-gauge wire.
- A set of jeweler's screwdrivers (flat and Phillips)—for opening toys with tiny screws.
- A Swiss Army knife.
- A pair of scissors.
- A cheap digital multimeter, capable of reading resistance, voltage, and current.
- Plastic electrical tape.

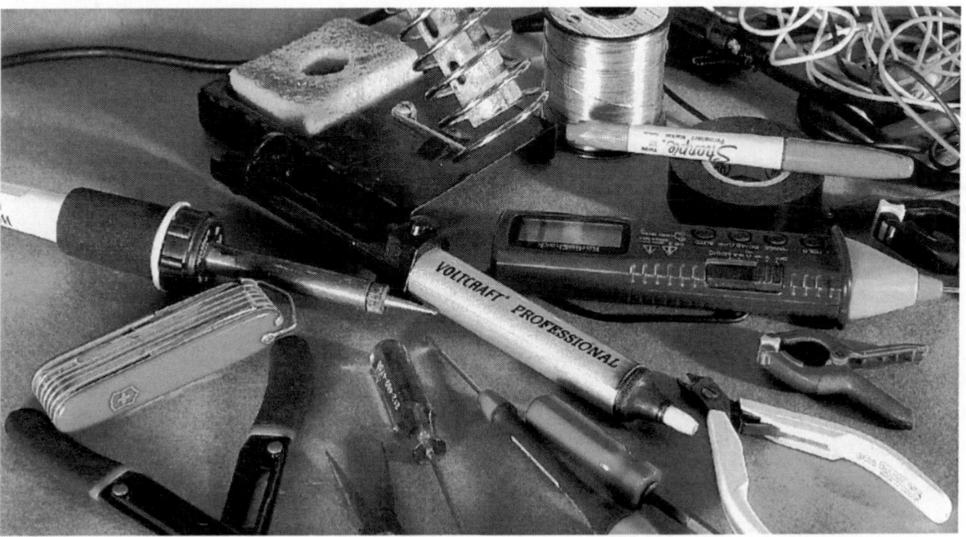

Figure 1.3 Some handy tools.

- Mini jumper cables with small alligator clips at each end, at least 20 of them—you can never have too many.
- A "Sharpie"-style fine-tipped permanent marker.
- Some small spring clamps or clothespins.
- A small vise or "third hand" device for holding things while you solder them.
- Basic shop tools—such as a small saw for metal and plastic, files, and an electric drill—are useful when you start to work on packaging.

PARTS

At the head of each chapter you'll find a list of the specific parts needed to complete the projects covered, and a complete inventory of parts for all projects is included in Appendix B. But here are some supplies you'll need most frequently, so you might as well pick them up early. Appendix B also lists online retail sources for most of this stuff. If any of these terms are unfamiliar or otherwise daunting, skip ahead and don't worry: each component will be explained in greater detail at the point that it is needed for a specific project. Likewise if the list at the start of any chapter is not self-explanatory just read on, since each new tool or part is discussed as it comes into use.

- Lightweight insulated hookup wire, 22–24 gauge, one roll stranded, one roll solid.
- Lightweight shielded audio hookup cable, single conductor plus shield.
- Assortment of standard value resistors, 1/8 or 1/4 watt. Sets are easily and inexpensively available from Radio Shack or mail order/Web retailers. If you want to make the minimum investment, the critical values we use are: 100 Ohm, 1 kOhm, 2.2 kOhm, 10 kOhm, 100 kOhm and 1 mOhm.
- Assortment of capacitors, in the range of 10 pf to 0.1 uf monolithic ceramic or metal film, and 1 uf to 47 uf electrolytic. These can also be bought in sets, but since they are a little more expensive than resistors you might prefer to purchase a handful of each of a few different values from across the full range, then replace or supplement them as needed. The most commonly used values in our projects are: 0.01 uf, 0.1 uf, 1.0 uf, 2.2 uf, 4.7 uf and 10 uf. Tiny capacitors (10 pf and 100 pf) will be handy when we start building distortion circuits in Chapter 22, and our power supplies (Chapter 29) will use 1000 uf or larger capacitors.
- Nine-volt battery clips—the things that snap onto the nipples at the end of a battery and terminate in lengths of wire. Get five or more.
- Assorted audio jacks and plugs to mate with other devices, i.e. your amplifier, MP3 player headphone jack, CD player, radio, toys, etc.

BATTERIES

Because of our core philosophy of avoiding unnecessary electrocution, we will be working exclusively with battery-powered devices. This means we will need a lot of batteries, for your amplifier, toys, radio, and the circuits you make. Please be *milieu vriendelijk* (a friend of the environment) as the Dutch say, and invest in some rechargeable ones if at all possible. Your groundwater will thank you.

ARCHITECTURE

You'll need a clean, well-lighted place. It should be well ventilated—soldering throws up some pretty unhealthy fumes (if you end up doing a lot of soldering you should consider investing in a compact bench-top "fume extractor"—your liver will thank you far more than your wallet will complain). You'll also want a fair amount of table space, since hacking has an unfortunate tendency to sprawl (see Figure 1.4). The table surface can be damaged by soldering, drilling, and filing, so no Boule inlay please. You'll need electrical power at the table for your soldering iron and a good strong desk light.

Ok, are you feeling ready to hack? First, a few rules to live by . . .

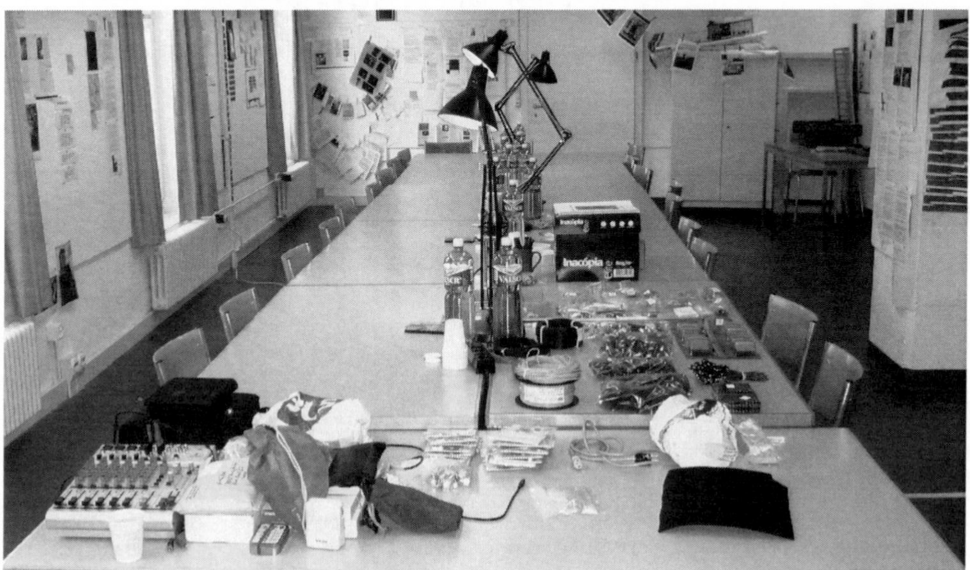

Figure 1.4 A typical worktable, before and during hacking.

CHAPTER 2

The Seven Basic Rules of Hacking: General Advice

Like boot camp or Candyland, this book is light on theory, but heavy on rules. Here are a few guidelines for keeping you healthy and happy:

Rule #1: Fear not!
Ignorance *is* bliss, anything worth doing is worth doing *wrong*, and two wrongs *can* make a right.

Rule #2: Don't take apart anything that plugs directly into the wall.
We will work almost exclusively with battery-powered circuitry. AC-powered things can kill you. AC adapters (wall-warts) may be used *only* after you have displayed proper understanding of the difference between insulation and electrocution.

Rule #3: It is easier to take something apart than put it back together.
Objects taken apart are unlikely to function normally after they are put back together, no matter how careful you are. Consider replacement cost before you open.

Rule #4: Make notes of what you are doing as you go along, not after.
Most wires look pretty much alike. As you take things apart make notes on which color wire goes to where on the circuit board, to what jack, etc. Especially important are the wires that go to the battery. Likewise, note what you add as you add it, what you change as you change it.

Rule #5: Avoid connecting the battery backwards.
This can damage a circuit.

Rule #6: Many hacks are like butterflies: beautiful but short-lived.
Many hacks you perform, especially early in your career, may destroy the circuit eventually. Accept this. If it sounds great, record it as soon as possible, and make note of what you've done to the circuit so you can try to recreate it later (see Rule #4).

Rule #7: In general, try to avoid short circuits.
Try to avoid making random connections between locations on a circuit board using plain wire or a screwdriver blade. This can damage a circuit—not *always*, but inevitably at the most inconvenient time.

Additional rules will emerge from time to time throughout the book, and are reiterated in Appendix C.

PART II

Listening

CHAPTER 3

Circuit Sniffing: Eavesdropping on Hidden Magnetic Music

You will need:

- A battery-powered AM radio or two.
- A battery-powered amplifier.
- An inductive telephone pickup coil or an electric guitar pickup.
- Optional: 100 feet of light-gauge insulated wire, an audio plug and two pieces of wood, approximately 1 inch × 2 inches × 5 feet.

RADIOS

Radios make the inaudible audible. Unlike microphones and amplifiers, which merely make very quiet *acoustic* sounds much louder, radios pick up electromagnetic waves that have no acoustic presence whatsoever and translate them into signals that can be heard through a loudspeaker. Radios are manufactured for listening to intentionally transmitted electromagnetic waves (i.e. those sent from radio stations) from which they extract music and speech through a process of *demodulation*—multiple stages of amplification, filtering, and frequency shifting. But radios can also be used to sniff out other types of electromagnetic signals, such as those emitted by lightning, sunspots, Aurora Borealis, meteorites, subway trains, and a gaggle of household appliances (see Art & Music 1 "Mortal Coils"). Generally speaking, AM radios (the cheaper the better) do a better job of picking up these "spurious" noises than FM radios. (In fact, the invention of FM technology was celebrated as the triumph of signal over noise, thanks to its superior rejection of exactly the kind of weird stuff we want to hear here.)

Put batteries in the radio and turn it on; if it has a *band select switch* set it initially to AM. Tune it to a dead spot. Try moving it around various electrical appliances: fluorescent lights, electric motors, computers, portable CD players, cell phones, MP3 players, infrared remote controls, and controllers for RC planes and cars are especially noisy. Fire off a camera flash next to it. Experiment with tuning the radio to different stations, in-between stations, and to the dead bands at either end of the dial.

As the FCC often warns you, certain electrical appliances can cause "radio interference." What this means is that, as a byproduct of whatever useful thing the appliances are doing, they emit lots of spurious electromagnetic radiation in the same frequency region as radio and TV broadcasts—they whistle while they work. As you tune the radio it becomes sensitive to specific frequency ranges (mostly very high) of electromagnetic waves; it shifts the modulated audio down into the range of our hearing, then demodulates and amplifies it. Compared to radio stations, these appliances put out very weak signals. The noise from a computer drops off rapidly as you move the radio a few feet away (hence the FCC advice), since electromagnetic waves follow what physicists have so elegantly dubbed "the inverse-square law": their strength fades by the square of the distance from the source.

If your radio has a FM band, try it as well. The technique of FM radio transmission and reception is designed to minimize interference, but strong periodic signals (like the clock frequency of a computer) can sometimes be tuned in, just like a proper radio station.

COILS

An alternate approach to picking up electromagnetic signals is to use a simple coil of wire and an amplifier. A *telephone pickup* consists of yards of thin copper wire wrapped around an iron slug (see Figure 3.1). Plugged into an amp, this coil acts like a radio antenna for frequencies low enough to be within the range of our hearing without the need for radio-style demodulation (i.e. between c. 20 Hz–20 kHz). Stuck on a telephone receiver (or held against any other loudspeaker), it picks up the electromagnetic field generated by the voice coil of the speaker, allowing you to record your landlord making unsavory threats.

Plug the tap coil into your portable amp and repeat the experiments we did with the radio. Sometimes you will hear different sounds from the same appliance. The coil

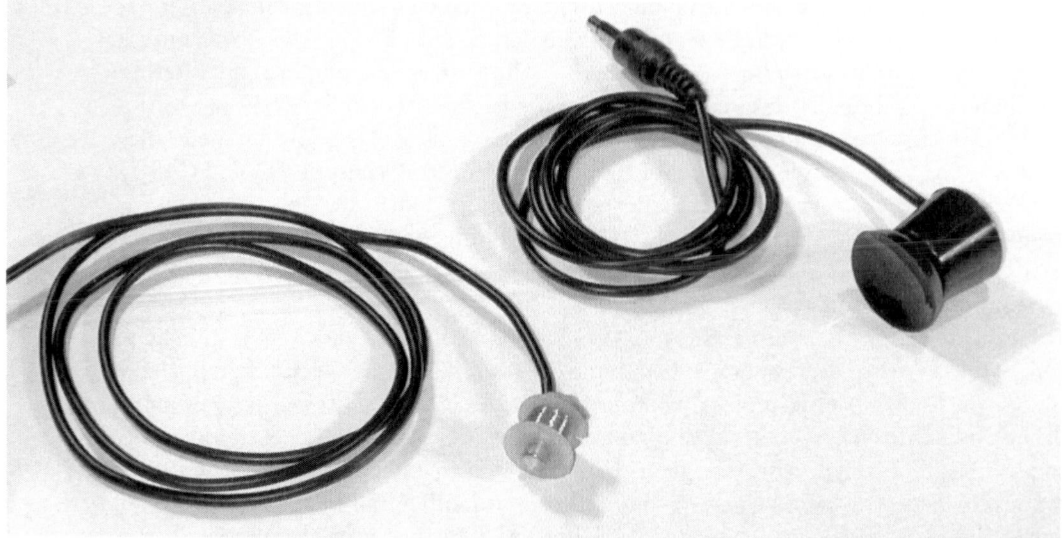

Figure 3.1 A telephone coil pickup, showing internal coil construction (left) and packaging (right). From the collection of Michael V. Hayden.

is small enough that you can move it close in to precise locations, like a stethoscope. Boot your laptop, and note the change in sound as you move the coil across its surface, from the CPU area to the RAM to the battery to the CDROM drive to the screen (see audio track 1). Stick it onto a portable CD player and notice the racket as you press "play," ">>|" and "|<<"—you're hearing the frantic electromagnetic twitching of all those little motors and servos that spin the disk and move and focus the laser. Eavesdrop on a cell phone as you initiate a call. Listen to small motors in fans, vibrators, and toys; notice the change in pitch as you change the motor speed. Cozy up to a neon sign.

Coil pickups are highly directional, like shotgun microphones. Wire up two of them to the left and right inputs of a portable audio recorder before you ride on a subway—the sounds of the motors and doors as you whisk in and out of stations acquire a vivid stereo presence (see audio track 2). You can assemble a handful of audio adaptors to connect two pickups to a stereo mini plug for use with your flash recorder or video camera. If you're feeling a bit more adventurous refer to Figure 7.12 in Chapter 7 for advice on soldering two coils directly to a stereo plug.

The stethoscope-like accuracy of the coil moving over a circuit board makes it a useful, non-destructive device for pinpointing the location of interesting sounds that we can later tap off directly with wires soldered to the board (see Chapter 17).

If you move the coil near the speaker of your amplifier it will begin to feed back with the coil that moves the speaker cone—if it doesn't squeal, turn it around so the other end of the coil faces the speaker. As with feedback between a microphone and speaker, the pitch is a function of the distance between the two parts, but here the pitch changes smoothly and linearly, without the odd jumps caused by the vagaries of acoustics, giving you a Theremin-like instrument. Try this with a full-size guitar amplifier for greater pitch range and more impressive volume. If you connect your amplifier to a raw speaker, you can place the speaker on its back like a candy dish, rest the coil inside, and turn up the gain: the coil should bounce around and change pitch as it feeds back.

Speaking of guitars, you can use a guitar pickup in place of a telephone tap—a guitar pickup is just a coil of wire, wrapped around a magnet, inside an expensive package (see Figure 3.2). You can repeat the above experiments with a whole guitar in your hands,

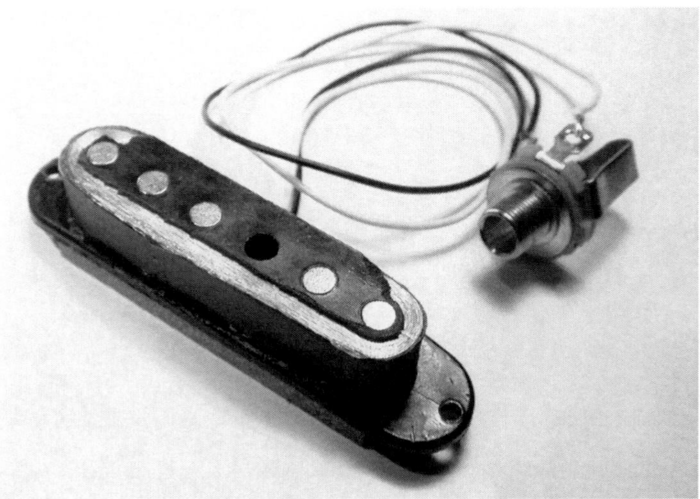

Figure 3.2
A guitar pickup with case broken away to reveal internal coil.

but a loose pickup is handier. At guitar repair shops you can usually buy cheaply the low-end pickups removed when better ones are installed. Jump ahead to Chapters 6 and 7 if you need advice on wiring the pickup to a cord and plug. As most guitarists know, "single coil" pickups are better at picking up hum and weird electromagnetic noise, while "humbuckers" are so called because they tend to be less sensitive to exactly the kind of garbage we want to hear. But work with whatever you can find.

Basically, almost any chunk of ferric metal (a magnet or piece of metal that is attracted to a magnet) with enough wire wrapped around it will pick up magnetic fields. You can solder a plug onto the wires coming out of any transformer (such as a wall-wart external power supply), relay coil, solenoid, or electric motor, connect it to your amp and listen. You can scrounge small coils called "inductors" from various online retailers and solder them up to an audio cable and plug—you'll have to experiment with various inductor values, but once you've found a part that sounds good you'll have reduced the cost of your pickup to well under a dollar (see Figure 3.3). Sometimes you can skip the metal core and just wrap your own coils of wire, which brings us to our next subject . . .

CULTS

The *length* of wire used in the coil affects its sensitivity to different frequencies (like the tuning dial on a radio.) Fans of what is known as VLF (Very Low Frequency) radio make big coils by wrapping yards of wire around big wooden crosses and then camp out on remote hilltops like hermit Klansmen. Get far enough from civilization's ubiquitous 60/50 Hz hum and you may be lucky enough to pick up the Aurora Borealis, "whistlers" induced by meteorites self-immolating as they enter the earth's atmosphere, the pipping of GPS satellites, or top-secret submarine radio communication.

If you want to experiment, take a 100 feet or so of ordinary insulated wire and wrap it around a wooden armature (nail two 5-foot pieces of 1 inch × 2 inch lumber together and notch the ends to keep the wire from slipping off). Solder one end of the wire to

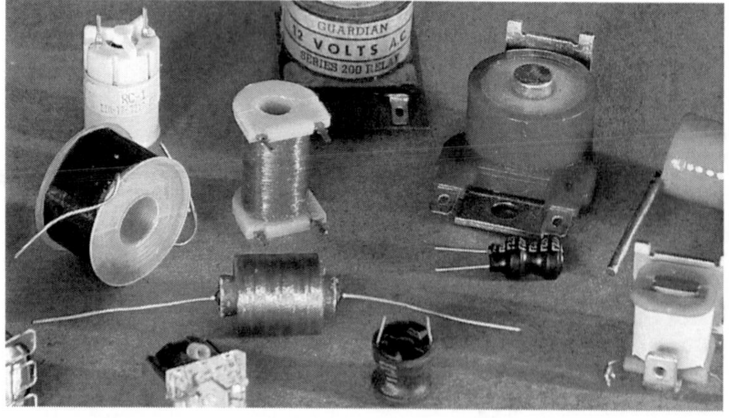

Figure 3.3 Assorted inductors and relay coils (left) and a coil pickup made by soldering an inductor to a shielded cable (right).

MORTAL COILS

In addition to constituting the basic mechanism of radios, microphones, and speakers, electromagnetic fields have spookier aspects that have been central to instrument design and artists' works for almost 100 years. The siren song of the Theremin (the earliest commercial electronic instrument, invented by Leon Theremin in 1924) resulted from two high-frequency radio signals beating like out-of-tune strings on a piano. Seventy years later Gert-Jan Prins (Netherlands) (see Figure 3.4) and David First (USA) (see audio track 3 and his video on the DVD) created music out of Theremin-like interference and feedback between radio receivers and transmitters. Some of the earliest realizations of computer music were heard through radios placed on top of the central processing units of mainframes: engineers would run programs with instruction cycles whose lengths were calculated to emit a composed sequence of radio frequencies, which were duly demodulated by the radios.

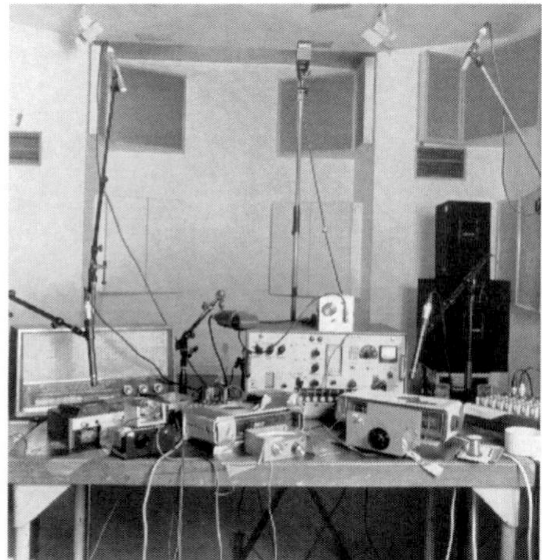

Figure 3.4 Gert-Jan Prins, installation for "Sub V" at STEIM, Amsterdam, The Netherlands, March 1996 (left), and detail of his radio transmitter and receiver feedback system (right).

Alvin Lucier's (USA) "Sferics" (1980) is a recording of electromagnetic "tweaks," "bonks," and "swishes" originating in the ionosphere, the result of self-immolating meteorites, the dawn chorus, and the Aurora Borealis. More recently Lauren Carter and Joe Grimm, in a Ben Franklin moment, sent a wire-wrapped kite up in the sky to record the same atmospheric sounds (see their video on the DVD). The squeal and chatter of mistuned shortwave receivers has been an inspiration to composers from Karlheinz Stockhausen (Germany), whose 1968 composition "Kurzwellen" used four receivers in live performance, to Disinformation (UK), who has made a career of recording and performing with radio signals from across the spectrum, emanating from both human activities (power stations, navigation satellites, submarine communication, camera flashes) and natural sources (sun spots, thunderstorms). Australian artist Joyce Hinterding has worked with enormous coils both in gallery settings and in the wild. Telephone taps have been used to pick up stray radio emissions from laptops, CD players, subway trains, and a host of other unexpected objects by artists such as Nathan Davis (USA), Haco (Japan), Andy Keep (UK) (see audio track 1), Rob Mullender (UK), Jérôme Noetinger (France), Yuri Spitsyn (Russia), and Sonia Yoon (USA).

Toward the end of the 1970s, German sound artist Christina Kubisch began using electromagnetic induction to transmit local sound fields that followed wires she arranged around rooms to form "sound labyrinths," to be heard over specially designed receivers, often embedded in headphones). In 2003 she began a series of site-specific urban "Electrical Walks" in which listeners don special headphones and follow maps that guide them through a series of specific sonic landmarks resulting from the electromagnetic signals emanating from the electrical grid, heavy machinery, security gates in stores, elevators, subways, ATMs, etc. (see Figure 3.5).

Figure 3.5 Christina Kubisch, "Electrical Walks," Birmingham, UK, 2006.

Recently there has been a revival of interest in historic crystal radio technology. Crystal radios receive signals through a seemingly impossible combination of bits of rock, diodes, coils of wire, and headphones—no batteries needed. Sawako Kato (Japan and USA) builds a radio from these raw materials in her séance-like performance work

"Ishi –Listening to Stones" (see Figure 3.6 and her video on the DVD). And by playing the 12 keys of his "FMkbrd," Portuguese musician Vasco Alvo can gate on and off 12 radios, tuned as he wants, to create a solo update of Cage's infamous "Imaginary Landscape No. 4" (see Figure 3.7 and his video on the DVD).

Of course, the electric guitar pickup just might be the single most important musical discovery of the twentieth century.

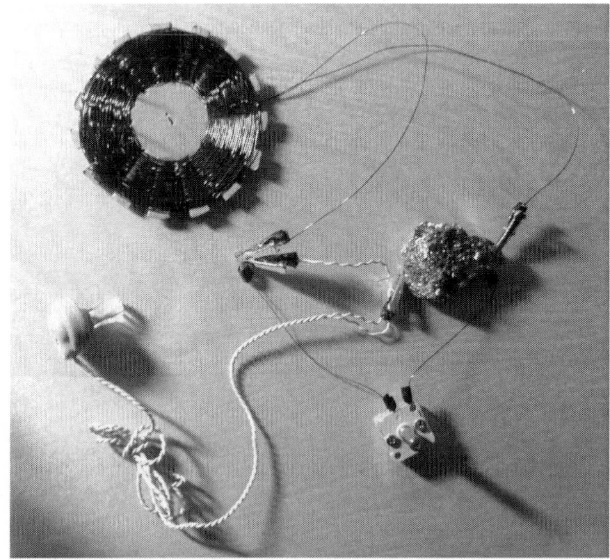

Figure 3.6
Sawako Kato, crystal radio made with pyrite.

Figure 3.7
Vasco Alvo, "FMkbrd."

the *tip* of a plug that fits into your amp or tape recorder, and solder the other end to the *sleeve*. Plug in, turn on, drop out.

DUELING RADIOS

In the process of receiving and demodulating transmissions, radios actually generate and send back out intermediary electromagnetic signals. These transmissions aren't very powerful, but evidently are strong enough that airline passengers are warned not to turn on radios in flight for fear of disrupting the navigation system (knowing just how weak these signals really are further diminishes one's faith in air travel safety). What one radio transmits another will receive: turn on two AM radios, tune them to the dead band at the end of the tuning range, and set them close together; by moving the radios and varying the tuning you should be able to produce Theremin-like whistling and interference patterns (see audio track 3 and David First's video on the DVD).

CHAPTER 4

In/Out: Speaker as Microphone, Microphone as Speaker—The Symmetry of it All

You will need:

- A battery-powered amplifier.
- A pair of headphones or a small speaker.
- A dynamic microphone (i.e. Shure SM58).
- A pair of jumper leads with alligator clips and a plug to fit the input jack of your amplifier.

ELECTROMAGNETISM

There is a beautiful symmetry to the principles of electricity that are most commonly used to translate acoustic sound into an electrical signal and then back into sound again. Inside every dynamic microphone (such as a typical PA mike) is a lightweight plastic membrane affixed to a coil of fine wire encircling a cylindrical magnet. Chris Martin sings, and his sound waves jiggle the membrane, which moves the coil in the field of the magnet, generating a very small electrical current. This current is amplified, equalized, flanged, reverberated, compressed, and finally amplified even more before being sent back out to a bigger coil wrapped around an even bigger magnet. Now this shimmering electromagnetic field pushes and pulls against the big magnet (think of the two magnetic Scotty dogs, forever trying to align themselves nose to tail), moving a paper cone back and forth, producing sound waves of . . . a louder, possibly improved, Mr. Martin.

A record player cartridge is basically a microphone with a needle where the diaphragm should be; and record cutting heads are beefy backwards phonograph cartridges. Headphones are tiny speakers. The telephone tap coils we used earlier are electromagnets with no additional moving parts, receiving or emitting electromagnetic waves rather than acoustical sound waves.

Not only is the same electromagnetic force at work in both input and output devices (microphones and speakers), but sometimes the gizmos themselves are interchangeable. Try plugging a pair of headphones into the input jack on your amp or cassette recorder; speak into it and listen—more than one band's demo tape was recorded this way. Use clip leads and a plug to connect a raw loudspeaker to the input of your amp or recorder.

According to legend, Motown engineers recorded kick drum and bass guitar with a large speaker placed in front of the drumhead or amp—a sort of "subwoofer mike."

These alternative microphones don't sound as generically "good" as a $5,000 Neumann tube mike, but (as Motown's sales have shown) for special applications they can be very useful. Non-standard microphones like these essentially "pre-produce" the sound of whatever you are recording or amplifying, by introducing idiosyncratic equalization and distortion that you might otherwise add later in the mixdown process—if you like the effect, why not get it over with at the start of your session and avoid the trouble of replicating it further down the line? A flat sound is not always the best sound, and delaying the fixing until the mixing is sometimes a sign of procrastination rather than perfectionism.

Likewise any dynamic microphone (i.e. based on a coil and magnet design, such as the ubiquitous Shure SM58, spat upon by singers in clubs around the world) can be used as a very quiet speaker or headphone. Use whatever chain of adaptors is needed to plug a dynamic mike into the headphone jack of your amplifier. Patch some audio source into the amp input (MP3, computer, CD player, etc.) and hold the mike up to your ear. Slowly turn up the volume until you can hear the music. In a pinch many a PA engineer has substituted a mike for missing headphones when tracing a suspected fault through a mixing board (although watching an engineer seemingly amplifying his earlobe can be disconcerting for nearby members of the audience).

Microphones have very delicate coil windings, however, and can be easily blown out, so BE CAREFUL. Also, condenser mikes (like the "plug-in-power" mikes for cassette recorders, or expensive studio mikes) use a different, not-so-easily reversible principle of translation, so:

IF THE MIKE USES A BATTERY OR PHANTOM POWER OR IS REALLY EXPENSIVE, DON'T USE IT BACKWARDS.

Which brings us, possibly pyrotechnically, to our next Rule of Hacking:

Rule #8: In electronics some things are reversible with interesting results, but some things are reversible only with irreversible results.

Some of you may recognize that the 8th Rule of Hacking is a sobering exception to the First Law of the Avant-Garde:

Do it backwards.

TELHARMONIUM LITE

A related experiment will introduce you to the fundamental operating principle of what is generally accepted to be the first synthesizer: the Telharmonium. Patented by Thaddeus Cahill in 1897, the Telharmonium weighed in at over 200 tons and resembled a power station more than a musical instrument. It generated sine tones by spinning the shafts of dynamos—essentially producing an AC current, like that running in your household wiring, but with variable frequency instead of fixed at 60 or 50 hz. The electrical output was carried on the recently installed telephone grid until it was banished from the network for overpowering conversations. We can mimic the effect, on a more modest scale, by clipping the terminals of a small DC motor to a plug connected to the input of an amplifier (see Figure 4.1). Turn on, turn up, spin the shaft and you should hear a whirring sound whose pitch is proportional to the speed of the shaft. Recently, California artist Lorin Edwin Parker has created a steam-powered synthesizer by connecting a DC motor to a homemade steam engine (see Figure 4.2 and his video on the DVD).

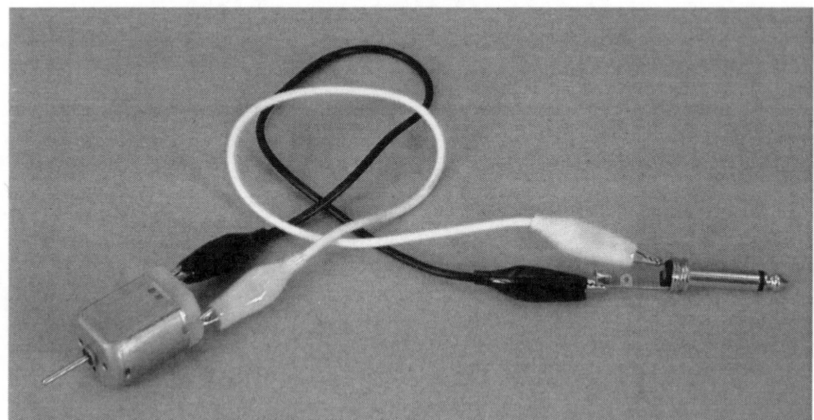

Figure 4.1 Motor-as-oscillator: listen as you spin the shaft.

Figure 4.2 Lorin Edwin Parker's steam-powered synthesizer.

CHAPTER 5

The Celebrated Jumping Speaker of Bowers County: Twitching Loudspeakers with Batteries

You will need:

- A few dispensable raw loudspeakers of any size (salvage them from old TVs, boomboxes, stereos, answering machines, etc.).
- A 9-volt battery and a few C or D batteries.
- Some jumper leads with alligator clips.
- Some hookup wire.
- Electrical tape or gaffing tape.
- A medium-size nail.
- A metal file.
- A sheet of copper, steel, or iron, or a chunk of some conductive metal—the more corroded or scratched the better.
- Pop-tabs from soda cans, paper clips, loose change, nuts and bolts, assorted metal scraps.
- A can whose diameter is slightly less than that of one of your speakers.
- Some rice, beans, or gravel.
- Plastic or metal bowls, larger than your speaker, or a toilet plunger.

Creative mistreatment of loudspeakers goes beyond Motown, and even precedes amplification as we now know it. British computer scientist and musician John Bowers has developed a beautiful electric instrument, evoking the spirit of nineteenth-century electrical experimentation (think twitching frogs legs and early telephones) out of nothing more than a speaker, some batteries, wire, and scrap metal.

Hook up the circuit shown in Figure 5.1. Clip one end of a test lead to one terminal of the speaker (it doesn't matter which). Clip the other end to the "+" or "–" terminal of the battery (again, it doesn't matter which one). Now tap the loose end of the second clip lead to the open terminal of the battery. The speaker should pop in or out from its position of repose. Disconnect the clip and the cone should pop back to its original position. Reversing the polarity of the battery will change an inny to an outy or vice versa: the cone that popped out should now suck in.

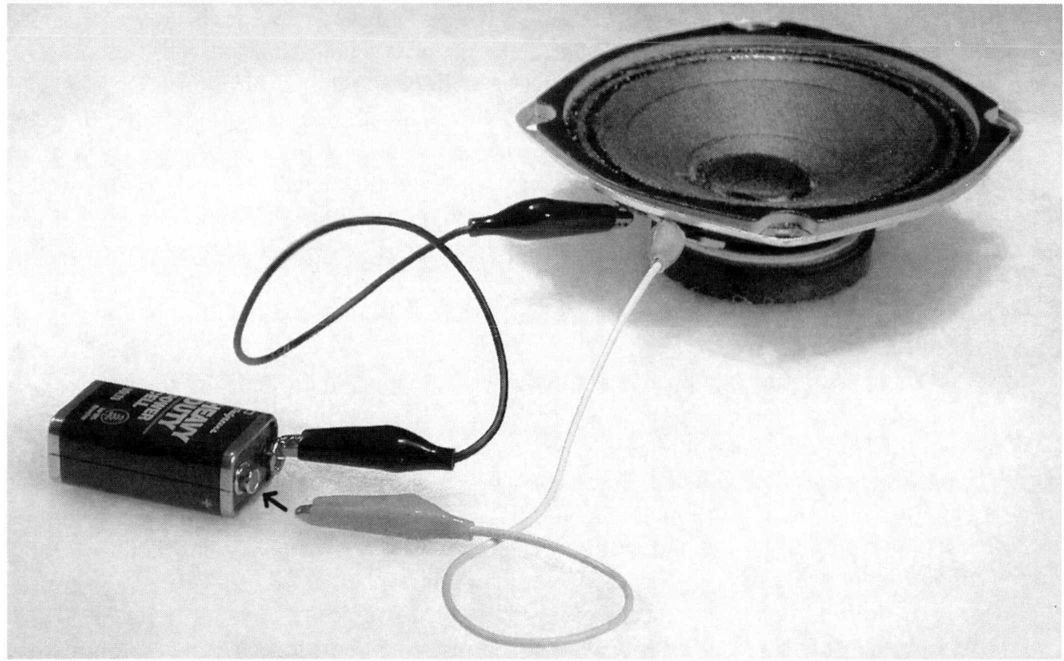

Figure 5.1 Twitching a speaker.

The cone doesn't pop as it should? Check that the battery is not dead: measure its voltage with a multimeter (see Chapter 14); put it in a functional toy or effect pedal and see if it works; or use the admittedly gross but nonetheless effective expedient of touching both of the battery terminals with the tip of your tongue—a good battery will electrically tickle your tastebuds, inducing a curious, salty sensation residing in that fuzzy region between pain and pleasure.

Battery OK? Then try another set of clip leads. I've bought mine from a variety of sources, and every batch seems to have about a 10–15 percent failure rate. A good worker never blames her tools, unless they end in alligator clips (or crocodile clips, for our British hackers). Still no luck? Try another speaker.

What's happening? Passing the battery current through the speaker coil (which is attached to the paper cone) creates an electromagnet that interacts with the speaker's fixed magnet (attached to the metal framework) and moves the cone in or out, depending on the polarity of the battery and resultant magnetic field (remember our Scotty dogs from the last chapter). Incidentally, the racetrack-like path of the current from the positive terminal of the battery, through the leads and speaker coil, to the minus terminal of the battery neatly demonstrates the etymological root of the electronic usage of the term "circuit."

Tapping the alligator clip against the battery terminal will produce a nice little percussive accent, at one and the same time drum-like and "speaker-ish," acoustic and electronic. But that's just the start. Keep one lead connected between the battery terminal and the speaker terminal as before. But this time, instead of connecting the second lead directly from the battery to the other speaker terminal, clip it between the speaker and a metal file or a chunk of some conductive metal: a pie tin, a cookie sheet, scrap copper flashing, an anvil, a piece of girder, a brake drum, a frying pan, etc.—the rougher or more corroded the metal surface, the better. Clip one end of a third jumper lead to the other terminal of the battery and the other end to a nail, bent paperclip, knife, or some other pointy piece of conductive metal (see Figure 5.2).

Touch the nail to the metal. When it contacts the metal, the nail completes the circuit, sending current through the speaker coil, and making the cone jump, as before. Now scrape the nail across the metal: as the contact is broken by irregularity of the surface, the speaker emits scratchy, percussive sounds whose character is quasi-controllable through hand movement. Drawing the nail across a file elicits sounds curiously like those of turntable scratching.

You may notice small sparks as the contact is made and broken, and the battery will probably get warm—the speaker coil is almost a short circuit, and sucks a lot of current from the battery. Avoid holding the nail on the metal for an extended period of time—loudspeakers get hot and bothered when presented with a steady DC voltage, so it's better to send them shorter pulses. Don't try this with your roommate's Bang & Olufsen, and don't plug a speaker directly into the wall. A 9-volt battery won't last very long under such treatment (see Chapter 17 for a little bit of battery science)—try substituting one or more C or D cells. You can buy holders for multiple batteries at Radio Shack or online retailers if you're fussy, or you can just tape them together in a column with a bit of bare wire held against each end, onto which you then clip the leads previously attached to the 9-volt battery terminals.

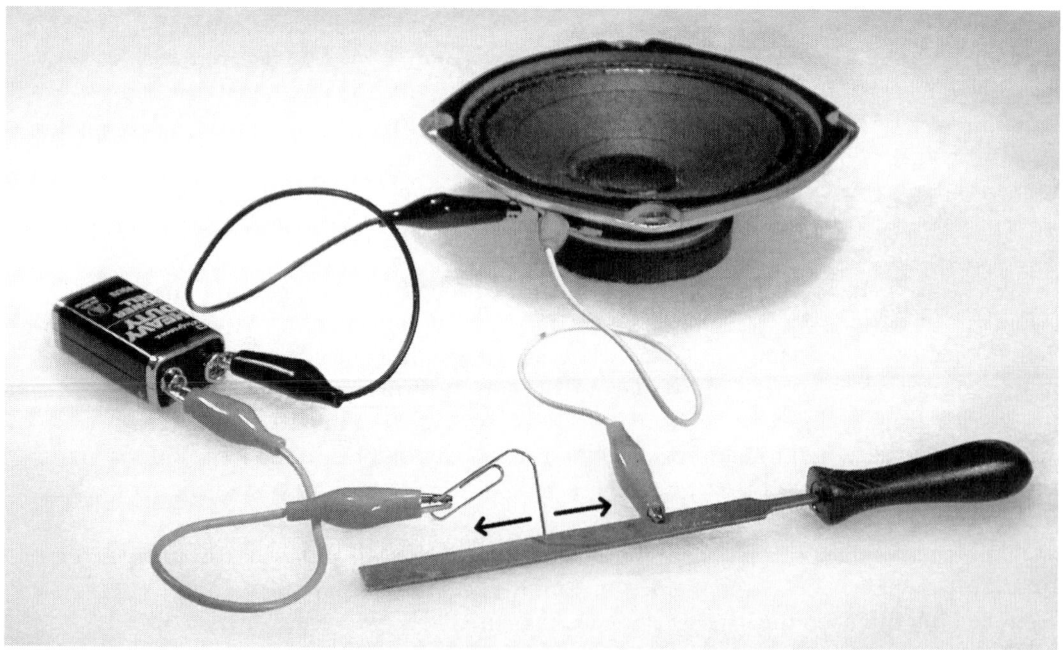

Figure 5.2 Scratching a speaker.

Instead of using the nail and file, you can clip the leads to two paperclips, washers, coins, aluminum pop-tabs, or loops of copper wire that you place inside the speaker cone while you hold the two leads near the clips. The cone jumps when contact is made, breaking the connection for a moment; then the metal bits fall against each other and the process starts all over—a mechanical oscillator and the beginning of what Bowers calls "The Victorian Synthesizer" (see Figure 5.3 and audio track 4). By holding the two contacts close together against the speaker cone as you vary your touch and the location on the cone, you can change the pitch, rhythm and timbre of the buzzing sounds.

You can line the cone with aluminum foil or apply metal tape (such as the kind sold in hardware stores or Radio Shack for preparing windows for burglar alarms), connect one lead to the foil or tape and the other to a flip-tab or other light metal fragment. The tab gets thrown up from the foil or tape, breaking and making contact as before. Or fill the cone with spare change, loops of wire or scrap metal: the mechanical interaction of all the conductive bits creates a kind of pre-computer algorithmic music.

Multiple speakers can be wired in series (like those frustrating Christmas lights from my childhood) or parallel, with contacts resting in each cone, so they interact to produce more complex rhythms. You can substitute a tilt-switch (see Chapter 16) for the aluminum tabs as another way of using the speaker's own movement to turn on and off the current.

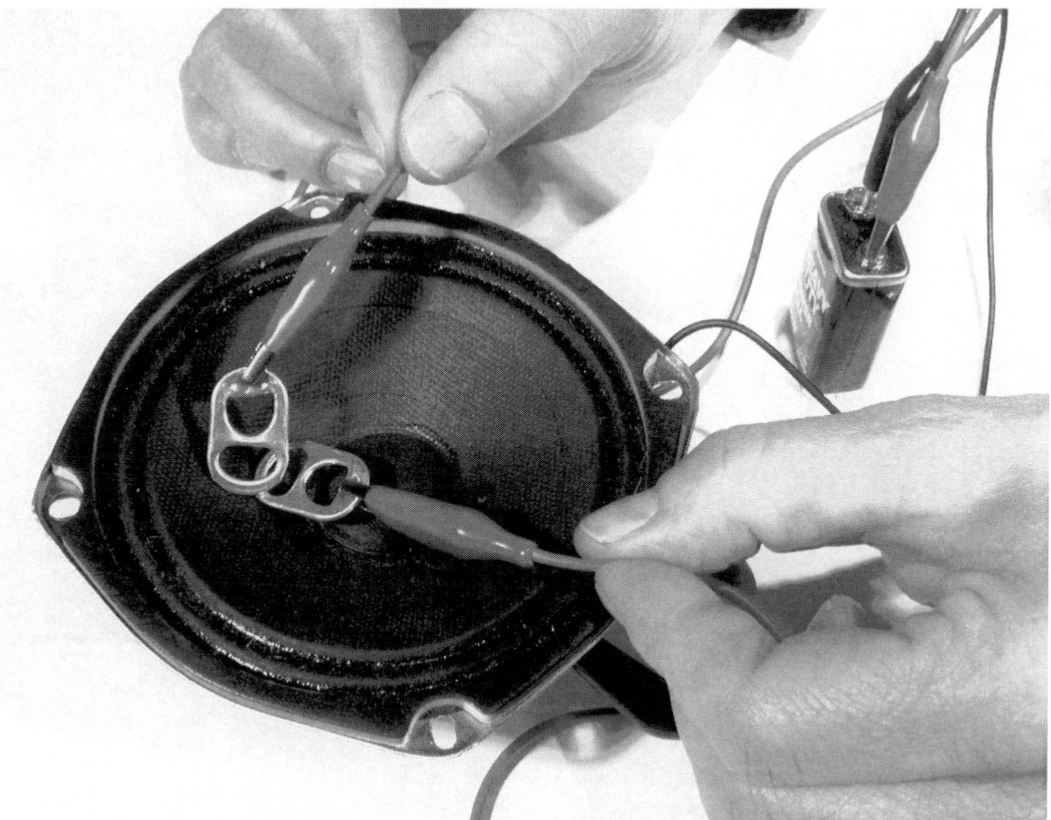

Figure 5.3 The Victorian Oscillator.

Sound doesn't end at the loudspeaker; it starts there. You can use your hands, bowls, or toilet plungers to mute and resonate the sound further. Put gravel, loose change, or dried lentils inside the cone for additional rhythmic accents. Place a can on the cone, open end down; clip one lead to the can and one to a metal washer placed on top of the can (see Figure 5.4). The speaker cone will jump, breaking and remaking the contact as before, but in addition, as the can jiggles it changes resonance like a trumpet mute; loose coins or beans placed on top of the can produce additional percussive accents. Alternatively, put some jangly things inside a small glass bottle/vial and place it inside a cone—*maracas de cristal*. Watch Aaron Zarzutski's video on the DVD to see a cymbal being played by a jumping speaker.

You'll notice that different speakers sound different, even in similar configurations. It's mostly a function of size, as with drums, but if you try these experiments with a speaker in an enclosure (such as one from a home stereo) you'll hear that it has considerably more bass presence—the box gives a woofer its woof.

You can further extend the sound world of the jumping speaker by placing a telephone tap (see Chapter 3) in the cone and connecting it to an amplifier. The sound will change as the signal is amplified into a second ("normal") speaker, and the bouncing of the coil inside the cone produces variations in the speaker's percussive snap.

Finally, there's a visual element. British artist Alex Baker tosses ping-pong balls with a variant of the Victorian Oscillator. You can fill the speaker cone with talcum powder or light sand and watch it make patterns as the cone jumps. For a touch of the old Fillmore light show, waterproof the speaker cone by painting it with enamel or rubber cement. Fill the cone with water or oil and turn down the lights; reflect a flashlight or laser pointer off the surface, and watch the resulting patterns on the wall or ceiling. Think Summer of Love. (In Chapter 24 we'll return to this subject.)

Figure 5.4
A "prepared speaker."

CHAPTER 6

How to Solder: An Essential Skill

You will need:

- A soldering iron with a fine tip.
- A small damp sponge (or, in a pinch, a folded wet paper towel).
- Rosin-core solder.
- Diagonal wire cutters.
- Wire strippers.
- Some light gauge insulated wire, solid or stranded.
- An audio jack or plug of some kind.

Soldering is one of the fundamental skills of hardware hacking. It is almost impossible to hack hardware *without* knowing how to solder. We've had fun with our clip leads, and some serious hackers make it a point of pride to use them exclusively, eschewing solder for the transient glory of an alligator's jaws. But the day will come when these minuscule members of the order *crocodilia* will fail you: plugging a Stratocaster into a Marshall stack, lowering a hydrophone into the Mariana Trench, or building with integrated circuits. As a skill soldering commands a lower hourly wage than Java or C++, but your friends and parents will be very impressed at your acquisition of such arcane knowledge (as if you had learned fire eating or Linear B).

Successful soldering, like fundamentalist Christian comedy performed in mid-winter by an L-Dopa patient, depends on cleanliness, heat, steady hands, and . . . timing!

Soldering is not a question of dropping melted solder onto a joint. Rather, one must first melt a thin layer of solder onto each surface, then let them cuddle up to one another while you heat both surfaces to re-melt the solder until it commingles to form a strong bond. The process is similar to gluing wood: the strongest bond comes from permeating the surfaces of both pieces of wood with a layer of glue before assembly, rather than just squeezing out a blob of glue and slapping them together.

We'll start by soldering wires together—high temperature knitting. Not very exciting, but a cheap way to learn (and much easier than trying to solder old license plates, which is how my father and I tried—and failed—to learn when I was 10).

1. Plug in the iron and place it somewhere where the tip will not make contact with flammable, meltable, or scorchable surfaces, or its own power cord (cute little wire

rests are sold for this purpose). Wait a long time for it to warm up. The iron is hot enough to use when solder touched to the tip melts.

2. Wipe the tip of the hot iron across a damp sponge. The tip must be smooth and clean enough that the solder flows evenly, leaving a shiny silver coating. If blobs of solder fall off and the tip remains grey and crusty even after sponging, unplug the iron and, after it has cooled down, polish the tip with steel wool, fine sandpaper, or a file, and try again (see Figure 6.1). If the tip of the iron is seriously pitted you will need to replace the tip (or, if it is a cheap iron with non-replaceable tip, the whole iron).
3. Use your wire strippers to remove about 1 inch of insulation from the ends of two pieces of wire. There are several styles of wire stripper, differing (proportionally) in cost and complexity of mechanism—I prefer the cheapest kind, which resembles squashed pliers with an orthodontic problem (see Figure 6.2). Use the adjustments on the strippers, or a fine sense of touch, to avoid cutting through the wire as you nick and slide back the insulation—connect with your inner female dog carrying her puppies. If the wire is stranded, twist the strands to eliminate frizzing. Hold the wires in something so that the tips are up in the air but don't wiggle around too much. You can use a fancy "third hand" gizmo (two articulated arms with alligator clips, affixed to a weighted metal base,) or a vise, or just weigh down the coil of wire under this book or a brick.
4. "Tin" the wires. Melt a small blob of solder on the tip of the iron. Hold this blob against one of the wires—this blob conducts the heat from the iron to the wire. Hold the tip of the solder roll against the *wire*, not the iron. After about 2 to 5 seconds the wire should be hot enough that the solder will melt, flowing around the wire to coat it evenly in a smooth layer; if not, apply a *tiny* bit more solder to the tip of the iron and try again (see Figure 6.3).

Figure 6.1 A happy soldering iron (top) and a sad soldering iron (bottom).

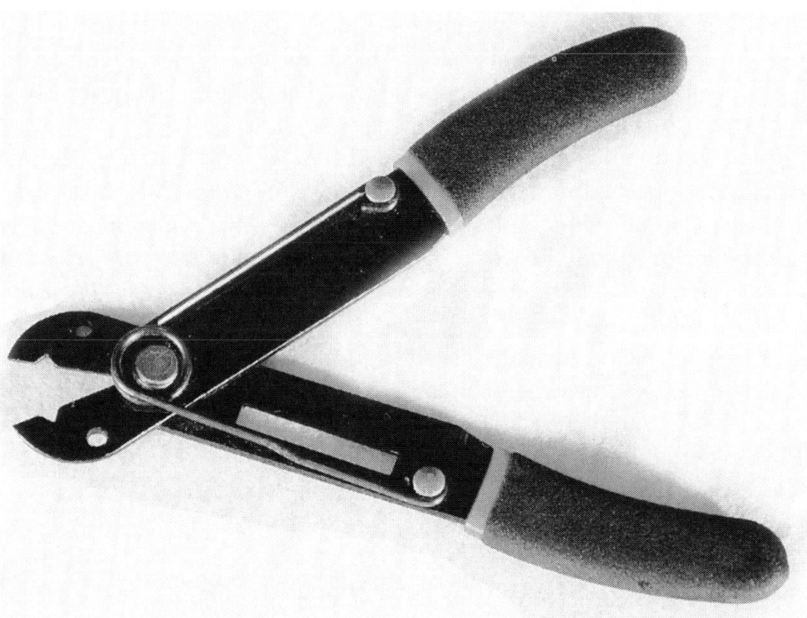

Figure 6.2 Simple wire strippers.

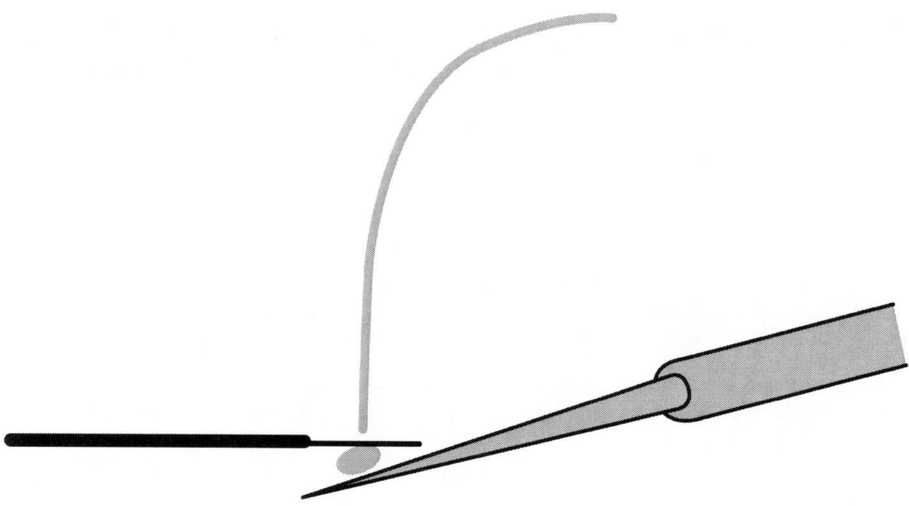

Figure 6.3 Tinning a wire.

Remove the iron from the wire. The solder should cool to a smooth, shiny silver; if it is rough and grey you did not get the wire hot enough—try again. Then go ahead and tin the second wire.

5. Twist the wires around one another like strands in rope. Again, apply a small blob to the iron and use the blob to conduct heat to the bundled wires. After a few seconds the tinned solder should re-melt and flow together; you may apply a little bit more solder to strengthen the joint, but only as much as can flow and distribute

itself smoothly—like a wax-impregnated candle wick. Wait several seconds *without wiggling* for the joint to cool and harden. Blobs of solder on the wire or dull grey solder are signs of a "cold solder joint" (see Figure 6.4). Such a joint is neither electrically nor mechanically strong. Do it again.

When tinning and soldering, be sure that you apply heat for the minimum amount of time needed to get the solder to flow, otherwise you may damage the components you are soldering (for example, melting the insulation off the wire).

6. Repeat this process until you get it right and feel comfortable with the "touch" of soldering—how much heat and solder to apply for how long, etc. It's a small step from here to cracking safes.
7. You can now move on to soldering wires to plugs and jacks—the next step down the road to making your own guitar cords. Tin the wire and jack terminals as before, then solder them together by holding the wire against the terminal as you heat them with the soldering iron. If the terminal lugs on the jack have wire-sized holes, you can make your life easier by bypassing the tinning, and simply looping the end of the wire through the hole to secure it before soldering.

When soldering circuit boards (such as a simple amplifier kit recommended in Chapter 1), use as fine a tip as possible. Keep it cleaned and tinned by frequent swipes across the sponge. Use solder sparingly to avoid blobs of excess solder unintentionally connecting separate pads on the circuit board (what are known as "solder bridges").

Be advised that cold solder joints sometimes sort of work, for a while, but will come back to haunt you at the most inauspicious moment (Amateur Night at the Apollo? *After* you get to Carnegie Hall? Grammy acceptance speech?), so it's worth getting soldering right before going on stage.

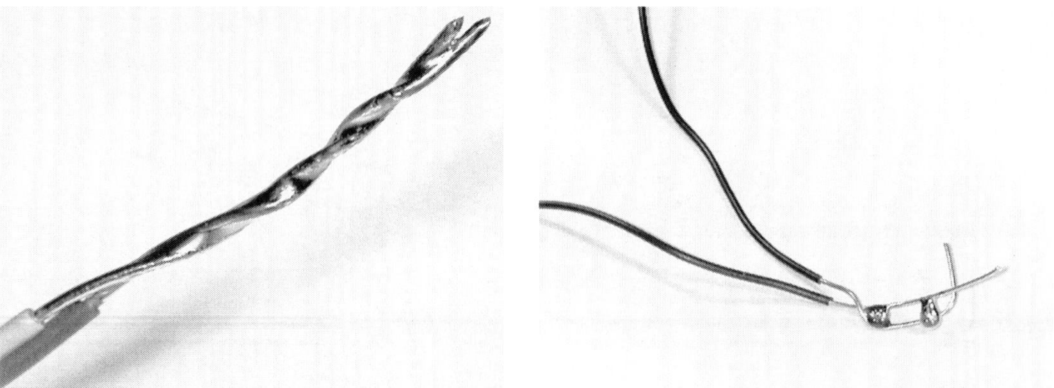

Figure 6.4 A happy solder joint (left) and a sad solder joint (right).

CHAPTER 7

How to Make a Contact Mike: Using Piezo Disks to Pick Up Tiny Sounds

You will need:

- A battery-powered amplifier.
- A piezoelectric disk or two.
- About 8 feet of lightweight shielded cable.
- A plug to match the input jack on your amp, recorder, or mixer.
- Plastic insulating electrical tape.
- A can of Plasti-Dip or similar rubberized plastic paint (sold in hardware stores for dipping tool handles).
- Small spring clamps.
- Molex-style terminal block.
- Hand tools, soldering iron, and electrical tape.
- Sparklers, small blowtorch, guitar strings, metal scrap, Slinky, springs, and condoms.

THE PIEZOELECTRIC EFFECT

In addition to electromagnetism, another common principle of reversible sound translation is the "piezoelectric effect," which depends on the electrical properties of *crystals*, rather than electromagnetism, as discussed in Chapter 4: bang a crystal with a hammer and it will generate a pretty sizeable electrical signal (enough to light a flashlight bulb); conversely, if you send an electrical current into a crystal it will twitch a bit.

Piezoelectric disks, made by bonding a thin layer of crystal to a thin sheet of brass, are everywhere today, inside almost everything that beeps: appliances, phones, toys, alarm clocks, computers, etc. (see Figure 7.1). Because they are manufactured in huge quantities, out of very few separate parts (many fewer than go into a traditional speaker), they are incredibly cheap. They also happen to make even better contact microphones than they make speakers. Drum triggers and commercial contact mikes are often made from piezo disks and sold at absurdly marked-up prices. As gratuitous markups are anathema to the hacker, making our own cheap contact mikes is an excellent way to while away an evening.

Figure 7.1
Assorted piezo disks.

HOW TO MAKE A PIEZO DISK CONTACT MIKE

1. Try to find a piezo disk that already has wires attached—soldering directly to the disk's surface is infuriatingly difficult. Better yet, get a few of them in case you break any. If you can only find wireless disks, skip ahead to step 6. You can salvage disks from all sorts of trashed electronic devices or buy them from any number of Web-based electronic surplus outlets (see Appendix A)—or, if you're feeling flush, at your local Radio Shack, at several times the online price.
2. The disk may be encased in a kind of plastic lollipop. If so, *carefully* pry open the case and remove the disk. Try not to impale yourself, but do not bend or scratch the disk, since this can result in the contact mike distorting. Prior work experience at the Oyster Bar or Clam Shack pays off here.
3. The disk may have a tiny circuit board attached. Snip off the connecting wires close to the circuit board, so that the wires attached to the disk are as long as possible. Remove the board.
4. Once removed, the disk should appear as a circle of gold- or silver-colored metal, with a smaller circle of whitish crystal within. Depending on the design, there will be two or three wires connected to the disk. One will always be connected to the *metal* portion, somewhere near the edge; this we will call the "ground" connection. One will connect to the *main* part of the inner crystal circle; this we will call the "hot" connection. In some cases there will be a narrow, tongue-like shape differentiated within the crystal, to which the third wire connects; this we will call the "curious but unnecessary" (CBU) connection.
5. Cut the connecting wires so that they protrude about 2 inches from the disk. Strip about 1/2 inch of insulation from the ends of the ground and hot wires; don't bother to strip the CBU wire—you may cut it off near the surface of the disk, or leave it dangling. Tin the stripped ends. If there are no wires attached to the disk prepare it by tinning a small spot on the surface of the crystal and one on the surrounding

perimeter of the brass disk; solder fast and be very careful to apply only a minimum of heat, since the crystal surface is easily damaged.

6. Shielded cable consists of stranded wire inside insulation, surrounded by a layer of braided or twisted wire, which is in turn covered by another layer of insulation. A cross-section looks like tree rings or a minimalist target. Shielded wire is used to protect an audio signal from hum and other electromagnetic interference. Shielded cable comes with any number of internal conductor wires, but for audio purposes most varieties have one or two internal conductors plus the shield. Unless otherwise specified, we only need cable with a single internal conductor plus shield; if your cable has two or more internal conductors (in addition to the shield) that's ok, but we'll only use one, so pick the color you like and cut off the others.

Rule #9: Use shielded cable to make all audio connections longer than 8 inches, unless they go between an amplifier and a speaker.

Cut a 5-foot section of shielded cable, the thinner and more flexible the better. Strip back 1 inch of the outside insulation. Unbraid the shielding and twist it into a single thick strand. Now strip back 1/2 inch of the inner insulation, and twist the center conductor into a neat strand. Keep the two strands separate. Tin both strands, quickly and carefully so as not to melt back the insulation (see Figure 7.2).

7. Twist together the hot wire from the center of the piezo disk and the inner wire from the shielded cable. Solder them together. Twist together the ground wire from the edge of the piezo disk and the shield from the shielded cable. Solder them together (see Figure 7.3). Wrap both joints separately with a little piece of electrical tape so that they cannot short out if they touch each other.
8. If you are using a piezo disk that did not arrive with wires already connected, you will have to solder the tinned ends of the shielded cable directly to the disk's surface. This is NOT easy. First tin a contact spot somewhere on the metal perimeter—this part is not so tough, just make sure the metal is clean and shiny (you can carefully

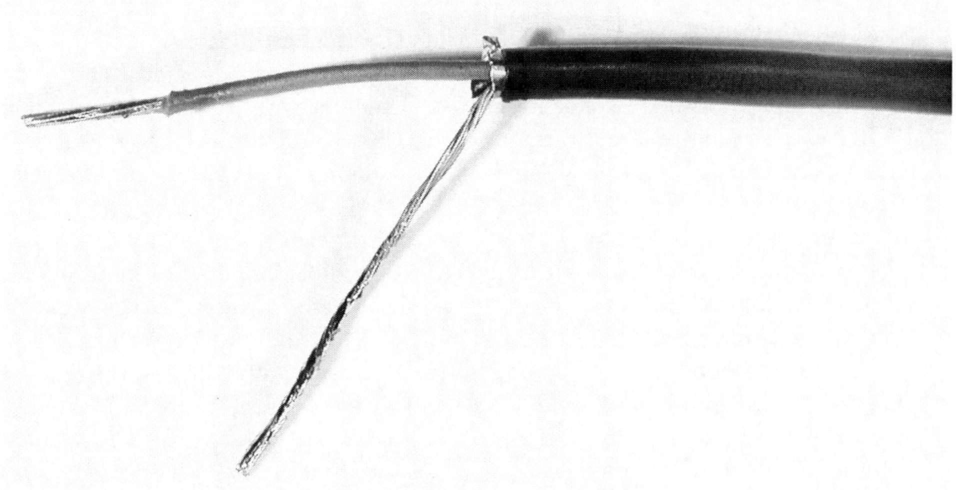

Figure 7.2 Shielded cable prepared for soldering.

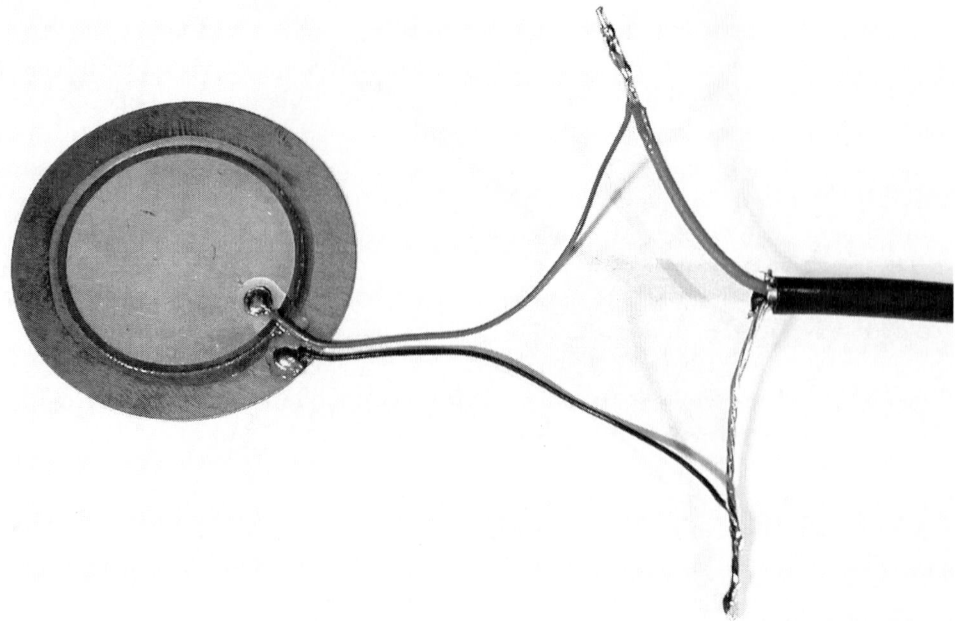

Figure 7.3 Piezo disk soldered to cable.

rough it up a bit with steel wool or fine sandpaper). Then tin a spot on the main part of the white crystal, by applying the heated iron *as briefly as possible* to melt a bit of solder into a small pad. Solder the end of the tinned shield to the tinned spot on the edge of the disk. Finally, quickly solder the tinned center conductor to the tinned pad on the crystal itself. If everything holds, congratulations—you've just earned the advanced soldering merit badge! If, on the other hand, you find this impossibly difficult, go back out, find a disk with wires attached, and start all over again.

9. Now move on to the other end of the shielded cable and strip it as in step 6: 1 inch outer insulation, twist shield, 1/2 inch inner insulation, twist conductor, tin the wires. If you are satisfied connecting to your amplifier with clip leads, skip to step 11. To attach a plug proceed to the next step, but not before you test the contact mike by connecting the stripped cable ends to your amp via a pair of clip leads and a loose plug.
10. Select the appropriate plug for whatever device you plan to use with your contact mike: a 1/4 inch "guitar plug" for connecting to a guitar amplifier or most mixers; a 1/8-inch plug to connect to most mini-disk or flash recorders. If you want to connect the contact mike to the XLR inputs of a microphone preamplifier please see Chapter 10 (Figure 10.3) for wiring instructions for this plug. Unscrew whatever plug you are using to connect the contact mike to your amplifier/recorder/mixer. Slip the barrel back over the shielded wire toward the disk so that the threaded portion faces the freshly tinned end. Unscrewing the barrel should reveal two solder tabs on the plug: the shorter one connects to the "tip" of the plug and the longer one connects to the "sleeve."

Solder the inner conductor of the shielded wire to the tip of the plug and the shield to the sleeve (see Figure 7.4). Sometimes there are small holes in the connector tabs that you can hook your wire onto so that it is held in place before soldering. Otherwise you will have to tin the tabs and hold each wire against its respective tab while soldering—a job for three hands, a vise, or a fearless buddy with a steady hand.

Now is as good a time as any to introduce the 10th Rule of Hacking, if it is not obvious already:

Rule #10: Every audio connection consists of two parts: the signal and a ground reference.

In the case of a contact mike the signal comes from the white part of the piezo disk, while the ground is the brassy bit; on the plug the tip carries the signal and the sleeve is the ground. In future chapters I may get a bit sloppy and only refer to the signal when describing connections—always assume that a ground connection must accompany every signal.

11. Plug into your amp and check that your new contact mike works—tapping the mike should make a solid thunking sound. If there is no sound, check the joints at both ends of the cable to make sure they are good and there are no shorts. If there is a lot of hum, you may have connected the hot and ground wires to the wrong conductor of the cable—de-solder and reverse them. If it works, screw the barrel down onto the plug and test again—sometimes squeezing the barrel down over a marginal solder joint will break or short it. A small piece of electrical tape can be used to isolate the connections if excess wire tends to short when the barrel is screwed down.

    Whoops! Did you forget to slide the barrel onto the wire before you soldered? If so, de-solder the plug, go back to step 10, but don't feel too stupid—everybody makes this mistake. But remember:

Rule #11: Don't drink and solder.

12. When you are sure you have an electrically functional contact mike, cover the ceramic side with a piece of electrical tape—you can trim it around the circumference with scissors or a knife, or you can wrap the edges over to the other side of the disk. Don't worry, it doesn't have to look pretty.

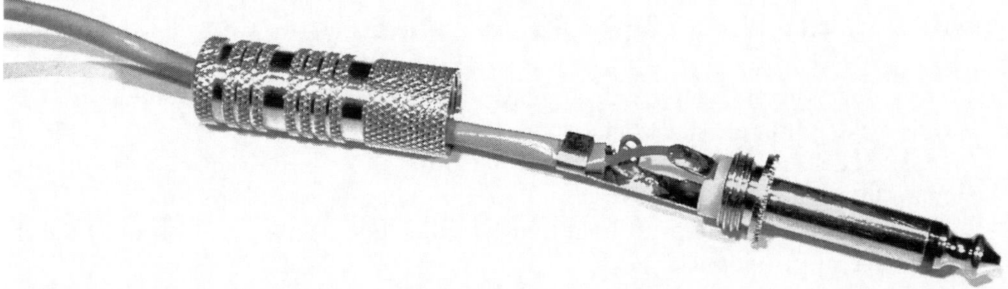

Figure 7.4 Cable soldered to plug.

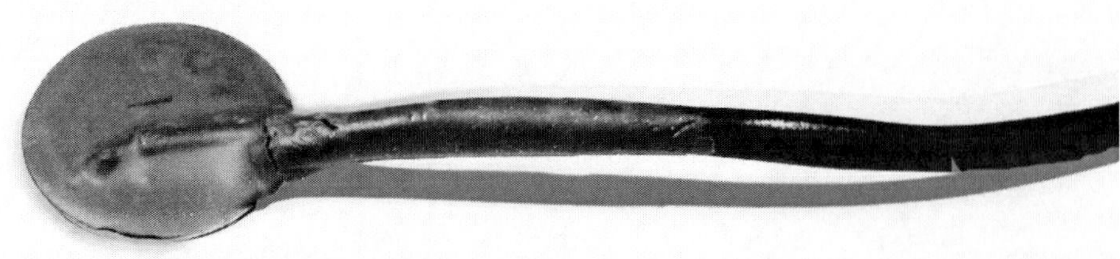

Figure 7.5 Contact mike encased in Plasti-Dip.

13. Buy a can of "Plasti-Dip" or similar rubber paint—it's sold in most US hardware stores as a coating to improve one's grip on tool handles. In Europe it's harder to find, but sometimes you can obtain "latex paint" that works as well. Find a well-ventilated space. Open up the can and stir. Slowly dip the contact mike end of your cable into the goop until you have covered the wire past the electrical tape (see Figure 7.5). Slowly withdraw it and hang it up (preferably outside) to dry. Go away and take a break—this stuff is stinky. You can dip a second layer after the first one dries thoroughly, which can take several hours. More than three layers tend to muffle the sound, so don't overdo it without listening carefully after each new layer.

The tape and Plasti-Dip treatment serves several functions:

a. It strengthens the connections between the wires and the piezo disk.
b. It insulates the disk from electrical shorts, and prevents hum when you touch it.
c. It waterproofs the contact mike, so you can use it to record underwater sounds, freeze it in ice-cubes, dangle it in a drink, swallow it, etc.
d. It deadens slightly the pronounced high-frequency resonance of the disk (similar to the effect of gaffing tape on the head of an unruly snare drum).
e. It looks really cool.
f. Dipping is an illicit-smelling treat after all that soldering.

The discovery of Plasti-Dip as the perfect contact mike sweater must be credited to the ingenious Robb Drinkwater of The School of the Art Institute of Chicago.

WHAT TO TRY WITH YOUR CONTACT MIKE

Contact mikes are great for greatly amplifying hidden sounds in everyday objects. The trick is making firm physical contact with the vibrating object.

Use double-stick tape or Blu-tak to attach the mike to the surface. Try: guitars, violins, drums, pots and pans, wrists and knees, foreheads, pinball machines.

Use small spring clamps to hold things to the contact mike. Try: strips of metal, gaffing tape, rulers, popsicle sticks.

"Terminal strips" are used to make electrical connections in lamps and other appliances. They can be salvaged from discarded appliances (Ikea loves to use them in

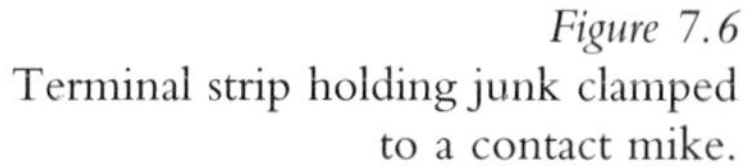

Figure 7.6
Terminal strip holding junk clamped to a contact mike.

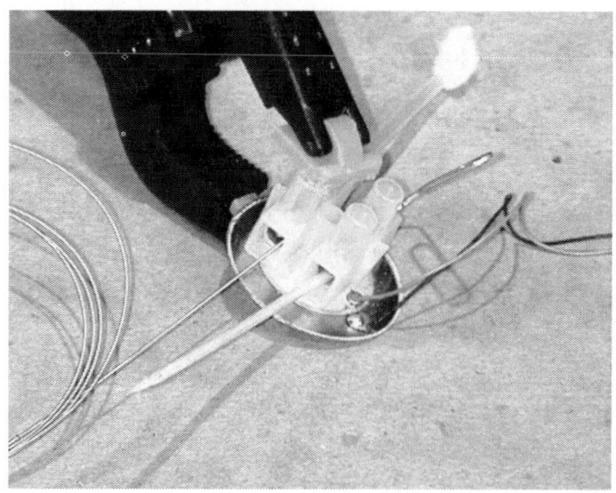

lighting fixtures) or bought from any number of sources, including Radio Shack, DIY centers like Home Depot, and online electronic surplus retailers. Cut the terminal strip into small sections and clamp them onto the mike with a spring clamp. Insert thin objects into the terminal openings and hold them in place by tightening the screws (see Figure 7.6). Try: Slinkies, springs, loose guitar strings, toothpicks, sate sticks, broom straws, porcupine quills, cactus needles.

This is an excellent way to replicate the old-fashioned phonograph cartridges used by John Cage in his visionary work of live electronic music, "Cartridge Music" (see Art & Music 2 "John Cage—The Father of Invention").

Many metals make unusual sounds as they heat and cool. Clamp a sparkler in the terminal block, light, and listen. Or clamp steel wire and heat with a torch (see audio track 7).

Although a naked piezo disk picks up very little airborne sound (try speaking into one), you can make unusual air-mikes by clamping the disk to lightweight, flexible material. Try drum heads (see Alex Baker's video on the DVD), thin sheets of metal, glass windows, plastic "clamshell" packaging from salad bars or toy purchases, Styrofoam cups, etc.

Connect the contact mike to an amplifier and wire a raw speaker to the amplifier output jack. Place the speaker on its back, like a candy dish. Rest the contact mike inside the cone and turn up the gain. The contact mike should jump up and down as it feeds back with the speaker—a slightly higher-tech variation on the jumping speaker in Chapter 5. (This is the underlying principle of Lesley Flanigan's "SpeakerSynth"—see her video on the DVD and Figure 30.4 in Chapter 30.)

Once waterproofed with the electrical tape and Plasti-Dip, contact mikes will also serve as affordable hydrophones and submersible mikes.

- Fill a plastic yogurt container with water, drop in the contact mike and pop it in the freezer. Listen as it freezes. Once frozen, remove the ice block from the container, float it in a bowl of hot water, and listen to it melt (see audio track 5).
- Drop it in the water next time you go fishing and check if the fish are really laughing at you (see audio track 6).

ART & MUSIC 2

JOHN CAGE—THE FATHER OF INVENTION

The influence of John Cage (1912–92) on American avant-garde music cannot be overstated. Given the breadth of his impact as a composer and theoretician, his significance in the rise of hacker electronic culture is sometimes overlooked. Throughout his career Cage had a passionate curiosity for new sounds and compositional strategies. Lacking institutional support in the form of orchestral commissions and the like, Cage, the son of an inventor, chose to develop new instruments from everyday technological and commonplace objects. At the beginning of his career he literally made do with rubbish: his early percussion music, such as "First Construction in Metal" (1939), used brake drums and other scrap iron from junkyards. In the 1940s, Cage (together with Lou Harrison) inserted screws, washers, rubber erasers, and other pocket detritus into the strings of pianos to create the gamelan-like sounds of the "prepared piano." "Imaginary Landscape No. 1," composed in 1939 for piano, bowed cymbal, and record player, was the first documented piece of music to feature the DJ as a musical performer, while "Imaginary Landscape No. 4" (1951) was composed for a dozen ordinary radios.

In "Cartridge Music" (1960) performers substitute springs, twigs, pipe cleaners, and other thin objects for the needles in cheap record player pickups; the surprising richness of these enormously amplified "micro-sounds" rivaled the more labor- (and capital-) intensive synthetic sonorities coming out of the European electronic music studios, and opened the ears of a generation of sound artists to the splendor of the contact mike. "Hpschd" (1967–69), composed in collaboration with Lejaren Hiller, was one of the first works of computer music.

Cage's later music, such as "Études Australes" (for piano, 1975), or "Ryoanji" (for mixed ensemble, 1983–85), reverted to more traditional instrumental resources. (Cage once told me, "If I don't write for these virtuosos they'll have to play music by even worse composers.") But the ethos of hacking lived on in his continually surprising methodology and in his persistent invention of new performance techniques.

- Hold the mike in your mouth while you drink or chew, but please observe safe sex practices: put an extra layer of protection between you and electricity by encasing the contact mike in a condom or balloon, and

NEVER CONNECT A WET DISK TO AN AC-POWERED AMPLIFIER OR RECORDER—ONLY TO BATTERY-POWERED EQUIPMENT.

A few hours spent with a contact mike and a fistful of junk should convince you of the significance of the Second Law of the Avant-Garde (and the First Law of Pop):

Make it louder, a lot.

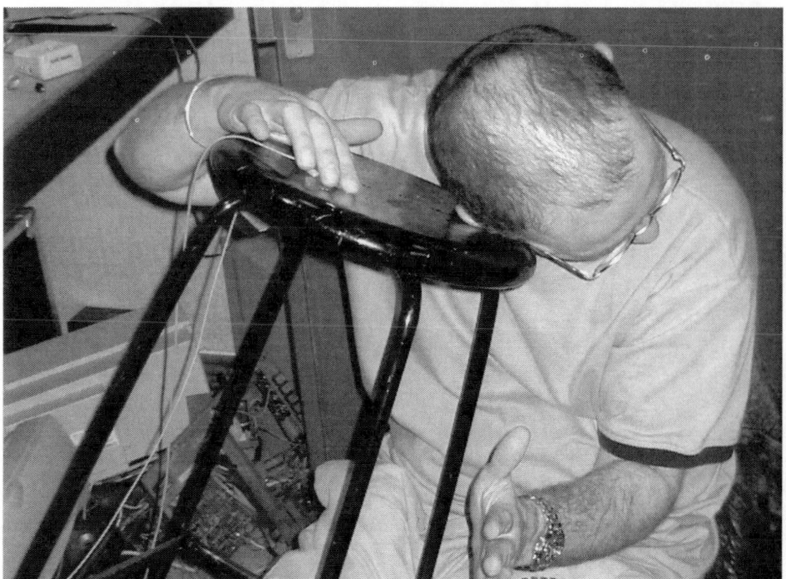

Figure 7.7
John Bowers enjoying his contact mike.

This is the credo of the "Piezo Music" movement that sprang up in the aftermath of Cage's experimentation and the fortuitous invention of the economical piezo disk (see Art & Music 3 "Piezo Music," and Figure 7.7).

HI FI

Ultrasonic transducers, such as those found in many motion-detecting alarm systems, contain very small piezo disks (see Figure 7.8). These disks usually have a slightly lower output level than the more common large brass ones (and are more expensive), but have a flatter frequency response. You will need to solder one connection to each side of the disk, since they have a different mechanical construction from the typical lollipop-encased beeper (see Figure 7.9).

Figure 7.8
Some ultrasonic transducers.

Figure 7.9
Element from ultrasonic transducer soldered to shielded cable for use as contact mike.

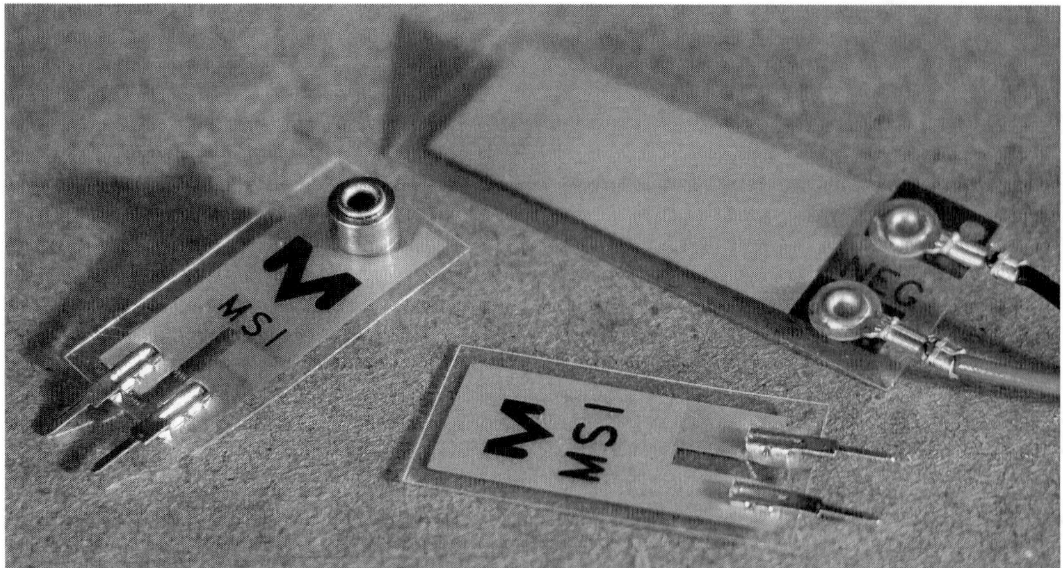

Figure 7.10 LDT vibration sensors from Measurement Specialties.

A few companies manufacture and sell piezoelectric material in the form of a flexible mylar-like film. Although somewhat more expensive than piezo disks bought on the surplus market, piezo film elements such as those by Measurement Specialties have a superior sound quality that justifies the cost in some cases (see Figure 7.10).

STEREO

Those of you running the contact mike into the 1/8 inch stereo microphone input of a video camcorder, flash recorder or mini-disk have a few wiring options to consider. If you solder the mike to a mono 1/8 inch plug (as discussed in step 10 above) the audio signal will usually record on the left channel only (the right will be silent). If you open a 1/8 inch stereo plug, you will notice that there are two short lugs in addition to the long sleeve—one connects to the tip of the plug and one to a ring between the tip and sleeve. You can solder the hot/center conductor to both the tip and ring connectors (as shown in Figure 7.11), in which case the mike's signal will record on both the left and right channel. Or you can make yourself two contact mikes; twist the shields of

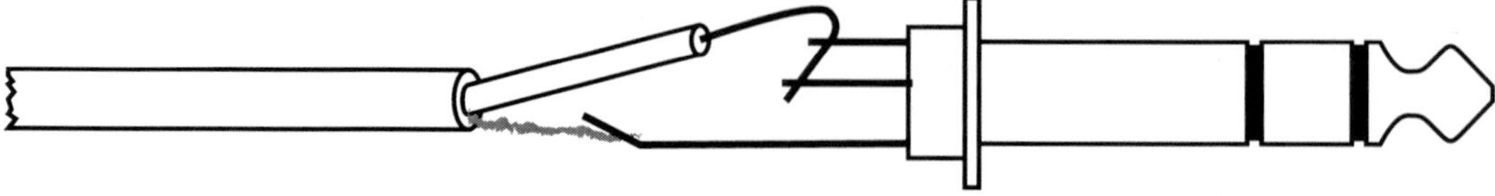

Figure 7.11 Hot lead of a single contact mike soldered to both the left and right connector tabs of a stereo 1/8 inch plug.

both together and solder to the long sleeve lug; then solder the inner "hot" conductor of each contact mike to one of the two short lugs, so that one mike connects to the tip and one to the ring (see Figure 7.12)—don't forget to slide the barrel over *both* mike cables before you start soldering. Voilá—stereo contact mikes.

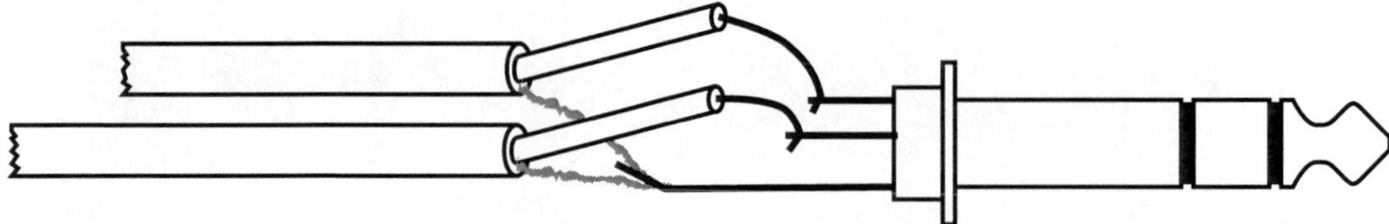

Figure 7.12 Wiring of a stereo contact mike pair to a stereo 1/8 inch plug.

PIEZO MUSIC

ART & MUSIC 3

In the aftermath of Cage's "Cartridge Music" many sound artists sought affordable techniques for amplifying mechanical vibration and microscopic sounds. Since the mid-1970s the proliferation of "Piezo Disks" in beeping appliances has effectively put contact mikes within reach of anyone with a soldering iron. Whether as pickups on bluegrass mandolins or as hydrophones for eavesdropping on whales, the disks have insinuated themselves into surprisingly diverse corners of our recorded soundscape, and have given rise to a genre of "Piezo Music." Hugh Davies (1943–2004) (UK) and Richard Lerman (USA) were two of the earliest innovators. Davies began inventing piezo-amplified instruments in the 1970s, the most poetic of which consists of a disk with short steel wires soldered around its rim. By plucking or blowing gently at these wires, he could elicit a wide range of surprisingly deep, marimba-like sounds, which he incorporated into composed and improvised work. Lerman (who has for many years maintained a wonderfully informative Web site with tips for working with piezo technology) uses similarly bewhiskered disks, but plays them with a small blowtorch: the whoosh of the gas creates an effect similar to that of bowing a cymbal, and while the wire heats and cools it snaps with gong-like solemnity (see Figure 7.13 and audio track 7). Eric Leonardson (USA) performs regularly on his "Springboard": a plank of wood festooned with springs, wires, and other bits of scrap metal that, when heard through the piezo pickup, evoke the sound world of a Balinese Gamelan (see Figure 7.14 and his video on the DVD). Adachi Tomomi's (Japan) "Tomoring" takes a different design angle on amplifying springs, rubber bands, and combs (see Figure 7.15 and his video on the DVD), and Ivan Palacky (CZ) creates similar textures by adding contact mikes to a 1970s-era knitting machine (see Figure 7.16 and his video on the DVD).

Recently several radical turntablists have taken the cue from Cage and extended their technique beyond the vinyl groove. Otomo Yoshihide (Japan) inserts bent wire and springs into the cartridge opening to "play" a cymbal placed on the turntable spindle—

Figure 7.13 Richard Lerman performing "Changing States 6."

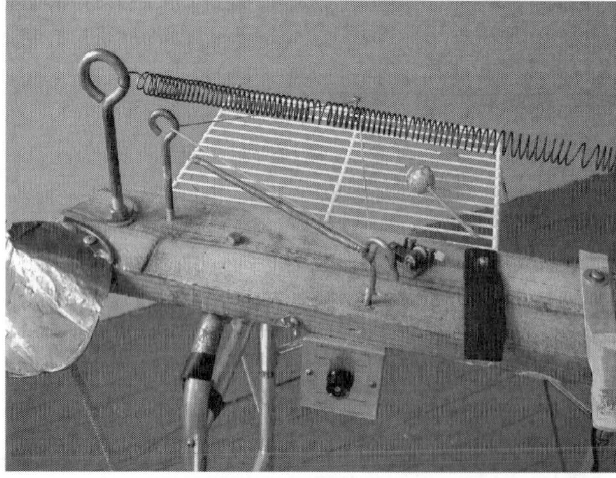

Figure 7.14 "Springboard," Eric Leonardson.

plugged into a guitar amp and turned up well past the point of feedback, the resulting din is one of the loudest things I have ever heard. By adding two extra tonearms to his turntable Janek Schaefer (England) can turn any record into a 3-voice round (see his video on the DVD). Using his "Stylus Pen," Vasco Alvo (Portugal) can play static records by drawing on them. Australian Michael Graeve assembles an impressive collection of cheap old hi-fis and builds thick, drone-like textures from feedback and the sound of the needles playing the rubber mats of the platters (see Figure 7.17).

Figure 7.15 "Tomoring," Adachi Tomoni.

Figure 7.16 Amplified Dopleta 160 knitting machine, Ivan Palacky.

Figure 7.17 0, 16, 33, 45, 78 (2004), Michael Graeve.

Many artists have used piezo contact mikes to record the inaudible or elusive soundscape. In the mid-1970s Australian neuroscientist Alan Lamb began composing music with recordings he made of the wind-driven sounds of abandoned telegraph wires (subsequently several artists, including Warren Burt, Chris Mann, and Jon Rose, began creating music by bowing and otherwise playing abandoned fences around the Australian countryside). Collin Olan (USA) froze waterproofed contact mikes in a block of ice and recorded the cracks and whistles of escaping air bubbles as it thawed (see audio track 5), while Peter Cusack (UK) used similar homemade hydrophones to record the breakup of ice on Siberia's Lake Baikal (see audio track 6).

INTERESTING HISTORICAL NOTE

The term "piezoelectric" originates in the Greek word *piezen* meaning "to press." The discovery of the effect is credited to Jacques Curie and his brother Pierre. The latter, along with his wife, Marie Curie, is better known for his work with another rather energetic, if less user-friendly, phenomenon: radioactivity.

CHAPTER 8

Turn Your Tiny Wall Into a Speaker: Resonating Objects with Piezo Disks, Transformers, Motors, and More

You will need:

- A battery-powered amplifier with output jack for an external speaker.
- A second battery-powered amplifier.
- Your contact mike from the previous chapter.
- Another piezoelectric disk.
- Two plugs to match the amp's external speaker connector.
- A female jack to match the plugs you are using.
- One small audio output transformer (Radio Shack 273–1380 or equivalent).
- About 8 feet of lightweight shielded cable.
- A few feet of lightweight speaker cable or stranded hookup wire.
- Electrical tape.
- A can of Plasti-Dip.
- Small spring clamps or clothespins.
- A sound source, such as an MP3 player, CD player or radio, and cable to connect it to the amp input.
- Small DC motors (from pagers, cell phones, vibrators, etc.).
- Small speaker.
- A wine cork.
- Glue or double-stick tape.

Piezo disks are used to make beeps because they do so very efficiently—they require very little current, and therefore are well suited to battery-operated devices. As loudspeakers they display a rather uneven, non-"hi-fi" response, but they can nonetheless be quite useful when coupled to other things to make *speaker objects*.

To get the most vibration out of a piezo disk speaker it is necessary to feed it a very high-voltage signal, albeit at a minuscule (and therefore harmless) current. A *transformer* is a chunk of iron wrapped in wire—basically two coil pickups (see Chapter 3) on steroids placed back-to-back. It functions as a kind of audio-lever that allows one to jack up the voltage of an electrical signal very easily, without batteries or what are known as "active" electronic components (transistors, integrated circuits, etc.). For this project we will wire up an *output transformer* backwards (see the First Law of the Avant-Garde) to step up the output voltage of a small amplifier from around 1 volt to over 100 volts. I learned this technique from Ralph Jones, a founding member of David Tudor's legendary ensemble, "Composers Inside Electronics" (see Art & Music 8 "Composing Inside Electronics," Chapter 14).

Output transformers were once commonly used in audio amplifiers: in most power amplifiers using tubes ("valves," in the UK), as well as in certain older transistor-based designs, these devices link the electronic circuitry to the loudspeaker. Chronic lower back pain in electric guitarists is an indication of the persistence of heavy iron transformers even in modern guitar amps. For our purposes we do not need to decommission our beloved Marshall, fortunately: for some inexplicable reason Radio Shack sells, for a very modest price, a small, lightweight output transformer that I suspect has not been bought since 1966 by anyone who has not read this book.

The transformer has a *primary* and *secondary* side. The primary will be designated as having an impedance of around 1,000 ohms (1 kOhm), and may have two or three wires. We will use the outer two wires—the center wire, if present, can be ignored. The *secondary* will usually have just two wires, and an impedance of 8 Ohms. In the case of the Radio Shack part, the outermost primary leads are blue and green and the secondaries are red and white.

1. Strip 1/2 inch of insulation off the ends of the primary and secondary wires and tin the ends.
2. Solder one of the *secondary* wires to the tip of a plug that mates with the output jack of your amplifier. Solder the other secondary wire to the sleeve of the plug. (If you are connecting to the amplifier via screw terminals or direct soldering, then omit the plug.) Polarity is irrelevant here—it does not matter which of the two secondary leads goes to the tip and which to the sleeve. If possible, slip the plug's barrel over the wires before you solder them up (as we did with the contact mike); there may not be sufficient wire length to allow this, in which case you can simply wrap some electrical tape around the connections later (after you have proven that the gizmo works, and being very careful not to squeeze the connections together to make a short).
3. Solder one of the *primary* wires to the tip connection of a female jack that mates with the plug you used on the contact mike you made in the previous chapter. Solder the other primary wire to the sleeve connection of the jack (see Figure 8.1). Once again, polarity is irrelevant here—it does not matter which of the two primary leads goes to which connection.

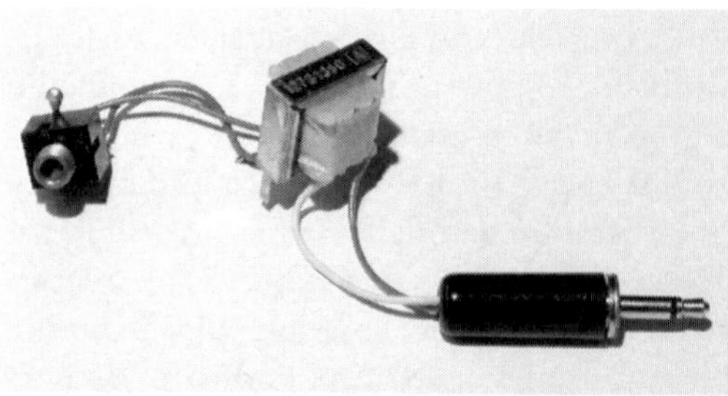

Figure 8.1
A transformer interface for driving a piezo disk.

4. Plug the plug-end of the transformer assembly into the external speaker jack of your amplifier or a boom box. You must use a connection designated for a *loudspeaker*, not a *headphone jack*, which cannot drive the very low 8-ohm load of the transformer. The small Radio Shack amplifier described in Chapter 1 (part number 277–1008) has such a jack—if you're in The Shack picking up an output transformer anyway why not make the salespeople a little happier by buying one of the amps? Very useful little things. If your amp has no proper external speaker jack (only a headphone jack) you will have to open it up, de-solder the wires going to the speaker, and use solder or clip leads to connect them directly to the *secondary* side of the transformer (instead of using the plug)—you can always re-solder them to the speaker when you're finished with this project. You might be tempted to scrounge up an old hi-fi amplifier with proper speaker terminals, but this presents two hazards: accidental electrocution (I'm trying to keep you away from the high voltage running through your walls) and the possibilities of melting the transformer with too much power (it can only handle about 1 watt, and most stereo amplifiers put out a lot more). Hold that thought until you have a bit more low-power hacking experience.
5. Plug your contact mike into the jack-end of the transformer assembly (the original primary side of the transformer).
6. Plug a sound source into the amplifier input: a microphone, MP3 player, CD player, computer, etc. Slowly raise the amp gain. The disk should start to radiate sound. If not, check your connections.

Use plastic spring clamps, clothespins, or tape to clamp the disk to different objects and resonate them with various sounds. Thin materials work better than thick ones: pie tins, etching plates, paper cups, tin cans, and balloons, rather than bricks, anchors, and baseball bats.

A quaint reverb unit can be made by sending signals into a spring or plate of metal using one piezo disk as a driver and picking them up with a second piezo contact mike. You'll have to solder up at least one more contact mike—go back to Chapter 7. This is similar to the technique used in early plate reverb units common in recording studios before digital reverb, and the principle behind David Tudor's "Rainforest" installation (see Art & Music 4 "David Tudor and 'Rainforest'"), which used sculptural objects to transform sound material. You can patch the amp/driver/contact mike assembly into your mixer just as you would a reverb or effect processor: connect a send bus output to your driver-amplifier input, and bring the contact mike back to any console input to amplify and mix in the "reverb" with the dry signal.

Often flexing or damping the object can affect the character of its filtering of the original sound—this is especially noticeable with that irritating semi-rigid clear plastic packaging used around toys and other goods, or the clamshell cases from salad bars; but you should also try loose guitar strings, Slinkies, balloons, plastic bags, bubble wrap, vinyl records, drumheads, old license plates, oil drums, buckets of water, bowls of Jell-O (see Figure 8.3). Whether as a reverb substitute in a mixdown situation or as part of a live performance setup, this is a cheap, easy, fun route to unusual signal processing.

With all due respect to the First Law of the Avant-Garde, this transformer assembly does *not* do a very effective job of increasing the gain of a piezo disk when it is used as a contact mike—it is effectively limited to boosting the efficiency of a piezo *driver*. But don't take my word—try it for yourself.

DAVID TUDOR AND "RAINFOREST"

ART & MUSIC 4

David Tudor (1926–96) began his career as a leading pianist of the avant-garde. By the early 1950s he was serving as pianist for the Merce Cunningham Dance Company, and assisting in the realizations of both Cage's piano music and his electronic works. Tudor gradually abandoned the piano and emerged as the first virtuoso of experimental electronic performance. Expanding on Cage's work with the "found" technology of radios and record players, Tudor embarked on the (at the time quite arduous) process of acquiring enough knowledge of circuit design and soldering to construct his own new instruments. He believed that new, object-specific, intrinsically *electronic*, musical material and forms would emerge as each instrument took shape: "I try to find out what's there—not to make it do what I want, but to release what's there. The object should teach you what it wants to hear." Although Tudor was not the first composer to make his own electronic instruments (he was inspired and assisted by pioneering composer/engineer Gordon Mumma), in no other composer's work is the ethos of music *implicit* in technology so fundamental and clear.

Beginning in 1968, Tudor composed a series of pieces under the title of "Rainforest," culminating in "Rainforest IV," developed in conjunction with a workshop in electronic performance that he gave in Chocurua, New Hampshire, in 1973 (see Figure 8.2). The principle underlying all the "Rainforest" works was similar: sounds are played through transducers affixed to solid objects; the objects filter, resonate, and otherwise transform the sounds; the processed sounds are directly radiated by the transduced objects, which serve as "sculptural speakers"; contact mikes on the objects pick up the vibrating surfaces of the objects, and these micro-sounds are amplified and heard through ordinary loudspeakers around the space. With an open-form score that encouraged experimentation in the design of sound generators and resonated objects, this work served as a creative catalyst for the workshop participants and, later, other young composers who were drawn to Tudor by word-of-mouth. They subsequently formed a loose collective ensemble called "Composers Inside Electronics." Over the next 28 years, this group served as a laboratory for artist-designed circuitry and experimental electronic performance, presenting dozens of installations of "Rainforest IV" worldwide, as well as performances of works by individual members of the ensemble, which over the years included John D. S. Adams, Nicolas Collins, Paul De Marinis, John Driscoll, Phil Edelstein, Linda Fisher, D'Arcy Philip Gray, Ralph Jones, Martin Kalve, Ron Kuivila, and Matt Rogalsky. (See Oscillatorial Binnage's recent realization of "Rainforest" on the DVD.)

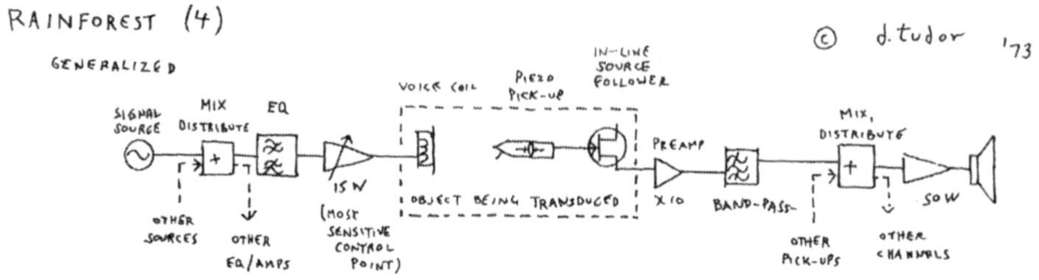

Figure 8.2 Generalized diagram for "Rainforest IV."

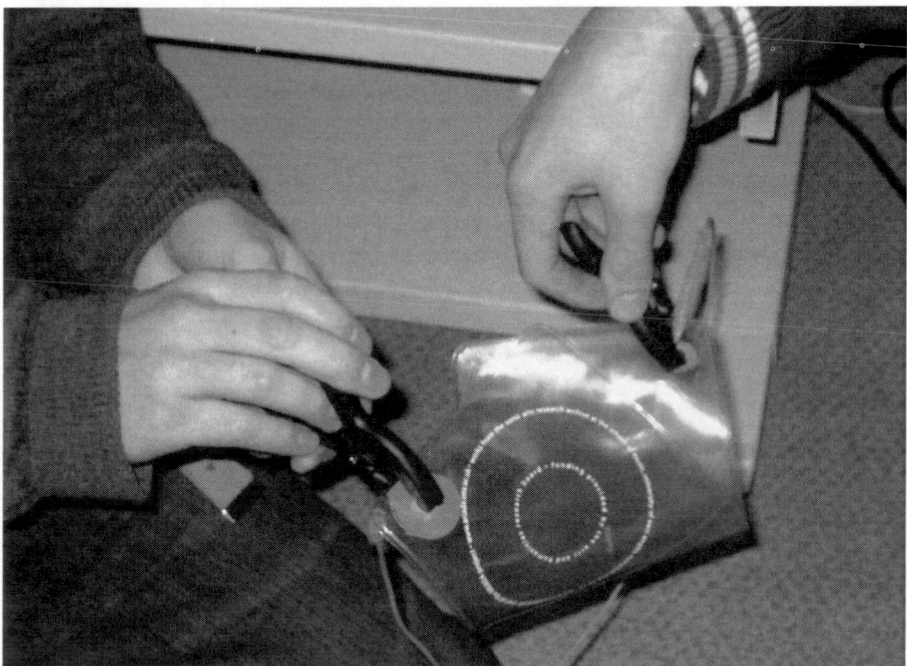

Figure 8.3 A piezo driver and contact mike pickup being used to filter a sound through a sheet of plastic.

You can create feedback by plugging a contact mike into the amp input and a piezo driver into the output, and attaching the two to the same object. Flexing or dampening the object can affect the feedback pitch, and turn a piece of garbage (such as the much-despised molded clear plastic packaging from toys) into a playable instrument—an electronic musical saw. You can configure several channels of amps, drivers, and contact mikes to send audio signals through a series of objects for multi-stage processing; using Y-cords you can branch off and mix after each resonator-object. Get together with your buddies, find a small garage, and form a piezo band. (See Chapter 26 for more information on matrix processing.)

Whereas size does not greatly affect the loudness of a contact mike made from a piezo disk, it makes a big difference when you are making a driver. If you have a choice, use the largest possible disk and you get a bigger sound out of whatever you are driving. You will notice that this is not a "high fidelity" device: the sound is often limited to high frequencies, displays the peaked resonance inherent to a piezo disk, and can be quite distorted—larger disks have a wider frequency response, as well as being louder and less distorted than small ones. You can insert some equalization, in the form of an inexpensive "stomp box" graphic EQ, to process the signal driving the piezo disk and/or the contact mike picking up the vibrating object. Substituting a more powerful amplifier, such as a boom box with external speaker jacks, for the Radio Shack mini test amplifier, will also give you more volume before distortion. The benefits of equalizers and stronger amplifiers will be especially noticeable when you set up the matrices and feedback systems described above.

The transformer dongle, equipped with jacks and plugs as we have built it, lets you adapt any contact mike as a driver. But you can save some time and money by soldering piezo disks directly to transformers (use lightweight speaker cable).

A word of caution: if you are tempted to use an amplifier of considerably more power (such as an old hi-fi receiver) you could blow out the little Radio Shack transformer, as I warned you earlier. Should you choose to ignore my advice you must acquire an output transformer with a power rating equivalent to your amplifier. For this I suggest you haunt a guitar specialty shop or other tube amplifier resource.

By the way, if you place your finger across an un-Dipped, bare piezo disk while it is being run as a driver, you may experience a mild, not entirely unpleasant electric shock. This demonstrates how high the voltage gets when jacked-up by the transformer. Electrical tape and Plasti-Dip will reduce distracting stimulation and protect the driver from damage. (Two considerably less-pleasant applications of transformer step-up technology are the Taser "non-lethal" stun gun and the "quaint" interviewing techniques popular at Abu Ghraib and Guantanamo.)

OTHER KINDS OF DRIVERS

In addition to piezo disks, small motors can be adapted as drivers. They are most effective for lower frequencies, and complement the rather tinny quality of the piezo—a sort of subwoofer-driver. Connect the two wires of any small DC motor (one that runs off batteries) directly to the speaker output of your mini-amplifier—this time *not* through the output transformer (see Figure 8.4). It should twitch in response to your sound source. Sometimes clamping the body of the motor directly to an object will be sufficient to transmit the vibration; sometimes you'll need to get clever with a cam on the motor shaft. Some motors work better than others: vibrators from pagers and cell phones tend to be good. You'll have to experiment. If you have a multimeter, you can use it to measure the resistance of the motor coil: touch the motor terminals with meter's probes while setting it to the lowest range for measuring ohms (Ω)—you're looking for motors that measure between 4 and 12 ohms (see Chapter 14 for advice on using a multimeter).

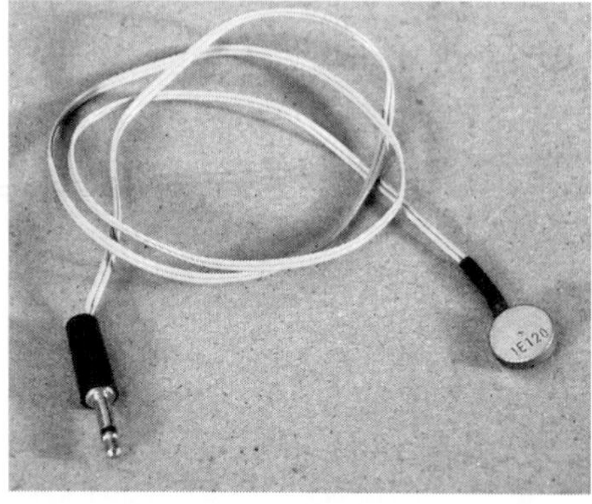

Figure 8.4 Cell-phone vibrator motor wired as driver (left) and connected as external speaker on alarm clock (right).

You can also make a pretty efficient driver by gluing a cork to the center of a small loudspeaker (see Figure 8.5). Connect any sound source through an amplifier to the corked speaker and hold the speaker against a sheet of metal, drumhead, cymbal, etc. The cork should vibrate the material and process the original signal. You may want to pick up the vibrating surface with a contact mike. The end of cork can be treated to further affect the sound: a thumbtack brightens it (like a honky-tonk piano), while a piece of felt softens it, and wood is somewhere in between.

A nice, simple spring reverb can be constructed by stretching a fat spring or Slinky between the center of a speaker cone and a contact mike. You can attach the Slinky to the cone with tape, glue, or a sandwich of self-adhesive Velcro.

There are commercially available transducers as well. Richtech Enterprises still manufactures the "Rolen-Star" wide-range driver originally used by Tudor for "Rainforest" (see Figure 8.6). Several companies make low-frequency drivers—essentially subwoofer transducers—for use in car sound systems, home cinemas, and to increase the realism in computer games. Aurasound's "Bass Shaker" bolts to the floor of your trunk and turns the whole back end of your car into a big subwoofer (see Figure 8.6). Similar drivers include the SmartDisk Tactile Sound Transducers, the Sonic Immersion I-Beam, and the delightfully named "ButtKicker." All of these require 10–30 watts of power, however, and cannot be driven from a mini test amp or boom box.

For more modest sound pressure levels, the "Soundbug" from FeOnic consists of a battery-powered amplifier and transducer package with a suction cup for affixing it to most smooth surfaces; it can be driven directly from the headphone output of a computer, portable CD player, MP3 player, etc. The SI-5 portable speakers by Sonic

Figure 8.5 A corked speaker.

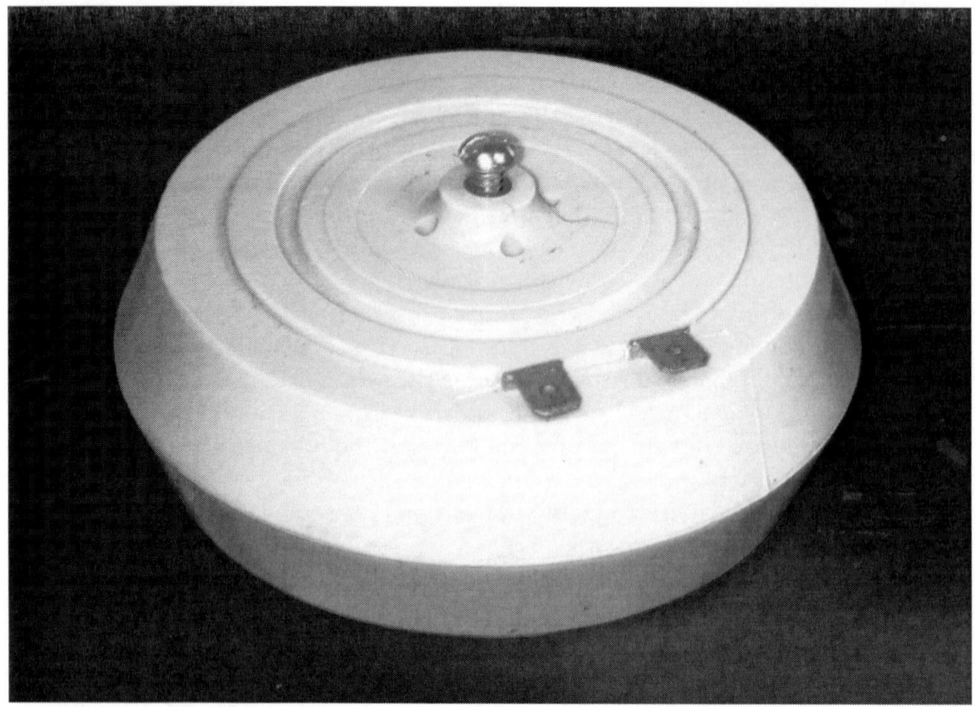

Figure 8.6 A Rolen-Star transducer.

Figure 8.7 A Bass Shaker.

Impact Technologies can be disassembled to extract a very effective wide-range transducer, powered by its own small battery amplifier (see Figure 8.8). The same company's Stick-On Speaker Sound Pad is a slightly higher-power version of the same transducer technology, packaged for direct adhesive application to any surface. Zelco recently released the "Outi," an earphone that clips to the outside of the ear and transmits sound by vibrating the cartilage—reviews mention that drum rolls are a problem for the ticklish, but this odd little gizmo might work fine for other, less biological applications.

It is possible to resonate guitar strings electromagnetically, using coils instead of vibrating transducers such as those described so far in this chapter. The EBow originated in 1967 experiments by the band Iron Butterfly: inside a small handheld device a pickup coil is amplified and fed into a driver coil; when placed over a string on an electric guitar the EBow produces local feedback to set the string resonating, with a bowing effect (hence the name). The EBow can be emulated by plugging a guitar into a small amplifier, attaching a low-impedance relay coil or other electromagnet to the speaker output, and holding this transducer near the strings. A simpler method involves just holding a battery-powered mini guitar amp (such as those pictured in Chapter 1) above the strings and raising the gain to the point of feedback. The guitar manufacturer Fernandes makes a sustaining device based on similar feedback principles—they market a few different guitars with the device built in, but one can also buy "Sustainer Kits" for retrofitting any existing instrument. Any of these systems can be hacked to resonate the strings with other signals besides feedback (see audio track 9 and the description of my "Backwards Guitar" in Art & Music 5 "Drivers").

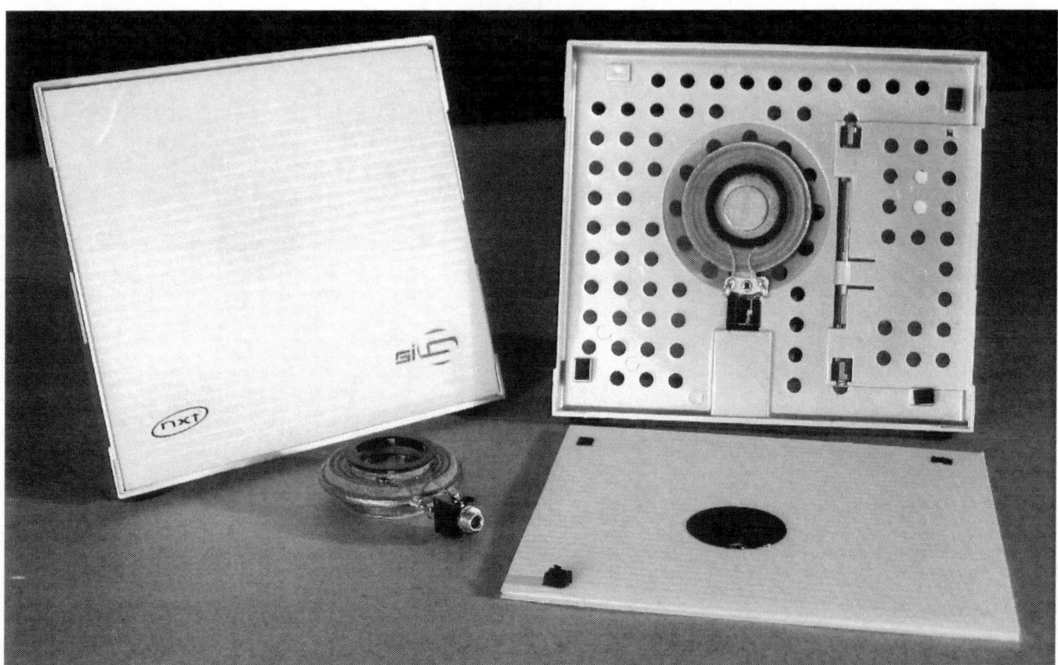

Figure 8.8 Sonic Impact SI-5 portable speaker and removed driver element.

DRIVERS

ART & MUSIC 5

David Tudor's "Rainforest" is one of the most conspicuous compositions to use transducers to resonate materials, but other sound artists have employed similar technology to different musical ends. In "Music For Solo Performer" (1965) Alvin Lucier places loudspeakers on the heads of drums and against cymbals, gongs, and other percussion instruments; the performer's amplified brainwaves are routed to the various speakers, and the low frequency (10–14 Hz) bursts play drum rolls on the percussion through the speaker cones. In later realizations Lucier attached solenoids to the outputs of some of the amplifiers, so they would tap against the bars of xylophones and gamelan instruments, and added corks to the speaker cones for similar effects. In Lucier's "The Queen of the South" (1972) sounds are played through a transducer affixed to a sheet of plywood (or other large plate) onto which fine powder has been sprinkled; as the pitch, timbre, and loudness of the sounds are changed, the powder forms different patterns according to the vibrational modes of the plate (Figure 8.9).

Berlin-based singer Ute Wassermann built her "Windy Gong" in 1995 (see Figure 8.10 and audio track 8). She sings into a mike, amplified through a small speaker with a cork attached to the center of the cone. The speaker is placed against the surface of a gong, which is in turn amplified with another microphone and contact mike. The vocal sounds are filtered and resonated by the gong, and the transformations can be manipulated by moving the speaker and microphone. In "Kupferscheibe" (1993–97) she extends long springs from speakers built into her clothing out to resonators encircling the performance area; her voice is processed by these spindly springs and heard, tin-can-telephone-style, through the megaphones.

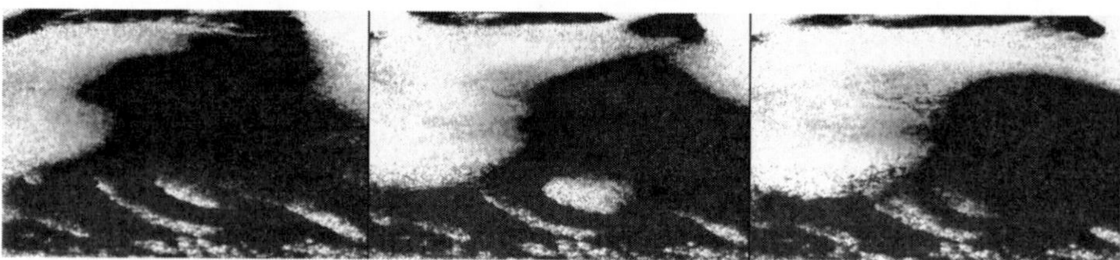

Figure 8.9 Three stills from "The Queen of the South," Alvin Lucier, performed by the SAIC Sinfonietta.

Figure 8.10 "Windy Gong," Ute Wassermann.

For "Tisch" (1994–95) German artist Jens Brand installed speakers inside a circular plastic café table. The surface of the table is oiled, and eight empty wine glasses are placed on top. Very low frequency sound is played through the speakers. Although barely audible, the vibration from the speakers causes the glasses to move very slowly across the surface. When they make contact, they ring against one another. The piece ends when the last of the glasses tips over the edge and crashes onto the floor. For "Mini-Fan Music" (1992), Brand and his collaborator Waldo Riedl placed handheld fans next to a dozen string instruments strewn around the performance space; the fan blades strum the strings until the batteries run down (typically 3–4 hours with cheap batteries), the droning sound field slowly changing as the fans slip along the floor and lose speed.

I have been working with "backwards electric guitars" since 1981 (see Figure 8.11 and audio track 9). In these instruments sounds are sent *into* guitar pickups or coils (scavenged from relays) whose fluctuating electromagnetic fields vibrate the strings of guitars and basses. The strings filter, resonate, and reverberate the original sounds (similar to the effect of shouting into a piano with the sustain pedal depressed), and are picked up, amplified, and further processed through distortion and other typical guitar effects. The filtering is "played" by fretting and dampening the strings, like a one-handed guitarist. This technique is an extension of the more familiar EBow technology, which uses an electromagnetic pickup and driver in a feedback loop to resonate guitar strings. Dan Wilson has used similar electromagnetic drivers to resonate everything from pitchforks to fences (see his video on the DVD and Chapter 30).

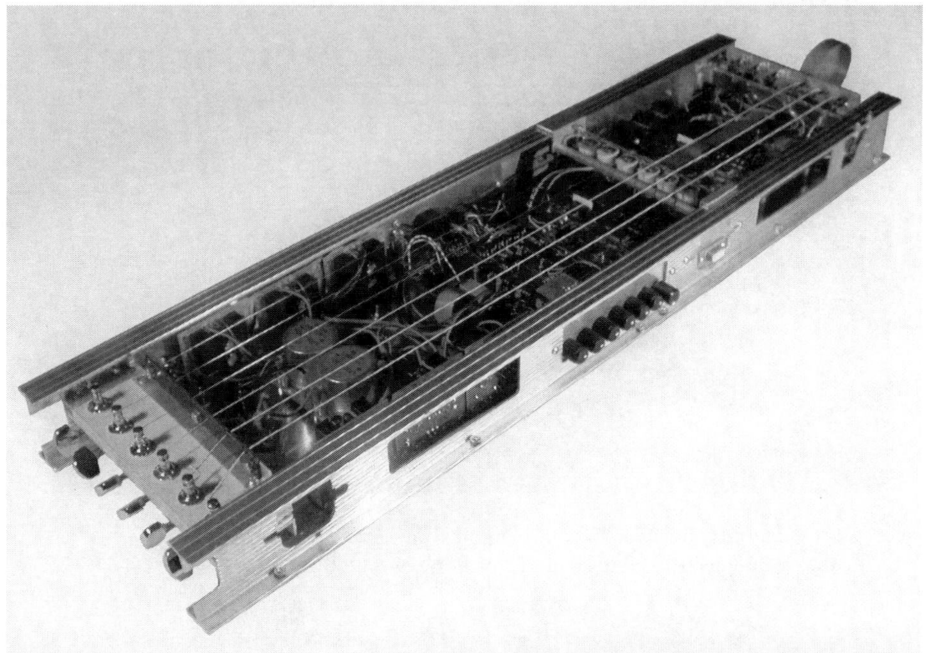

Figure 8.11 Backwards electric Hawaiian guitar, Nicolas Collins.

Dutch sound artist Felix Hess has created beautiful large-scale installations with tiny circuits that drive piezo disks directly (see Figure 8.12). Pressed against small sheets of balsa wood by the weight of a stone, the piezos produce a cicada-like chirping of astonishing intensity.

Recently Chicago-based artist Jesse Seay updated some of the fundamentals of "Rainforest" in her installation "Untitled (Resonant Objects)" (2006): a curtain of small, seemingly silent transducers hangs from the ceiling above a line of hollow objects (bottles, plastic yogurt containers, etc.); visitors are free to press the containers against the transducers, at which point her soundtrack miraculously radiates from the objects (see Figure 8.13 and her video on the DVD). Similarly, in his installation "Listening to the Reflection of Points" Japanese artist Toshiya Tsunoda arranges a series of objects on top of small speakers through which white noise is played; each object resonated and radiated the sound differently. The New Zealand duo of Chris Black and Christine White create performances in which they filter sounds through materials ranging from guitar strings to baking pans, using only small loudspeakers and contact mikes (see Figure 8.14 and their video on the DVD). Canadian Frederic Brummer's "Speaker Drum" is just that (see his video on the DVD).

Figure 8.12 "Luftsdruckschwankungen" circuit from installation of "It's In The Air," Felix Hess, for the Stroomgeest exhibition at Groot Bentveld, Netherlands, 1996.

Figure 8.13 "Untitled (Resonant Objects)" (2006), Jesse Seay.

Figure 8.14
Detail of performance of "Crude Awakening" by Chris Black and Christine White.

Finally, the truly adventurous among you might try building your own loudspeakers with magnets, coils of wire, and plastic cups—just follow the lead of Daniel Rodak (see www.hope.edu/academic/physics/REU/2000/students/DR_use.html).

As the "Drivers" box shows, many artists have made creative use of the kind of "physical filtering" that can be accomplished by sending sounds through objects via all sorts of drivers.

CHAPTER 9

Tape Heads: Playing Your Credit Cards

You will need:

- An expendable tape recorder/player, or a loose tape head.
- Some magnetic media: cassettes, reel-to-reel tape, transit cards, credit cards, etc.
- A battery powered mini-amplifier with considerable gain.
- An additional sound source, such as a CD or tape player.
- Optional: a surplus credit card reader.

Even in the age of iPods and CDs there's a lot of data sitting around in magnetic particles: music and phone messages on cassette tapes, personal data on your credit card, virtual money on transit cards. Whereas many of us can still remember what a cassette sounded like, it's not often we get to hear the information on other magnetic storage media. But all it takes is a tape head and an amplifier.

A tape recorder translates audio signals into a fluctuating electromagnetic field through a *tape head*, the small metal Brancusi-esque object you can see inside a cassette player or answering machine (see Figure 9.1). The tape head's undulating magnetism in turn aligns tiny magnetic domains in the iron-like powder covering one surface of the recording tape, as if they were midget compass needles. When the tape is played back the process reverses: the varying magnetic orientation retained by the mini-magnets on the tape now induces current flow inside the tape head which, when amplified, resembles pretty closely what went into the tape recorder earlier—another instance of the *reversibility* of electromagnetism discussed in Chapter 4. It's not so different from translating sound vibrations into grooves cut into a record's surface, to be traced later by a needle whose wiggling is re-translated back into sound waves—only with tape it's magnetic fluctuations instead of shimmying grooves. Digital tape recordings—such as Mini-DV tapes from camcorders, hard drives or credit card stripes—are like cassette tape only simpler: the magnetic domains just flop back and forth between two states, 0 and 1, instead of tracing the nuanced contour of an analog waveform.

PREPARATION

The easiest place to find a tape head is inside a broken or otherwise unwanted answering machine or cassette player. (If you have a functional boom box or other device that

Figure 9.1 Tape heads.

records as well as plays back tape, skip ahead to the "Recording" section below; if you have a working Walkman or other cassette playback-only device, skip to "Playback.") Many Web-based electronic surplus stores sell individual tape heads or credit card data readers at reasonable prices. The advantage of Aztecking a tape head (ripping it out of a still warm electronic body) is that audio wiring is attached, often in the form of a short shielded cable; in this case just cut the cable so as to leave as long a section as possible attached to the tape head. Other times the head will be connected to a circuit board through a translucent tape-like band, which we will discard. The answering machine will yield a simple mono tape head, while the Walkman will probably be stereo, but for the purposes of this experiment stereo is not very important, and you can get away with wiring up just one channel if you're feeling lazy.

The back-side of a stereo tape head will have four connections (as shown in the examples in Figure 9.1); a mono head will have two. If the head has a cable attached, each pin will probably be attached to a separate fine wire in a multi-conductor shielded cable, and the shield will be affixed to the metal shell of the head. Trace the free ends of the wires back to the pins on the head (often the individual wires will be color-coded, simplifying your job.) When wiring this cable from a *stereo* head to a *stereo* plug, solder "A" to the tip of your plug, "C" to the ring, and "B" and "D" and the shield to the sleeve; when wiring a *stereo* head to a *mono* plug, solder "A" and "C" to the tip of your plug and "B" and "D" and the shield to the sleeve (you can refer back to Figure 7.12 in Chapter 7 for a picture of this configuration). When wiring a *mono* head to a *mono* plug, connect "A" to the tip, "B" and shield to the sleeve (see Figure 9.2).

Figure 9.2 Wiring orientation for tape heads.

If the tape head arrives unwired, solder directly from the connecting pins on the head, to the sleeve and tip of the jack, following the above routing instructions. To minimize hum always use shielded cable, and solder an additional connection between the shield and the metal shell of the head. But bear in mind that tape heads are very hummy things by nature, and some noise is inevitable (in fact, you can substitute a tape head for a telephone tap coil to pick up electromagnetic fields, as described in Chapter 3).

PLAYBACK

If your tape head is inside a functional tape player, remove it from the player *without* disconnecting the cable if possible. You will probably need to extend its wiring with a foot or so of shielded cable—enough that you have room to move the head freely. Either cut the existing wires in half and splice in some additional cable, or de-solder the head's wiring at the circuit board (make careful note of which wire goes where!) and solder the extension cable between the board and the pigtails attached to the head. In either case make sure you re-connect the matching ends.

If the player has no built-in speaker, plug in headphones, or patch it into an amplifier; press the "play" button. If you are working with a loose tape head, plug it into a high gain amplifier, such as the Radio Shack mini test amplifier, a guitar amp, or the microphone input of a mixer.

Now rub the head over some recorded media: transit cards and credit cards, eviscerated cassette tapes, computer disks. If you're using cassette tape, it helps to stretch the audiotape across a sheet of cardboard, a tabletop, or some other flat surface and fasten it down with the sticky kind of tape at either end (or apply double-stick tape or spray adhesive to the back side). You will notice that one side of the tape (the emulsion side) will be MUCH louder than the other (backing). Digital data (credit cards, transit cards) tends to make a much louder sound than audiotape, and one that often sounds curiously like turntable scratching (see "Card Readers," on p. 63).

Sometimes the hum will increase when you touch the head—you can minimize this by wrapping it in plastic electrical tape. If you find it awkward to handle the tiny tape head you can lash it to the end of a popsicle stick or pencil with some electrical or gaffing tape, or solder it to a metal fingerpick (see Figure 9.5).

TAPE

Although invented for straightforward recording and playback of speeches in the service of the Third Reich, and largely known today in its more benign role as a trustworthy musical amanuensis, magnetic tape has proven to be a wonderfully flexible *performance* medium in itself. Composers such as Alvin Lucier ("I Am Sitting in a Room," 1970), Steve Reich ("Come Out," 1966), Pauline Oliveros ("1 of 4," 1966) and Terry Riley ("Rainbow in Curved Air," 1969) have all made pieces derived from the properties of tape loops and tape delays. When the tape is taken off the reels it becomes surprisingly instrumental. In 1963 Nam June Paik, on the threshold of his transformation from composer to video artist, attached dozens of strips of prerecorded tape to the wall of a gallery in Wuppertal, Germany, and invited the visitors to play it back via handheld tape heads. According to legend, John Cage once did a similar thing in reverse. He fully covered a tabletop with blank tape, invited the public to scribble across it with tape heads attached to pencils through which electronic sound was playing; at the end of the evening the tape was wound onto a reel and played back for all to hear.

Laurie Anderson's "Tape Bow Violin" (built in 1977 in collaboration with Bob Bielecki—see Art & Music 11 "The Luthiers," Chapter 28) substitutes a tape head for the bridge, and a strip of tape for the hair of the bow; the tape contains a recording that Anderson plays backwards and forwards as she draws the bow across the head (see Figure 9.3). "I began to work with audio palindromes, words that produced different words when reversed. Audio palindromes are not predictable like written palindromes ('god' is always 'dog' spelled backwards). With a lot of experimentation I produced songs for 'The Tape Bow Violin' that could be played forwards and backwards."

Figure 9.3 "Tape Bow Violin," Laurie Anderson.

Years later, César Eugenio Dávila-Irizarry glued recordings of percussion instruments onto the body of the gourd typically used to make a *güiro* (a percussion instrument consisting of a gourd scribed with notches that are scraped rhythmically with a comb-like *raspa*); his new instrument is played with a hand-held tape head (see Figure 9.4). In the installation version of Mark Trayle's *"¢apital magnetics"* (1999) visitors insert their credit cards into what appears to be an ordinary ATM; Trayle has programmed an internal computer to generate short musical compositions based on the data on each card, heard through speakers embedded in the ATM.

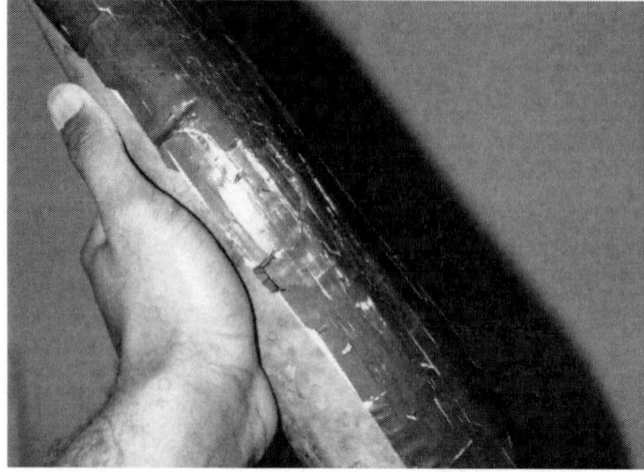

Figure 9.4
"Tape Guiro,"
César Eugenio Dávila-Irizarry.

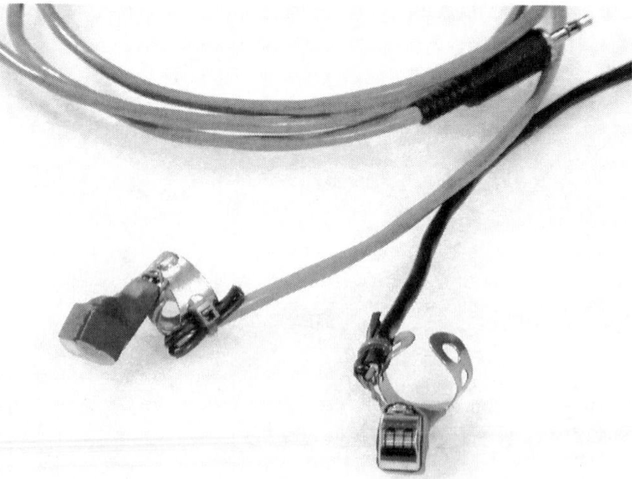

Figure 9.5
Tape heads mounted on fingerpicks, Nicolas Collins.

RECORDING

You can try *recording* with a hand-held tape head as well. Stretch cassette or reel-to-reel tape over a tabletop as above, making sure the emulsion side is up. If you are working with a loose tape head, plug an MP3 or CD player into the *input* of a mini-amplifier; plug the tape head into its external speaker *output*. While playing the sound source, move the tape head over the tape surface—keep the head in close contact with the tape. After a while stop recording and try playing back the tape—either by amplifying the head while moving it by hand across the surface, or you can reload the tape into a cassette or onto a reel and play it back on a tape recorder. Sometimes this works and sometimes it doesn't, so don't be too disappointed if you are unsuccessful.

(The student ID cards at The School of the Art Institute, where I teach, also serve as debit cards for use in on-campus vending machines. One day I came back to my Hacking class after a break to discover two of my students trying to "copy" money from one card to another: they carefully moved two tape heads along the magnetic stripes on the two cards, one connected to the input of a small amplifier, one to its output. Sadly, the experiment was not successful.)

You can get much better results if you start with a functional boom box or cassette recorder that you're willing to sacrifice on the altar of the weird. Disassemble the recorder to the point that you can carefully remove the record head from its mount in the cassette well. You will probably need to extend its wiring, as we described earlier in the "Playback" section. Mount some scrap cassette or reel-to-reel tape to a flat surface as we did above. Connect a signal to the boom box inputs, or tune in its radio. Insert a blank cassette in the well, or press that little prong at the back of the cassette well with your pinkie, in order to enable the record function, and press the "record" and "play" keys to start recording. Move the head smoothly across the scrap tape. Press "stop," then "play," turn up the boom box volume, and retrace your movements over the tape—you should hear the original signal, altered by the inconsistencies in speed and smoothness between your two passes. You can speed up, slow down, and reverse your original sounds by changing the speed and direction of your playback motion. Or you can reload the tape into a cassette shell or onto a reel.

Although a bit awkward, this method of recording and playing back tape through a kind of linear "scratching" offers a cheap, highly performable alternative to the DJ's practice of cutting dub plates for custom turntable work.

CARD READERS

Surplus outlets often sell "card readers" from ATM machines, public telephones, etc. The reader consists of a tape head inside a housing that guides the card smoothly past it, along with circuitry needed to decode the digital data (see Figure 9.6). Stealing credit card data is *advanced* hacking (see Art & Music 6 on "Tape") but for our immediate purposes you can discard the digital circuitry, wire the head up as shown in Figure 9.2, plug into an amp, and end up with a very nice instrument for "scratching" cards.

Figure 9.6
The "Scratchmaster" card reader (by Nicolas Collins, from the collection of Ted Collins).

CHAPTER 10

A Simple Air Mike: Cheap Condenser Mike Elements Make Great Microphones

You will need:

- An electret microphone element.
- 8 feet of lightweight shielded cable.
- An amplifier, audio recorder, camcorder, or mixer.
- A plug to match the jack on your amp, recorder, or mixer.
- 9-volt battery and battery hook-up clip.
- A 2.2 kOhm resistor.
- A capacitor in the range of 0.1uf.
- Packaging supplies as needed: heat shrink tubing, soda straw, Altoid tin, along with a few additional connectors.
- Hand tools, soldering iron, and electrical tape.

We've saved the most "normal" form of microphone for last: after coils, backwards speakers, contact mikes, and tape heads we finally get around to your basic hear-my-song mike. From any number of sources (Radio Shack, Web retailers, electronic surplus outlets) one can buy, quite cheaply, high quality electret condenser microphone elements (see Figure 10.1). You can also scrounge them from telephone answering machines, boomboxes, and even some toys. These are the basic building blocks of recording microphones that can sell for several hundred dollars. All that stands between your $2.00 purchase and a pretty good mike is a handful of cheap components, a soldering iron, and some ingenuity.

If you have any choice when you go to buy an element (and they are available from a number of online retailers), get a hold of some data sheets first and look for the models with highest signal-to-noise ratio and a flat, extended frequency response. If you have a choice between *cardioid* (directional) and *omnidirectional* pickup pattern bear in mind that omnidirectional microphones usually have a smoother response curve—the quality of sound often more than compensates for the lack of directionality (many purists in the field of classical music recording use omnidirectional microphones exclusively).

You may recall that when we transformed a speaker into a microphone in Chapter 4 all we had to do was connect its speaker terminals to the amplifier input—electromagnetism did the rest, miraculously converting acoustic sound waves into electronic waves. This is how a *dynamic* microphone, such as a typical PA mike, works.

Figure 10.1 Some electret condenser microphone elements.

Electret microphones, on the other hand, do not contain magnets and coils of wire; they operate based on even spookier *electrostatic* principles whose theory I'd prefer not to have to explain (check the sources in Notes and References before you nag me on this one.) The critical thing for you, the hacker, to understand at this point is that electret microphones require a small amount of externally applied voltage (typically from a battery) in order to function. Most electret elements use some form of what is known as "phantom powering" to send this voltage to the mike on the same wires that the mike uses to send its signal back to your mixer, recorder, amplifier, etc.

Figure 10.2 shows the connections for a typical 2-wire electret microphone. You will notice that the capsule has two pads to which wires can be soldered: one, marked "signal, + voltage" in the diagram, connects to both the positive terminal of the battery and your audio input (via the tip of a plug); the other, marked "ground," connects to both the negative terminal of the battery and the shield of the audio plug. The pads on the actual electret element are not labeled, but you can identify the ground terminal by the presence of a thin trace connecting it to the metal shell of the capsule.

The positive terminal of a 9-volt battery connects to the mike through a resistor—typically a value of 2.2 kOhm. A capacitor (usually around 0.1 uf) blocks this voltage from entering your amplifier or mixer. You can buy these two parts very cheaply at Radio Shack if you don't have any kicking around (we'll discuss more about how these parts work in later chapters—for now you can take them on faith). An optional switch turns the battery on and off—the microphone drains very little current, but the switch will help your battery last for years rather than months. It's best to turn the mike on or off when it is *not* connected to your recorder or amp to avoid big thunks. Solder on whatever plug matches your recorder, amplifier, or mixer—I've included in Figure 10.3 the rather odd connections needed if you want to use an XLR connector to plug into the microphone inputs on a mixer or a professional field recorder. (Please note: if you connect this XLR-equipped electret design to a mixer's microphone input make sure the mixer's internal phantom power is switched *off*; powering an electret capsule

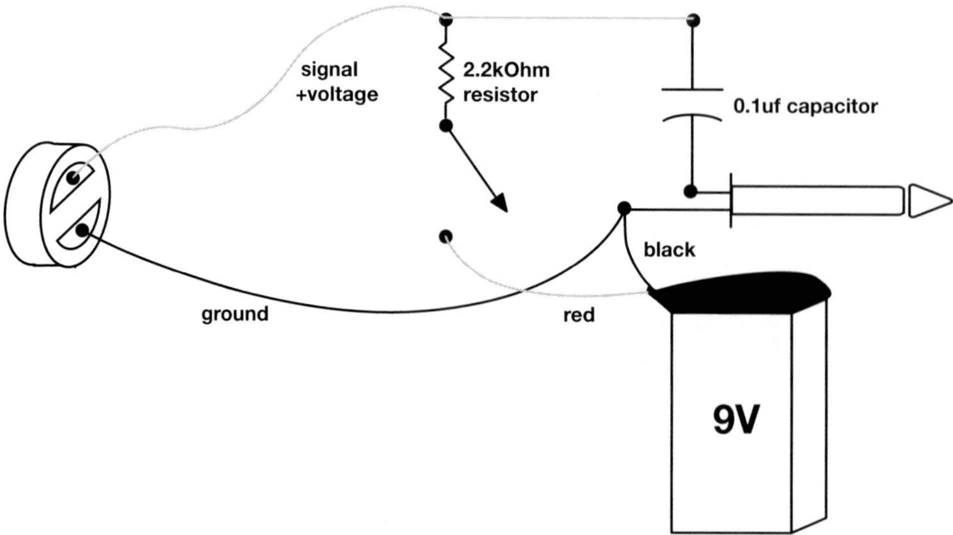

Figure 10.2 Basic electret microphone wiring.

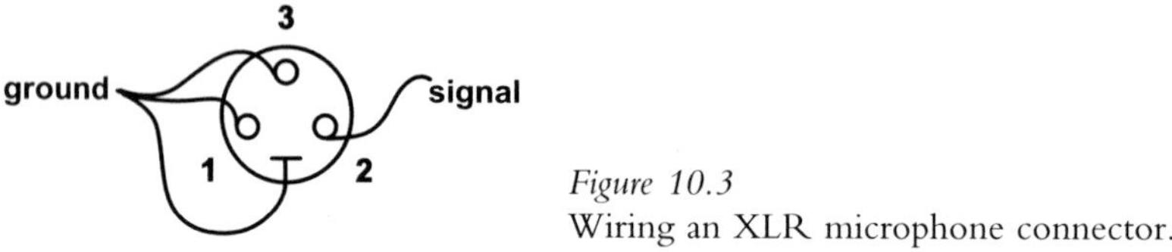

Figure 10.3
Wiring an XLR microphone connector.

off a mixer's phantom power—usually rather high voltage—requires a rather different, more complicated circuit.)

Electret microphone elements can usually be powered by a wide range of voltages. I've indicated a 9-volt battery in the figures, but two 1.5-volt batteries are usually sufficient—you can use AA, AAA, or even those tiny, absurdly overpriced button cells (see Chapter 17 for advice on battery substitution). Sometimes a single 1.5-volt battery will suffice. If a data sheet is available from the manufacturer, it will tell you the safe voltage range; otherwise you'll have to experiment.

Occasionally one finds electret elements with three wires instead of two. Instead of combining signal and battery power on a single wire, one wire will be designated as the signal, and will connect directly to the tip of the plug or pin 2 of the XLR; another will be labeled "power" and will connect to the "+" of the battery; the third (invariably the shield) connects to both the connector ground and the battery's "–" terminal (see Figure 10.4). This microphone element requires shielded cable with *two* internal conductors (plus the shield) rather than one.

The diagrams above show you how to make the electrical connections needed to get sound out of the mike, but leave you holding an unwieldy rat's nest of wire, batteries, and loose components. Mechanical packaging can be trickier than the basic electronic connections. Where do you put the various parts?

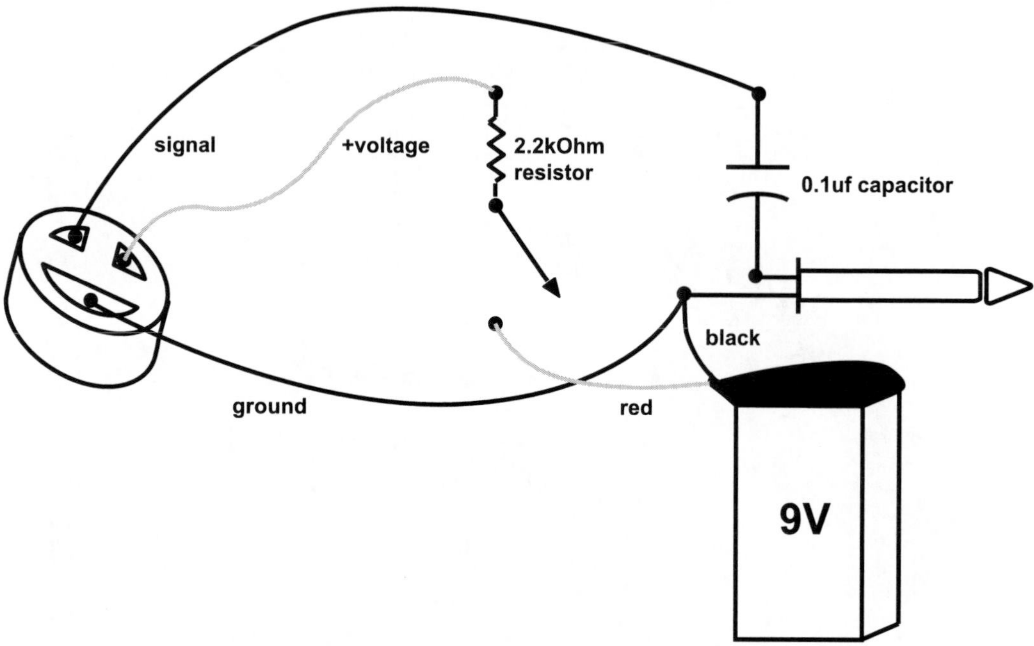

Figure 10.4 Wiring a three-wire electret microphone.

One solution is to run some shielded cable from the electret element to a remote power supply, in the form of a small box (another perfect application for an abandoned Altoid tin) containing the battery and related components; then run another cable from this box to whatever plug you are using to connect to the rest of your audio system. You can simply pass the wires through holes in the box, or you can use an additional set of plugs and jack (see Figure 10.5).

The electret capsule itself can be packaged in a number of ways:

- Glue it into the end of a plastic drinking straw, with the cable running through the straw and out the other end.
- Remove the hood covering the alligator clip in a busted test lead (I'm sure you have a pile of them by now) and slip it over the electret element (see Figure 10.6).
- Encase it in heat-shrink tubing (as shown in Figures 10.5 and 10.7): heat shrink comes in a range of diameters, and can be bought at electronic shops such as Radio Shack, as well as hardware stores and DIY shops; select a diameter a little bit wider than the capsule and cut a piece an inch or so long; slip it over the mike so that one end is flush with the front surface of the capsule; aim a blow dryer or heat gun at the tubing and it should shrink down to fit as snugly as the sweater on a Littleneck clam (if it shrinks down over the front of the mike you can trim it with a razor blade).
- Mount it in the same box as the power supply: drill a hole the diameter of the capsule; secure the capsule with epoxy or silicon caulking; build in the remaining components; run a cable out of the box to a plug. This is a good way to approximate a "Pressure Zone Microphone" (PZM).

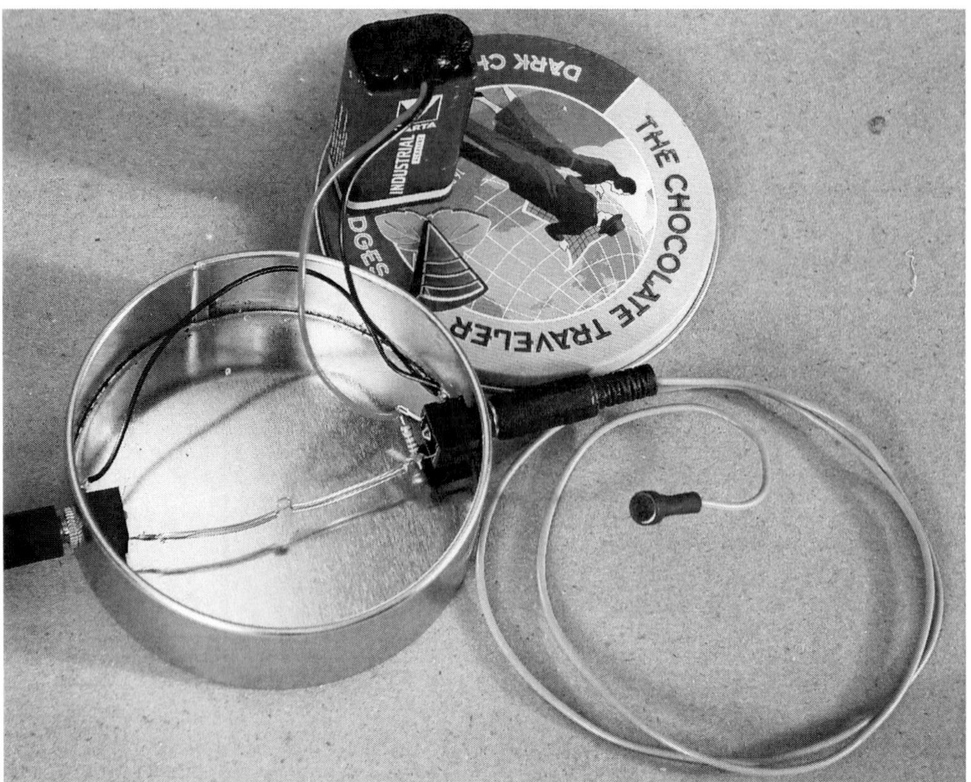

Figure 10.5 Remote power supply for electret mike.

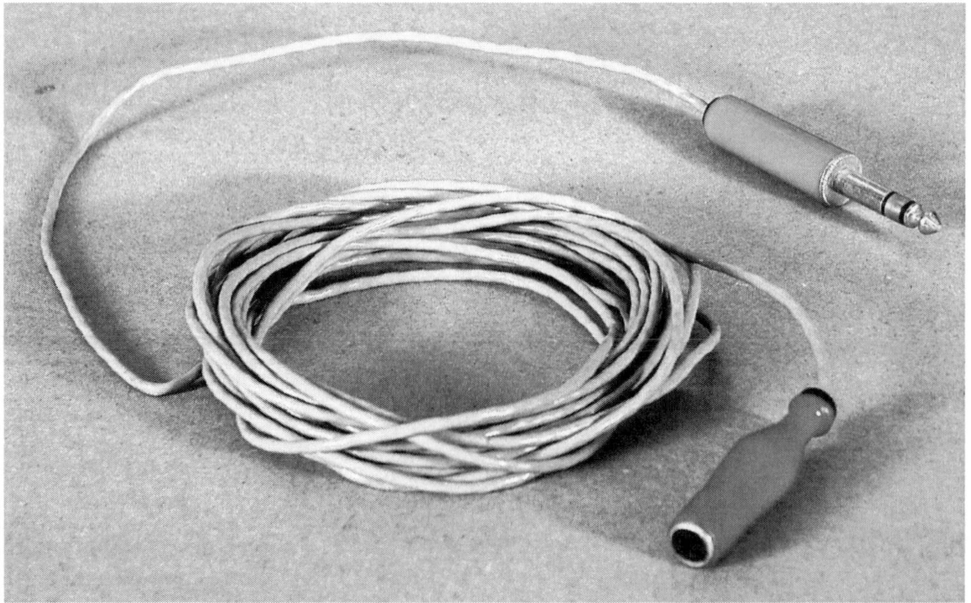

Figure 10.6 Electret capsule in clip lead hood.

- Use small PVC plumbing parts to make something that resembles a handheld mike, and build the power supply right inside.
- Use your imagination.

The issue of packaging is not merely aesthetic: the shape and mass of the material can affect the frequency response and directionality of the microphone. You should "ear-test" various options. The challenges of the mechanical design process may increase your tolerance of Neumann's mark-up, but it's worth it.

"PLUG-IN-POWER" AND STEREO MICROPHONES

As mentioned above, the 48-volt phantom power available on many audio mixers is too high for these little mikes to use directly, but a lot of semi-professional audio video equipment (such as mini-disk recorders, flash recorders, and camcorders) provide a form of 5-volt phantom power on the 1/8 inch stereo jacks used for connecting external stereo microphones. "plug-in-power," as it is known, can be used in lieu of the remote power supply we discussed above—Figure 10.7 shows two electret elements connected to a single 1/8 inch stereo plug to make an inexpensive stereo microphone for use with this low-voltage phantom power (refer to Figure 7.12 in Chapter 7 for a clear view of the wiring). The power supply shown in Figure 10.8 emulates plug-in-power if you want to use this mike design (or any commercially available stereo mike compatible with plug-in-power) with devices that do *not* provide the 5-volt phantom. Note that this power supply requires no on/off switch, since the battery drains only when the mikes are plugged in. "+ Voltage" is unspecified on the schematic: a 9-volt battery will work for most electret elements; you can use three AA or AAAs if you prefer to conform to the 5 volts "industry standard."

APPLICATIONS

"Binaural recording" is a miking technique optimized for playback over headphones: two small microphones are mounted in the ear openings of a specially-designed dummy

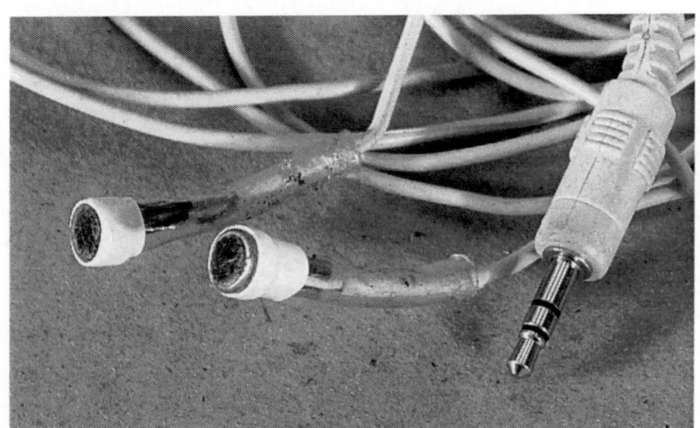

Figure 10.7
Two electret capsules wired as a stereo pair for use with plug-in-power.

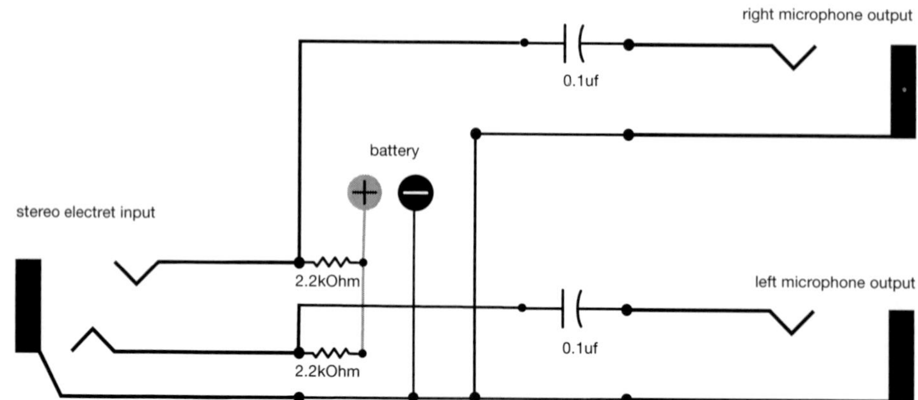

Figure 10.8 Power supply providing Plug-In-Power phantom voltage for any compatible microphone.

head, or worn like headphones against the outer ear of the recording engineer; when the recording is played back over headphones there is great spatial realism, because the headphones are in the same spatial relationship as the mikes were. Since more and more music is being heard through ear-buds in the age of the iPod, you should consider experimenting with this recording technique. You can glue two electret mikes a to a pair of ear-buds or headphones (but look out for feedback if you monitor through the headphones while recording). Alternatively if you can find the head from a mannequin you can try a home cochlear implant—drill holes where the ear canals would be and glue in two electrets (be prepared for odd looks when you carry the head into a concert hall or bird sanctuary . . .).

Mount a microphone at the focus of a parabolic reflector (a satellite dish or kid's snow-saucer work fine) for a hyper-directional mike for wildlife sound recording. These mikes are cheap enough to embed recklessly in musical instruments—they're great inside accordions and melodicas, and in mutes for trumpets and trombones; New Zealand musician David Watson dropped one inside his bagpipes (see his clip on the DVD). You don't even have to blow into the instrument: drop one into the mouth hole on a flute, aim the instrument at a speaker, wiggle your fingers on the keys, and let feedback do the rest. Glue electrets and small speakers at opposite ends of PVC pipes of different lengths, patch them through small amplifiers, and build a feedback organ (see the clip by Valve/Membrance on the DVD). Hang one in your shower, put a speaker in the sink, and turn your bathroom into a reverb chamber (shades of Phil Spector).

Even the least expensive of electret capsule has a surprisingly extended bass response—often reaching down to 20 Hz or lower. They are very susceptible to wind noise and breath pops; you'll want a windscreen of some kind (scrap foam rubber will do) and you may need to roll off the bass to avoid overloading your recorder input under certain circumstances. But their low frequency response also makes them excellent for picking up subsonic pressure waves, such as those produced by opening and closing doors, wind gusts and some barometric changes—roll off everything *above* 30 Hz and you can record the weather. Felix Hess has made recordings of such very low frequency barometric activity and sped them up into the audible range for our listening.

PART III

Touching

CHAPTER 11

Laying of Hands: Transforming a Portable Radio into a Synthesizer by Making Your Skin Part of the Circuit

You will need:

- A battery-powered radio.
- Batteries for the radio.
- Small screwdrivers, flat and/or Phillips, as required to disassemble the radio.
- Plastic electrical tape and some stranded hookup wire may be needed.
- Optional: cigar box, double-stick foam tape.

HOW TO CHOOSE A RADIO

It should be cheap enough that you won't be too angry if it never works again. The AM band is more important than the FM, but it doesn't matter if the radio picks up both. It should have *analog* tuning (i.e. a dial) rather than digital presets or scan buttons. Older radios are usually better than newer ones. Larger radios are easier to work with than tiny ones, and often produce a wider range of sounds. Boom boxes are great, and you can use the tape head for other experiments (see Chapter 9). It's better if it has a built-in speaker, not just a headphone jack, but a headphone jack *in addition to* a speaker can be useful. And most importantly: IT MUST BE BATTERY POWERED! Beware: an alarm clock radio with a built-in "backup battery" is *not* suitable, since it requires AC power to function as a radio.

LAYING OF HANDS

Install the batteries and confirm that the radio works *prior* to disassembly; if not functional, return it to the store. If it works, remove the batteries.

Remove the screws holding radio together. Put them somewhere safe (like a cup, *not* loose on top of your table), taking care to make a note of location if they are of different sizes. Some screws may be hidden beneath stickers or under the batteries. Gently separate the halves of the radio. If plastic wedge-fasteners are used you may need to twist a thin flat screwdriver or clam shucker along the seams. Don't force it—check for hidden screws if it resists. Avoid tearing wires. Once open, make note of any wires connecting the two halves of the radio or the circuitry to the speaker, battery, antenna, etc., in case they get torn later (see the Fourth Rule of Hacking).

Locate and remove the screws holding the circuit board to the radio housing. Carefully remove the circuit board from the chassis. Sometimes adhesive may be used as well as screws. Knobs and switches may intrude into slots in the case and require bending or cutting the plastic to release the board; alternatively, you can remove the knobs (they usually attach with small screws). Circuit boards, especially cheap ones, can be very brittle, so don't bend the board!

The side of the board with most of the little bumpy colorful things (resistors, capacitors, chips, etc.) is called the "component side"; the side that consists mostly of

Figure 11.1 Laying of hands.

wiggly lines (usually silver or copper colored, sometimes under a translucent green wash) is the "solder side." Turn the board so that the solder side is accessible. If it has a telescoping antenna this may need to be disconnected in order to expose the circuit board—you probably won't need to reconnect it. Remove the volume and tuning knobs if they are large enough to cover over parts of the circuit board, leaving short nubbins by which you should still be able to adjust settings. Replace the batteries; depending on the construction of the case you may have to hold the batteries in place using plastic electrical tape, or extend the battery leads from the discarded case with extra wire.

Turn on the radio and tune it to a "dead spot" between stations or at the end of the dial. Lick your fingertips, like a safecracker in a film noir. Touch the circuit board lightly in different places with your fingers until you find a location that affects the radio's sound (see Figure 11.1). Search for touch points that cause the radio to start to whistle, squeal, or "motorboat." Tune the radio across the band, and continue to experiment with finger placement. Try several fingers at once. The moisture on your fingertips increases conductivity, and makes your touch more sensitive, but I suggest you do not lick the circuit board directly. And, observing Seventh Rule of Hacking, avoid shorting points on the circuit board with screwdriver tips, bare wire, or full immersion in drool.

Don't worry if you don't get new sounds immediately—it's a bit like trying to make your first sound on a trumpet or flute, or learning to ride a bike. Sometimes you have to work a while before you find a sweet spot, but then you'll lock in and form a tight feedback relationship with the instrument, and the sounds should pour forth (see audio track 10). It can take up to an hour to make your first squeal. If you can't get anything after an hour, try another day or another radio.

WHAT'S HAPPENING?

As Ol' Sparky has demonstrated on far too many occasions, flesh is an excellent conductor of electricity. By bridging different locations on the board with your fingers you are effectively—if haphazardly—adding free-range resistors and capacitors to the existing circuit. Your body literally becomes part of the circuit. Varying the pressure (or dampness) of your finger changes the values of these components. Depending on the location and pressure, you may end up merely re-tuning the radio, or affecting its loudness, but you may change the radio into a very different kind of circuit, like an oscillator. This happens when the output of a gain stage (such as an amplifier) flows back through your skin into an input—voilá, feedback, the musician's friend!

A radio contains most of the basic modules of a classic analog music synthesizer: oscillators, noise generators, amplifiers, filters, ring modulators—in addition to the world's largest sample library, courtesy your local radio stations. Your skin re-tunes these modules, patches them together, and adds feedback paths. Moving your fingers and changing the volume, tuning, and band selections reconfigure this synthesizer to make different sounds—a whistling oscillator, modulated white noise, a signal processor chopping fragments of radio broadcasts, etc.

You may not know exactly what you are doing (like plugging patchcords and twiddling knobs in the dark), but you should soon acquire a sense of touch: what points work best, how does pressing harder affect the sound, etc. This is a very direct,

ART & MUSIC 7

THE CRACKLEBOX

Legend has it that the inspiration for the Cracklebox (*Kraakdoos*, as it is known in Dutch) springs from the out-of-body experience of an adolescent Michel Waisvisz (1949–2008) after he attempted to play his father's shortwave receiver by laying his hands on its 240 volt-powered circuit board. He recovered, the radio was nailed against the wall of their house in Delft to prevent further mishap, and Waisvisz's vision of an electronic instrument that could be played by intuitive touch was eventually safely realized in the late-1960s in collaboration with the engineer Geert Hamelberg. After building a number of different keyboard-sized instruments, in 1975 Waisvisz worked with engineers at STEIM, a music research foundation in Amsterdam, to design the very portable and affordable Cracklebox. Four thousand Crackleboxes were sold, and the original instrument was reissued a few years ago and can be bought online (www.steim.org/steim/cracklebox.php). Touch circuits had been employed in the expressive keyboard controllers of the maverick synthesizer designers Donald Buchla in 1965 and Serge Tcherepnin in the early 1970s, but the Cracklebox was the first mass-produced electronic musical instrument that incorporated the player's skin as the primary variable component in a sound-generating circuit.

Figure 11.2
The Cracklebox.

interactive sense of control similar to that which a "real" instrumentalist, such as a violinist, uses to articulate and intonate notes. This principle of direct contact with circuitry is relatively rare among commercial electronic instruments, but has often been exploited by experimentalists (and is the soul of the infamous Cracklebox).

In the upcoming chapters we will modify circuits by replacing your flesh with specific "knowable" components—the effect may be more predictable and stable, but the sense of touch will be diminished. In the future if things start to sound *too* controlled, remember you can always add your body to the circuit. And if your eviscerated radio becomes too predictable, try a friend's or open another—different radios respond differently. (Listen to the Bent Radio Orchestra in Duncan Chapman and Stewart Collinson's video on the DVD.)

Older-style radios sometimes have tuning coils whose colorful slotted tops are just asking for the twist of a screwdriver. Doing so may diminish or disable the radio's ability to pick up stations, but can add whooshy noise and rhythmic motorboating to your instrument's palette (see Michael Bullock's video on the DVD).

Generally speaking, the older the radio, the greater the range of sounds you can coax out of it. Modern radios cluster more functions onto each chip, minimizing the number of interconnections you can make with your skin. Older designs use more components on larger, finger-friendlier circuit boards; they often have bigger speakers, with more extended bass response as well. When scavenging flea markets, thrift shops and eBay, think 1960s (like the radio in the lower center of Figure 11.1), rather than twenty-first century. The additional weight and mass in your touring luggage will be more than offset by its sonic splendor.

When you are through experimenting you may want to reassemble the radio—this is the safest way to carry it around, and to insure its future functionality as a radio. But if you are so enamored of your electronic Ouija board that you cannot bear to seal it up again, welcome to the most hardware part of hacking: finding a box. Cigar boxes work great: using double-stick tape, you can stick down the circuit board (solder side up), speaker, and related parts (see Figure 11.3). Close the lid to transport, open it to play. Don't do this with metal boxes, as they may short out the circuit, but wood or plastic are fine.

Figure 11.3 "Laying of Hands" radio mounted inside a cigar box, Nicolas Collins.

A tickled radio often swoops over a very wide frequency range, but if the built-in speaker is small you might never hear the bass end. Try placing a telephone pickup or a guitar pickup on the speaker and plug it into an amplifier of some sort—a mini-amp, or a guitar amp, or a mixer and speakers. The coil will pick up lower frequencies than a small speaker itself will reproduce. Alternatively, drop an amplified contact mike onto the speaker: listen as it bounces around, adding a percussive edge to the radio's squeals—like the bottle caps around the calabash of an mbira.

If your radio has a headphone jack you can listen to it over headphones, or connect it to a *battery-powered* amplifier—if this amp has a larger speaker than the radio it should give you a louder, fuller range signal. Do *NOT* connect from the headphone jack into any amplifier, mixer or recorder that connects to AC (Mains) power at the wall.

EXTREMELY IMPORTANT NOTE: DON'T EVEN *THINK* ABOUT "LAYING HANDS" ON ANYTHING THAT PLUGS INTO THE WALL!! AND NEVER PLUG YOUR RADIO'S HEADPHONE JACK INTO AN AC-POWERED MIXER OR AMPLIFIER UNLESS YOUR ARE 101 PERCENT CERTAIN THAT THERE IS NO POSSIBILITY OF A GROUND FAULT (I.E. NEVER)!

Another important caveat: solder usually contains lead—not a pleasant element to have thumping about in your bloodstream. If you plan to lay your hands on circuits with some regularity, please substitute a fingerbowl for your tongue—your mother will be happier, on the grounds of both health and etiquette. This advice is doubly critical if you are working in a country whose microbes are hostile to your digestive system, or if you are playing radio four-hand duets with multiple partners.

VARIATIONS

The primary electronic function needed to transform the radio into an oscillator is *amplification*. Although the radio's filtering, frequency shifting, and chopped noise add considerable character, you can get many of the more oscillator-like effects by laying damp fingers upon a simple battery-powered amplifier circuit instead, such as those described in Chapter 1. When playing a Walkman or boombox this way one can control tape speed as well as circuit feedback (see Figure 11.4 and Seth Cluett's video on the DVD).

Very spooky things start to happen when you link two or more separate radio circuits with your fingers—spread a few open radios in front of you like a set of Tarot cards and try it (see Figure 11.5).

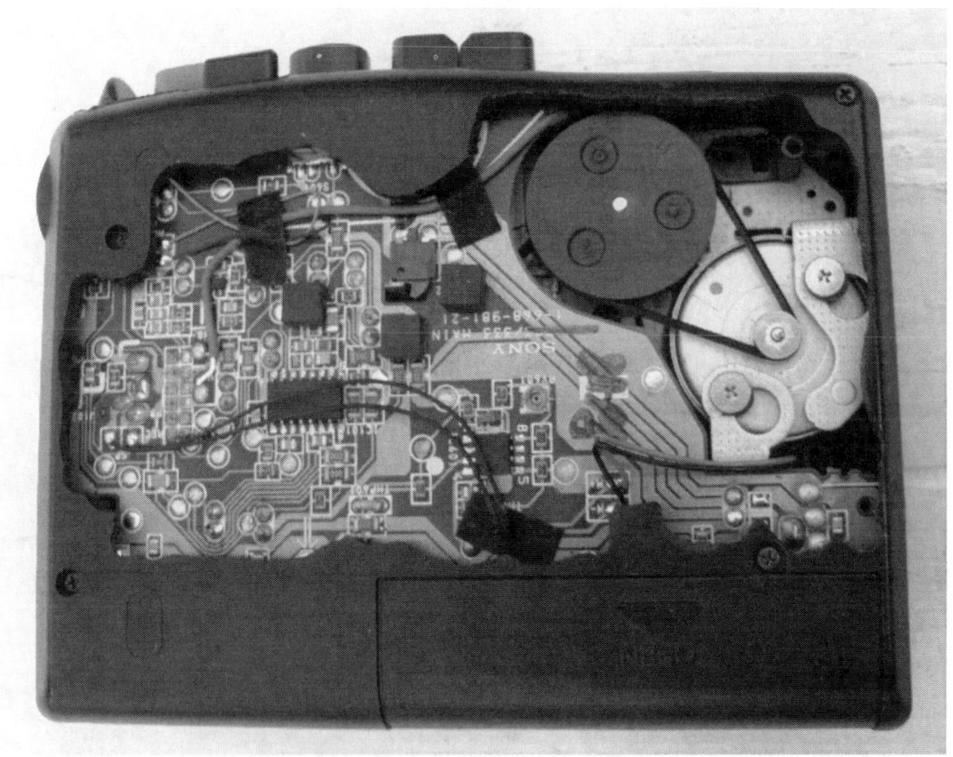

Figure 11.4 Open-back cassette player by Seth Cluett, played with hands on the circuit board.

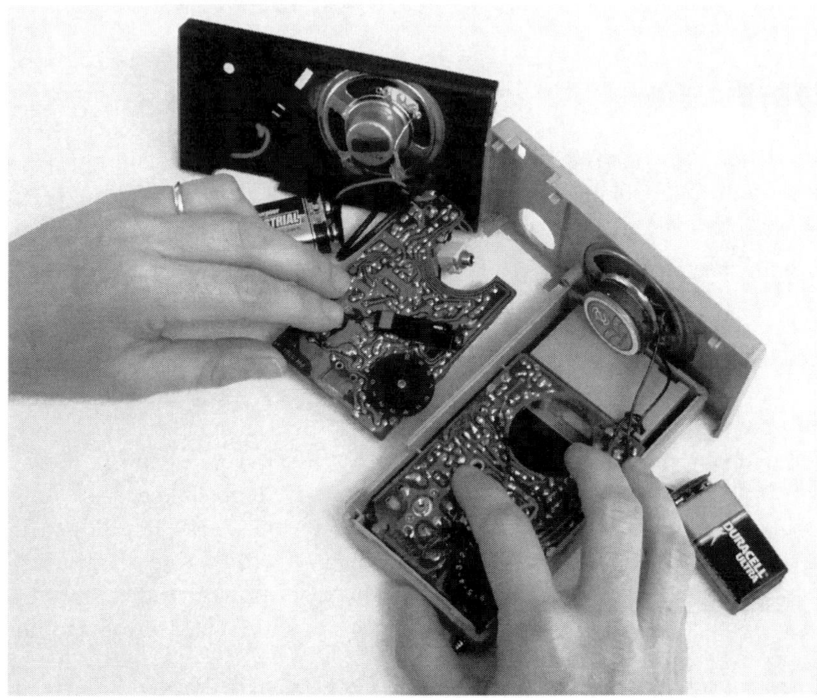

Figure 11.5 Susan Stenger does Rick Wakeman—spooky interaction between two radios.

CHAPTER 12

Tickle the Clock: Finding the Clock Circuit in Toys

You will need:

- An electronic toy.
- Small screwdrivers.
- A Sharpie-style fine-tip permanent marker.
- Optional: two test leads with alligator clips and a resistor in the range of 1 kOhm–4 kOhm.

Hacking is a lot like hot-rodding your car: you don't need to be able to build a car from scratch to swap in a four-barrel carburetor, but it helps to know what a carburetor looks like before you get too creative with the wrench. We'll use a simple but useful hack as a step toward identifying basic electronic components, and introduce some electronic axioms along the way.

HOW TO CHOOSE A TOY

As with the radio, select a toy that is expendable, not too tiny, and has a built-in speaker. A toy that makes sound is preferable to a mute one, and sampled sounds (like voices, animal sounds, or instruments) are more rewarding than simple beeps. The more buttons and switches the better, generally speaking. Keyboards are a gamble: some cheap Yamahas hack magnificently (the PSS-140 is especially satisfying), while others have curiously limited potential for interesting modification. Cheaper is usually better, especially for our first experiments—the more expensive toys, and almost all that put out video, often use crystal clocks, which are more difficult to hack (these include sophisticated handheld games such as Gameboys, robot toys such as Furbies, and more sophisticated music keyboards). However, lately there has also been an increase in the number of super-cheap toys coming from China that are really impossible to hack (this will be explained very soon), so beware. To the degree that one can distinguish such detail when trawling through thrift shops, toys from the 1990s hack more easily than more recent ones, and usually have a richer sound palette than most from the 1980s or earlier (with a few notable exceptions: the venerable "Speak and Spell," introduced in 1978, is worshipped by Circuit Benders).

And, of course: **THE TOY MUST BE BATTERY POWERED!**

CLOCKS

The majority of electronic toys manufactured since the early 1990s are essentially simple computers dedicated to running one program. In most, a crude clock circuit determines the pitch of the sounds and the speed of its blinking lights, graphics and/or program sequence. If you can locate the clock circuit and substitute one component, you can transform a monotonous bauble into an economical source of surprisingly malleable sound material.

WHAT'S UNDER THE HOOD?

Open up the toy, carefully noting wire connections in case one breaks. Keep the batteries connected, as we did with the radio in the previous chapter (and, similarly, this might require some ingenuity, clip leads and electrical tape). Expose both sides of the circuit board.

Study the board and try to identify the following types of components:

- Resistors: little cylinders encircled by colorful 1960s retro stripes (see Figure 12.1).
- Capacitors, in two basic forms (see Figure 12.2):
  - small disks of dull earth tones, or colorful squares;
  - cylinders, upright or on their side, fatter than resistors, with one stripe at most.
- Transistors: three wire legs supporting a small black plastic blob or metal can (see Figure 12.3).
- Diodes: glass or plastic cylinders, smaller and less colorful than resistors, usually marked with one stripe (see Figure 12.4).
- Integrated circuits (ICs): usually black or grey, resembling rectangular bugs with legs along one or more sides, or malignant looking black circular blobs oozing up from the circuit board (see Figure 12.5).

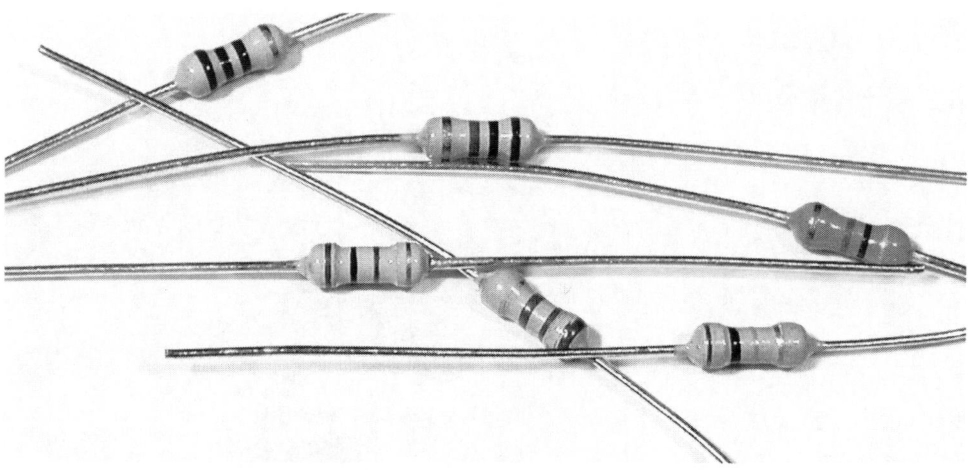

Figure 12.1 Some resistors.

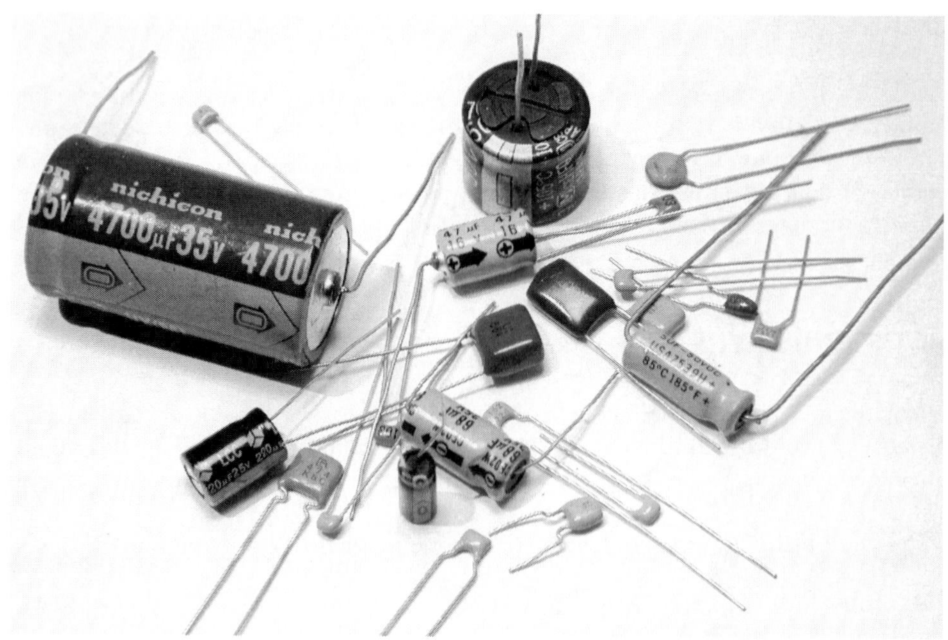

Figure 12.2
Some capacitors.

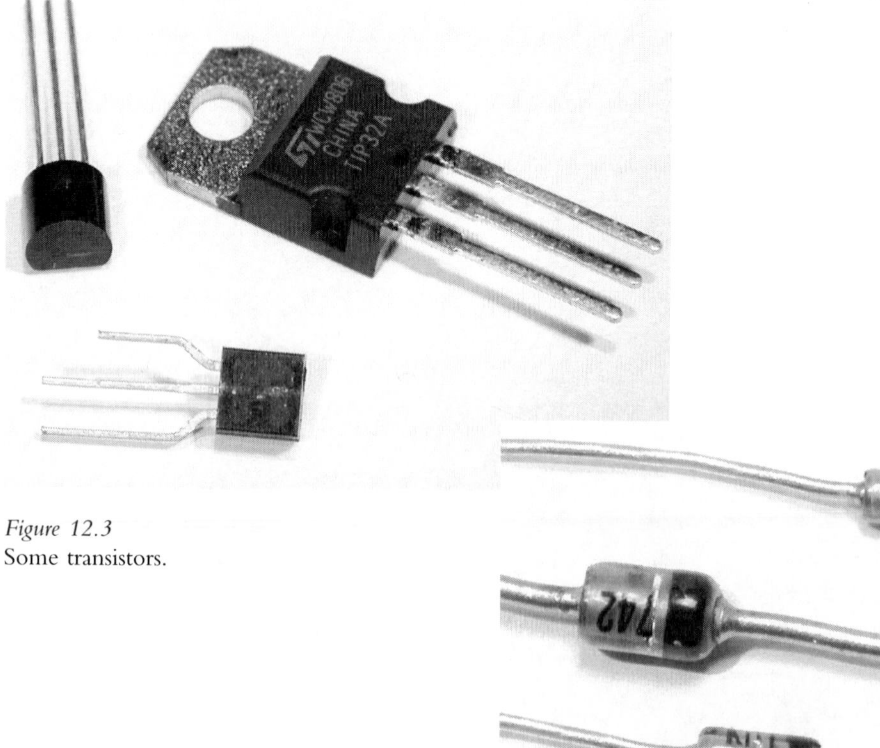

Figure 12.3
Some transistors.

Figure 12.4
Some diodes.

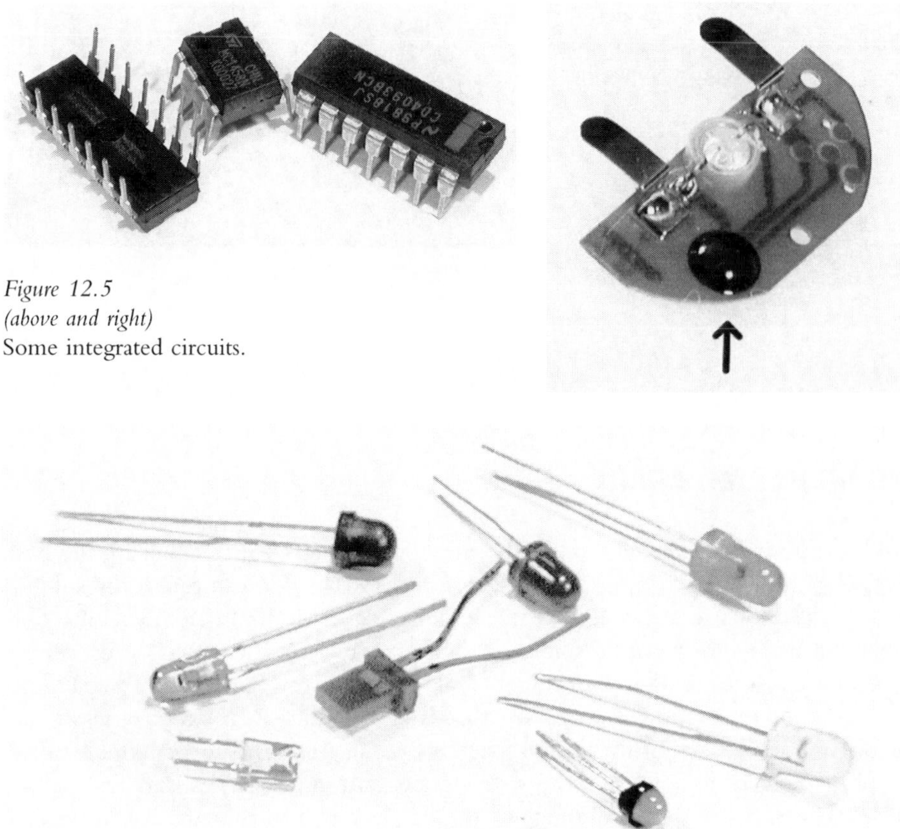

Figure 12.5
(above and right)
Some integrated circuits.

Figure 12.6 Some LEDs.

- LEDs (Light Emitting Diodes): colorful sources of light (see Figure 12.6).
- Other things you'll learn about later.

More and more toys are being made these days with "surface mount devices" (SMDs)—insanely tiny, rectangular versions of the above building blocks (see Figure 12.7). Until you gain some hands-on experience with them you can despair of distinguishing the various different types of components, and decoding and hacking these toys will be a doubly foggy experience. If you have a choice, start your experiments on a toy with the larger, more traditional and more easily identifiable components described above (another reason to scavenge an older, used toy in a thrift shop, rather than buy something new at the mall).

We're looking for *resistors*, especially those lying near an IC, flanked by a disk or square capacitor. The frequency of the simple, easily-hackable clocks in most toys is controlled by a resistor and a capacitor working together (explanation follows—be patient). If the toy was really cheap and its circuit board appears to have one shiny black blob and not a single resistor, you may be holding one of the unhackable new Chinese toys: the clock is hidden inside that blob, where you can't get at it, so you might as well hand the toy on to a grateful child and start over in a thrift shop somewhere.

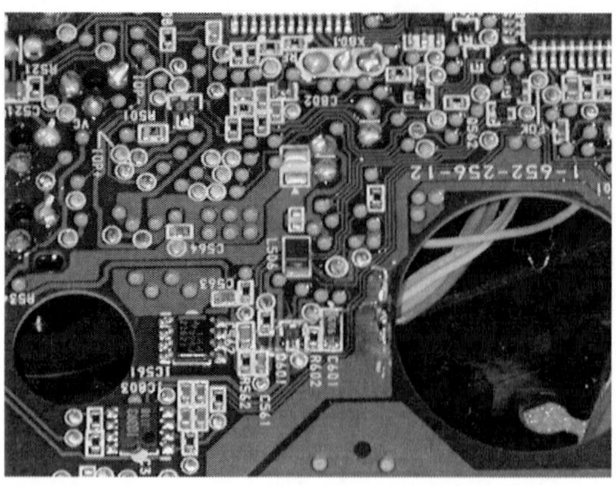

Figure 12.7
Circuit board with disturbingly small surface mount components.

LAYING OF HANDS, AGAIN

As with the radio experiment we did in the previous chapter, your fingers are the best tool here. Get the circuit making sound. Position it so that you can touch the solder-side of the circuit board, if possible while looking at the component side. Lick one fingertip and place it across various solder pads; in particular try to connect across points at either end of a resistor, so that your finger parallels the resistor's connection (see Figure 12.8). When your finger bridges a resistor that is part of the clock circuit you should hear the pitch slide up a bit, or the tempo speed up. If the circuit has lots of connections, and you are having trouble finding the spot, concentrate on those resistors lying close by small capacitors, usually near the biggest IC on the board. If the circuit is too small for your fingers, clip a test lead to each end of a resistor in the range of 1–4 kOhm, and touch the free ends of the leads to the ends of various resistors on the circuit board until you hear the pitch go up.

When you think you've found a hot spot, mark it on the circuit board with a Sharpie.

If the circuit incorporates the above-mentioned SMDs, most of the components *and* connections will be on the same side of the board, and it may be difficult to distinguish the capacitors from resistors. Sometimes resistors are identified by the letter "r" followed by a number, in a minuscule font, either on the component itself or (more

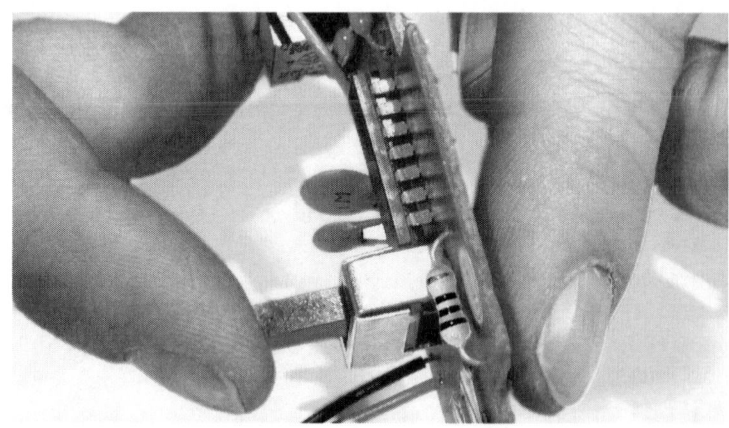

Figure 12.8
Finding the clock resistor (finger on right).

likely) immediately adjacent to it on the circuit board; likewise capacitors are sometimes labeled with a "c" plus a number. Go after the blips with *two* shiny solder blobs at either end, rather than three or more, and you're more likely to hit one of the timing components. Good luck—it can be very frustrating, and at a certain point you may want to give up, go out, and find yourself another (older) toy with bigger, more recognizable components.

If you have no success finding the spot, the toy might be one of the new, super-cheap Chinese models described above. Or it may use a crystal for the clock timing, rather than a simple circuit with a resistor and capacitor (as mentioned earlier, this is especially true for more advanced musical keyboards, toys with video output, and complex or expensive toys). Crystals often take the form of a small shiny metal cylinder or a three-legged epoxy-dipped blob, usually brightly colored. If you suspect a crystal is at work you'd best put the toy aside and try another. On the other hand, sometimes you'll get lucky and won't even need to lay a finger on the circuit board to find the clock resistor: some toys, such as the "Microjammer" guitars, include a pitch control knob or slider.

WHAT'S HAPPENING?

Electric current flows through wire like water through a fat pipe. Resistors are like skinny pipes, or the rust-laden risers of NYC loft buildings: the higher the resistance (measured in chantworthy Ohms), the less current flows. Capacitors also resist the flow of electric current, but resist it more at some frequencies than others, in a manner that defies a simple liquid analogy. Capacitance is measured in soukable Farads, usually in small enough amounts to be called "microfarads" or "picofarads." (Yes, the vocabulary of hardware is much cuter than that of software.)

Many oscillator designs rely on *feedback*: a speaker feeding back into a microphone is essentially an oscillator; when your damp finger bridges the right contacts on a radio circuit it produces feedback around an amplifier stage and sets the circuit oscillating. A clock circuit, whether in a cheap toy or an expensive computer, is just an oscillator, designed to run at a specific, carefully chosen frequency. In the simple clock circuit used in most toys a resistor and a capacitor are built into the feedback loop of a kind of amplifier. With enough gain this circuit starts to feed back, just like a mike and speaker, at a frequency determined by the values of the resistor and capacitor. Make the resistor or capacitor *smaller* and the frequency goes up; make either *larger* and the frequency goes down. When the frequency gets too high to hear, it enters the range of a useful clock rate for a computer or a digital toy (or a way to summon your dog).

When you place one resistor in parallel with another you *lower* the overall resistance (think of it as adding an additional pipe for the current to flow through.) Your skin is a resistor (as we demonstrated in the previous chapter)—when you press your finger across the circuit board contacts you effectively add a second pipe alongside the resistor on the other side of the board, decreasing the net resistance. More current flows and the pitch goes higher.

If the explanation is confusing, don't worry about it. For better or for worse, we'll revisit the secret lives of resistors and capacitors in a more clinical fashion when we get to building our own oscillators in Chapter 18. Until then just let your fingers do the thinking.

Got it? Good. But what do we do if we want to make the pitch *lower*? Read on . . . .

CHAPTER 13

Hack the Clock: Changing the Clock Speed for Cool New Noises

You will need:

- The electronic toy from the previous experiment.
- Some hookup wire, stranded and solid, 22–24 gauge.
- Test leads with alligator clips.
- A few resistors of different values.
- A potentiometer, 1 megOhm or greater in value.
- Soldering iron, solder, and hand tools.

Wet fingers are fine for making the clock go faster. But we all know that the Third Law of the Avant-Garde states:

Slow it down, a lot.

Lower and slower always sounds cooler. To slow it down we need to make the resistance *larger* instead of smaller. Which means removing the clock resistor from the circuit board (once you are sure which one it is) and replacing it with a larger one, instead of bridging it with your finger as we did in the last chapter, which, by lowering the resistance, can only make the pitch go up.

1. Locate the clock resistor you identified in the previous experiment. Wedge a small flat-bladed screwdriver under it. Melt the solder on the underside of the circuit board at one end of the resistor, and lever the screwdriver to lift that end free from the solder connection (see Figure 13.1). Do this quickly but gently, so as not to damage the circuit board or components. Now grip the resistor with a pair of pliers and pull it free of the board as you melt the other solder joint. Put it somewhere safe and don't lose it! If the circuit already has some kind of pitch-control potentiometer ("pot"), de-solder and remove it entirely.
2. Strip and tin the ends of two pieces of hookup wire (approximately 3–6 inches long). Press the end of one into one of the holes left after removing the resistor from the component side of the circuit board. It's easier to do this with *solid* hookup wire, but in the long term it's better if you use thin stranded wire for this, so you

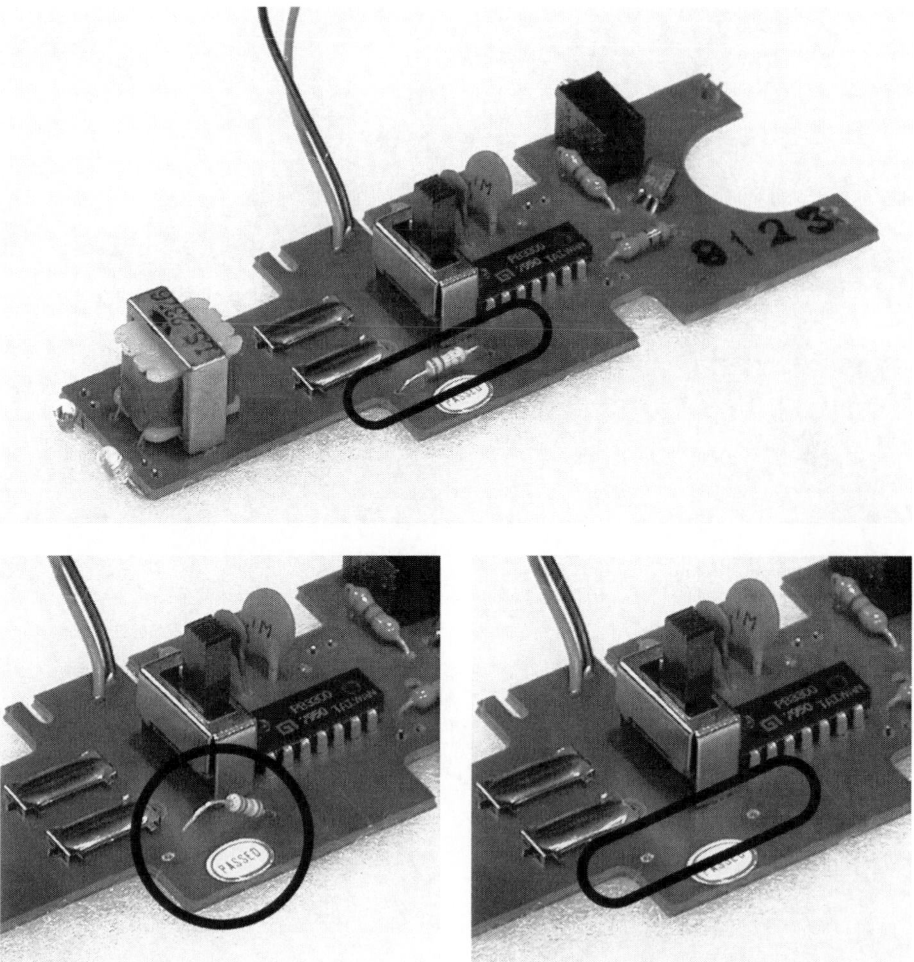

Figure 13.1 Removing a resistor.

might want to give it a try (twist the stranded wire before tinning and it will be more likely to pass neatly into the hole). Melt the solder on the solder side of the board as you press the end of the wire through the hole. Touch up the solder joint with a bit of fresh solder to make sure it is solid, but avoid excess solder "bridging" the gap between traces that are supposed to be separate. Repeat with the second wire into the other hole. Your circuit board should now have two colorful whiskers sprouting from among the other, vertically challenged components (see Figure 13.2). If you removed a pot from the board, there may be more than two holes; solder a wire to each of them. If you have problems fitting the wires through the holes you may have to clear out the old solder first—the best tool for this is the eerily cosmetological-looking "solder sucker" (see Figure 28.11 in Chapter 28), but an artfully wielded straight pin can work too.

The copper traces on printed circuit boards can be very delicate, and the twisting of the wires as you go through the following experiments can tear the trace,

Figure 13.2 Resistor whiskers.

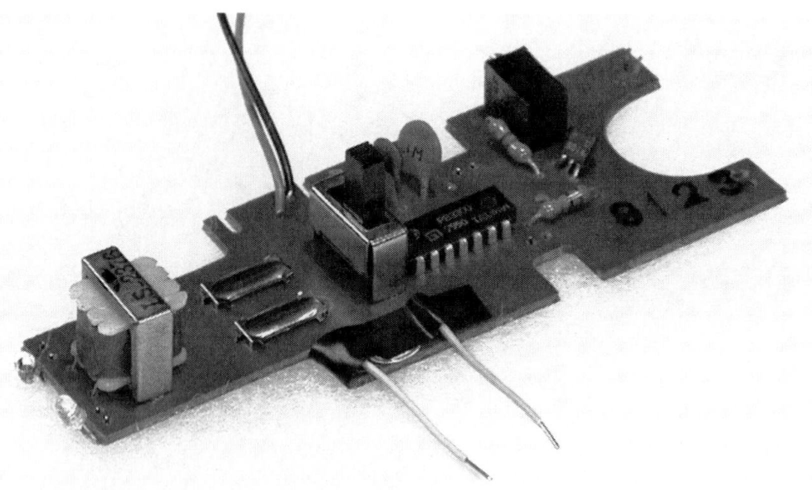

Figure 13.3 Whisker strain relief.

sometimes irreparably. Which is why the greater flexibility of stranded wire offsets the difficulty of forcing it though the holes. It is also a good idea to provide some kind of "strain relief" for your whiskers. The easiest method is to bend them gently so they lie flat against the board, and then tape them down with electrical tape to prevent them from moving at the point they pass through the board (see Figure 13.3).

3. Attach a clip lead to the free end of each of the wires. Clip the resistor you removed between the other ends of the clip leads, effectively re-inserting it to the circuit (see Figure 13.4). If you didn't damage anything in de-soldering, the circuit should behave as it did before the operation. If it doesn't, you may need to "restart" the toy by removing and reinstalling the batteries (see Rule #12, on p. 90).

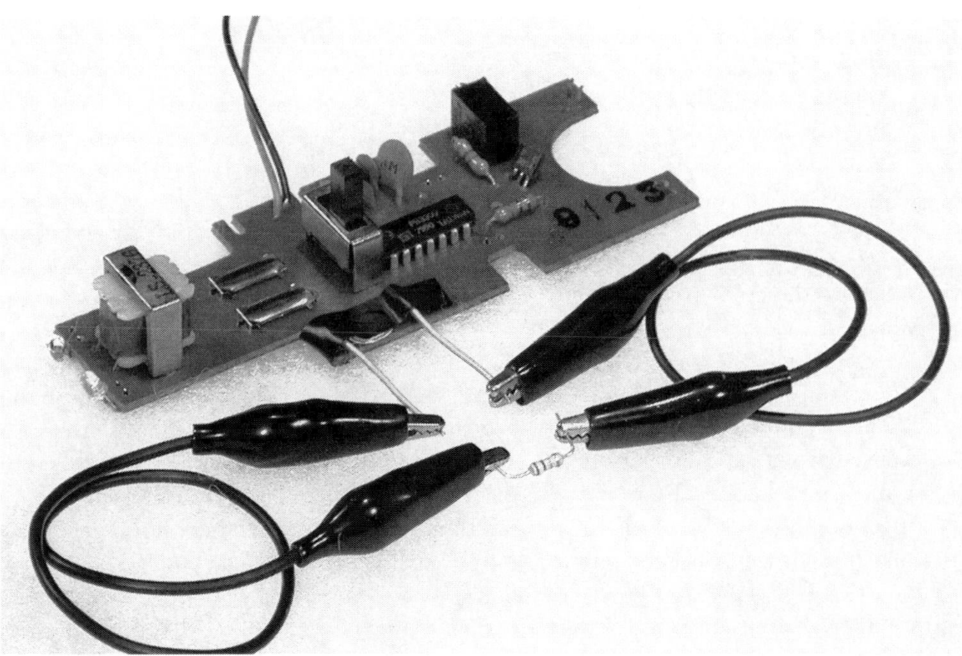

Figure 13.4 A remote resistor.

Table 13.1 Resistor color codes

Colour	*Value*	*Multiplier*
Black	0	1
Brown	1	10
Red	2	100
Orange	3	1,000
Yellow	4	10,000
Green	5	100,000
Blue	6	1,000,000
Violet	7	10,000,000
Gray	8	100,000,000
White	9	1,000,000,000

4. Those colorful stripes around the resistor indicate its value. Look at the decoder chart in Table 13.1 above.

   Study your resistor. The first two stripes represent number values directly, the third is a multiplier, and a final gold or silver band the tolerance. So if the bands go: brown, black, yellow, silver:

Brown = 1
Black = 0
Yellow = multiply by 10,000
Silver = ±10 percent tolerance
So we get: 10 × 10,000=100,000 Ohms (or 100kOhms) ±10 percent.

Another example: orange (3) orange (3) red (× 100) gold = 3,300 ±5 percent. Get it?

5. What are the color bands of the resistor you removed? __________
   What is its value? __________

6. Go to your resistor assortment and find a resistor at least twice as big, and one about 1/2 the value. Clip the larger one into the circuit and the pitch should go *down*. Replace it with the smaller one and the pitch should go *up*. If either one does not work it may be so extreme a value that the circuit shuts down, so replace it with one whose value is somewhere between the original resistor and the non-functional one. In the event of such a crash, observe the 12th Rule of Hacking:

Rule #12: After a hacked circuit crashes you may need to disconnect and reconnect the batteries before it will run again.
(Count to five before replacing them.)

Substituting resistors should give you a good idea of what values produce what kind of sound, but you will probably want to vary the pitch/speed more fluidly. A potentiometer is a continuously variable resistor. In order to extend the pitch downward you need a pot whose maximum value is *greater* than the resistor you removed. Since most clock circuits use rather large resistors (100 kOhm or larger) you will probably need a pot whose maximum value is 1 megOhm (1,000,000 Ohms) or greater.

Pots have three terminals—two "ears" and one "nose"—which are labeled A, B, and C in Figure 13.5. The resistance between the outer two ears (A and C) is fixed at the designated value of the pot, which is the pot's absolute maximum resistance (i.e. 1 megOhm). As you rotate the shaft of the pot clockwise the resistance between the center terminal (nose) B and the outer terminal A goes *up* from 0 Ohms to the maximum value, while the resistance between B and the other outer terminal C goes *down* from the maximum to 0—the two values change in contrary motion, like the ends of a seesaw. Reversing the pot's rotation tips the seesaw back the other way.

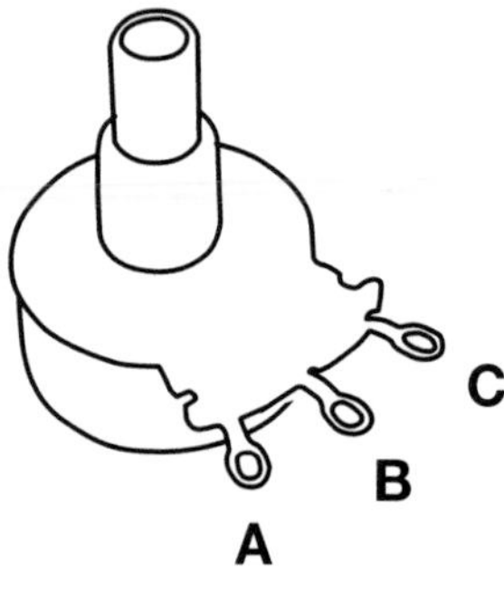

Figure 13.5
The three terminals of a potentiometer

Remove the resistor from the clip leads attached to the whiskers on your circuit board. Clip the free end of one of the leads to the center terminal (B) of your pot, and clip the other to the end terminal C (see Figure 13.6). Rotate the pot and listen. The circuit will probably crash if you raise the pitch past a certain point and you'll have to restart it (see Rule #12, above). But as long as you stay below that point in the pot's rotation you should be able to coax a pretty wide range of sounds out of your circuit. If the toy appears to shut down when the clock is at its slowest, but restarts on its own when the pitch is raised again, the problem may be simply that the sound is going too low to be heard on the built-in speaker; try putting a telephone pickup on the speaker and amplify it through a bigger speaker (as suggested for the radio in Chapter 11). Of course you can always use the center terminal B and the other ear, A, but in this case the response of the pot will be backwards: the pitch will go *down* as you turn it further clockwise, rather than up. Which is fine, except slightly counterintuitive if you have worked with most commercial electronic music devices—there's no time like now to start a revolution (or at least change its direction).

If there is no appreciable change in pitch or tempo you may have picked the wrong resistor as the clock timing component. Solder the part back into the board and start over at step 1 probing for the hot spot with your wet finger. Other common causes of failure include torn traces (as mentioned in step 2 above), or sloppy soldering accidentally joining points on the circuit board that should remain separated.

If you removed a pitch-varying pot (instead of a fixed resistor) from the circuit you will have to experiment with connecting the terminals of your new pot to various combinations of leads from the circuit board before you find the correct hookup—you can start by matching up the nose and ears of your replacement pot with those of the original part in the toy. Substituting a pot of larger value than the built-in one should give you a wider range of pitch/speed variation.

In case you were wondering, yes, you could also change the clock frequency by varying the *capacitor* in the clock circuit, rather than the resistor. But it is difficult to make a capacitor continuously variable over a wide range, and therefore this is a less practical approach to the problem of "playing" the clock. Later in the book we will substitute different size capacitors to set the frequency *range* of an oscillator, while a resistance is varied for continuous pitch change, but for now we'll limit ourselves to experiments with resistors of various kinds.

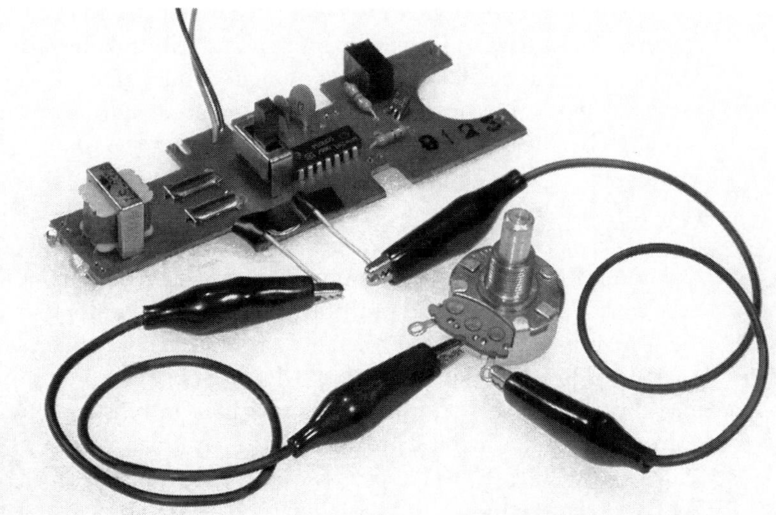

Figure 13.6
Pot substitute for a clock resistor.

CHAPTER 14

Ohm's Law for Dummies: How to Understand Resistors

You will need:

- The electronic toy from the previous experiment, or a new one.
- Some stranded and solid hookup wire.
- Test leads with alligator clips.
- An assortment of resistors.
- A potentiometer (1 megOhm or greater in value).
- A multimeter.
- Soldering iron, solder, and hand tools.

Now it's time for a smidgen of theory—sorry.

MEASURING RESISTANCE

A multimeter is a device for measuring various electronic properties, such as testing the voltage of a battery to see if it's dead or alive, or checking the value of a resistor. Meters come with analog readouts (a wiggling needle) or digital displays—simple digital meters are cheap these days and generally more useful. Most meters have a multi-position rotary switch for selecting different measurement modes (DC voltage, AC voltage, current, resistance, etc.) and the range of values measured and displayed within each mode.

Grab a meter, turn it on, and select the resistance/Ohm setting (Ω). Measure some resistors to confirm your prediction of value from the color code (as learned in the last chapter), and to acquaint yourself with this new tool. Most meters have a few ranges for resistance—experiment and see how changing the range affects the readout. You will notice that a handful of "identical" resistors (i.e. all the ones marked orange-orange-yellow for 330 kOhm) will probably indicate slightly different values—an indication of the 5 percent or 10 percent tolerance, as indicated by the gold or silver last band (and mentioned in the previous chapter). Measure between the nose and ears of a pot as you turn the shaft, and observe the seesaw change in resistance (you may need to use clip leads between the meter's probes and the terminals of the pot unless you have three hands or are skilled with chopsticks). Check the resistance of your skin—you'll need to use the highest setting in the Ohm range. Notice that the resistance decreases the harder you press the probes into your skin (ouch!), the closer together you place them, or the wetter your fingers are (digital confirmation of the experiments you've done in previous chapters).

SERIES AND PARALLEL (OHM'S LAW)

You may recall that a finger pressed on a circuit puts your mom-made resistor in *parallel* to the existing components, *lowering* the net resistance, *increasing* the speed of the clock and raising the pitch of the toy. In order to lower the pitch you had to remove the on-board resistor and substitute a pot of larger value. This demonstrates an aspect of Ohm's Law (a fundamental underpinning of electronic theory, worshipped by engineers of many lands) so essential to hacking that we will appropriate it:

Rule #13 (Ohm's Law for Dummies): The net value of two resistors connected in parallel is a little bit less than the smaller of the two resistors; the net value of two resistors connected in series is the sum of the two resistors.

To make a clock *slower* than it already is you must add a pot in *series* with the resistor on the board, making the net resistance *larger*: de-solder one end of the resistor and connect the pot between this loose end of the resistor and the hole out of which it came (see Figure 14.1). Because the toy will never run faster than its "stock" speed, this configuration minimizes the risk of freeze and crash.

To make the clock only go *faster* you connect the pot in *parallel* to the on-board resistor (essentially substituting it for your finger in the test we did in Chapter 12): leave the resistor in place in the circuit, and solder wires from the two tabs of the pot to the two pads at the ends of the resistor (see Figure 14.2).

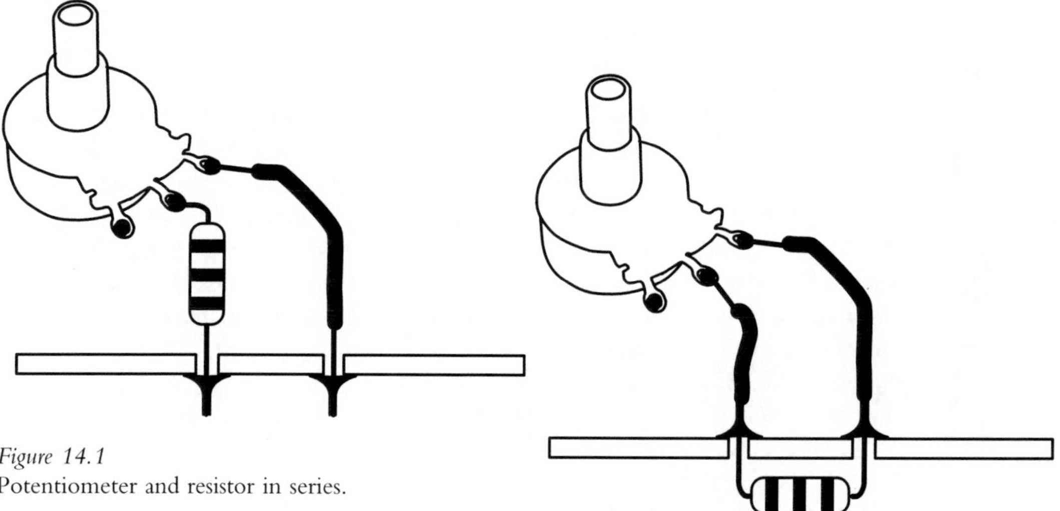

Figure 14.1
Potentiometer and resistor in series.

Figure 14.2
Potentiometer and resistor in parallel.

To make the clock go slower *and* faster you remove the on-board resistor entirely and connect a pot of *larger* value than the removed resistor in its place (see Figure 14.3).

Finally, if you want the toy to go slower and faster but never crash, put the original resistor in series with the pot with clip leads (as in Figure 14.1), then substitute progressively smaller resistors for your original value until you find one that lets the circuit run at the maximum speed without crashing, with the pot in the fully clockwise (i.e. 0 Ohm) position.

If all this doesn't make sense in the abstract, check it out. Use the meter to measure some series and parallel combinations of fixed resistors. Then try some of these pot and resistor variations on a toy clock circuit until you feel comfortable with The Law.

Theory class is over. Take a moment to read about some of the first artists to puzzle over Ohm's law as they struggled to make their own electronic instruments (see Art & Music 8 "Composing Inside Electronics"), then get back to work.

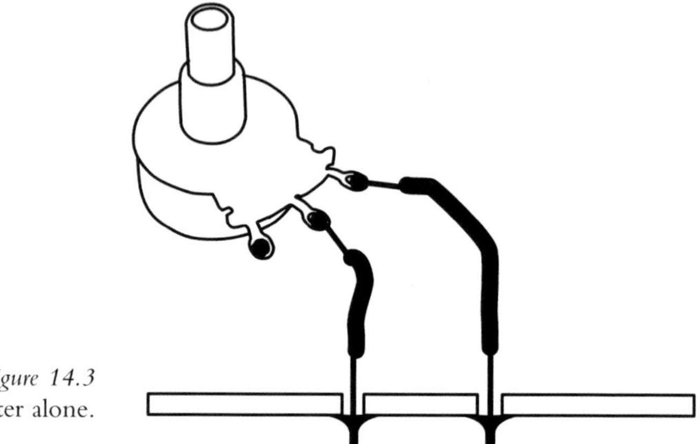

Figure 14.3
Potentiometer alone.

ART & MUSIC 8

COMPOSING INSIDE ELECTRONICS

The 1970s were a pivotal time in the evolution of the technology and culture of electronic music. Synthesizers were still impractically expensive for young musicians, but integrated circuits—the guts of those costly machines—were getting cheaper in inverse proportion to their sophistication. New chips contained 90 percent of a functional circuit designed by someone who really knew what he was doing; the remaining 10 percent could be filled in by someone relatively clueless. The trick was finding the right chips: in the days before the World Wide Web, information was much more segregated, with precious few leaks. When data did trickle down from engineers to amateurs, through magazines with titles like *Popular Electronics* or *Wireless World*, it was often passed from hand to hand like samizdat literature.

A musical community formed around this exchange of information. It included the "Composers Inside Electronics" who worked with David Tudor (see Art & Music 4 "David Tudor and 'Rainforest'," Chapter 8,) students of David Behrman (see audio track 11) and Robert Ashley at Mills College in Oakland, California (including Kenneth Atchley,

Ben Azarm, John Bischoff, Chris Brown, Laetitia de Compiegne Sonami, Scot Gresham-Lancaster, Frankie Mann, Tim Perkis, Brian Reinbolt, and Mark Trayle), students of Alvin Lucier at Wesleyan University in Middletown, Connecticut (Nicolas Collins and Ron Kuivila), of Serge Tcherepnin at California Institute of The Arts in Valencia, California (Rich Gold), and other musicians and artists scattered throughout the United States and (more thinly) Europe. Some participants were mere muddlers, who built beautiful, oddball circuits seemingly out of pure ignorance and good luck. Others became astonishingly talented, if idiosyncratic, designers. The prolific Paul De Marinis included bits of vegetables as electrical components so his circuits would undergo a natural aging process ("CKT," 1974), incorporated sensors that responded to the weak electronic field emanated by the human body ("Pygmy Gamelan," 1973; see Figure 14.4), and built automatic music-composing circuits that anticipated later trends in computer music ("Great Masters of Melody," 1975)—one of which could be played by a bird ("Parrot Pleaser," 1974.)

The European electronic music scene of the time was much more stratified—there was a well-established state-funded tradition of collaboration between composers and professional engineers, and homemade music circuitry never caught on there to the degree that it did in the United States (I have never seen a photograph of Stockhausen holding a soldering iron). There were notable exceptions, however. Andy Guhl and Norbert Möslang (Switzerland) formed "Voice Crack" in 1972, and over the next 30 years honed their skills at "cracking" everyday electronics; they became virtuoso performers with their new instruments, including circuits for extracting sound from blinking lights (see audio track 19), radio-controlled cars, radio interference, and obsolete Dictaphones (see Figure 14.5 and Andy Guhl's video on the DVD). Christian Terstegge (Germany) has been making elegant sound installations and performances with homemade circuitry since the early 1980s. In his 1986 work, "Ohrenbrennen" ("Ear-burn"; see Figure 14.6 and audio track 12) four oscillators are controlled by photoresistors inside small altar-like boxes containing candles; the pitches of the oscillators rise in imperfect unison, punctuated by swoops that trace the sputtering of the candles as they burn down over the course of a dozen minutes.

Figure 14.4
"Pygmy Gamelan" (1973), electronic circuit composition, Paul De Marinis.

Toward the end of the 1970s the first affordable microcomputers came on the market. Cajoled by the visionary Jim Horton (USA), a handful of musicians invested in the Kim-1—a single A4-sized circuit board that resembled an autoharp with a calculator glued on top. Programming this thing in machine language (and storing the program as fax-like tones on a finicky cassette tape recorder) was an arduous, counterintuitive, headache-inducing process, but coding offered one great advantage over building circuits: it was easier to correct mistakes by reprogramming than by re-soldering. Over the next ten years Apple, Commodore, Atari, Radio Shack, and others introduced increasingly sophisticated machines (and eventually disk drives!) which gradually reduced the angst-factor of programming, and homemade circuits faded into anachronism—until the anti-computer backlash of "Circuit Bending," as proselytized by Reed Ghazala (see Art & Music 9 Chapter 15), brought "chipetry" back into fashion.

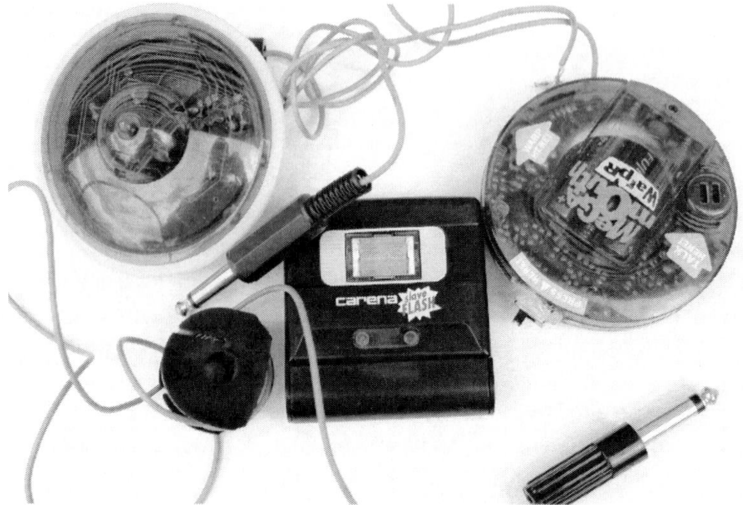

Figure 14.5 Circuitry by Norbert Möslang.

Figure 14.6 "Ohrenbrennen" (Ear-burn), Christian Terstegge.

CHAPTER 15

Beyond the Pot: Photoresistors, Pressure Pads, and Other Ways to Control and Play Your Toy

You will need:

- Electronic toys and radios from the previous experiments.
- Some hookup wire.
- Test leads with alligator clips.
- An assortment of resistors and pots.
- A few different photoresistors.
- A flashlight.
- A soda straw (opaque if possible), or a short piece of heat shrink tubing.
- Some loose change.
- Anti-static foam from packaging integrated circuits.
- A small sheet of corroded conductive metal, such as iron, copper, or aluminum.
- A lead from a mechanical pencil.
- Some paper and a soft pencil.
- Some fruit and or/vegetables.
- A telephone pickup coil and small amplifier.
- A multimeter.
- Soldering iron, solder, and hand tools.

You've opened a toy, tickled the clock, replaced its timing resistor with a potentiometer, and learned a bit of theory about swapping resistors—what's next in the way of toy hacks?

PHOTORESISTORS

A photoresistor (or photocell, as it is sometimes called) is a device that changes its value in response to light level: the resistance gets *smaller* when it is exposed to a bright light, and gets *larger* in the dark. It takes the form of a small disk, whose diameter can range from 1/8 inch to 1 inch; two wires come out of one side, and the other side displays a pleasing zigzag pattern of fine lines (see Figure 15.1). The side with lines is more sensitive to light than the other, but the back is translucent enough that light striking the back will affect the resistance as well. The lowest resistance in bright light is anywhere from

100 to 2,000 Ohms, depending on the kind of photoresistor; the "dark resistance" is very large, typically around 10 megOhms. Because this is higher than most pots, and because most clock circuits use pretty large resistors, photoresistors are a convenient variable resistor for slowing down toys a lot.

Photoresistors are pretty cheap, especially when bought from "surplus" outlets online. Some retail sources provide data on the range of resistance, sometimes not. In addition to different "light" (minimum) and "dark" (maximum) resistances, different photoresistors will respond at different speeds to changes in light level—some are more sluggish than others. All these factors affect how they perform in a musical circuit. You can test them with a multimeter, but ultimately your ear is the best guide to picking a good photoresistor for your circuit. Don't be disappointed if it takes a while to find the perfect one.

Select a photoresistor. Remove the pot from the clock circuit of the toy you've been working with for the past few chapters, or find and remove the clock resistor in another toy. Using clip leads, attach the two leads of the photoresistor where the pot tabs were connected, or solder the photoresistor directly in to the holes left when you removed the resistor (see Figure 15.2). Turn on the toy and listen to how the circuit

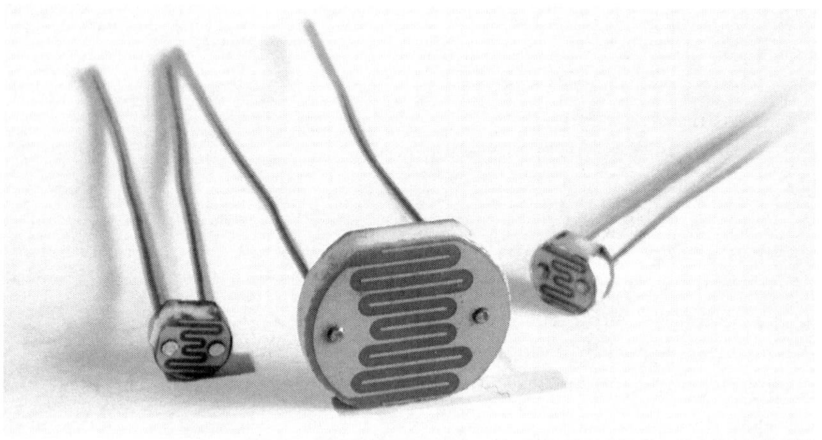

Figure 15.1
Some photoresistors.

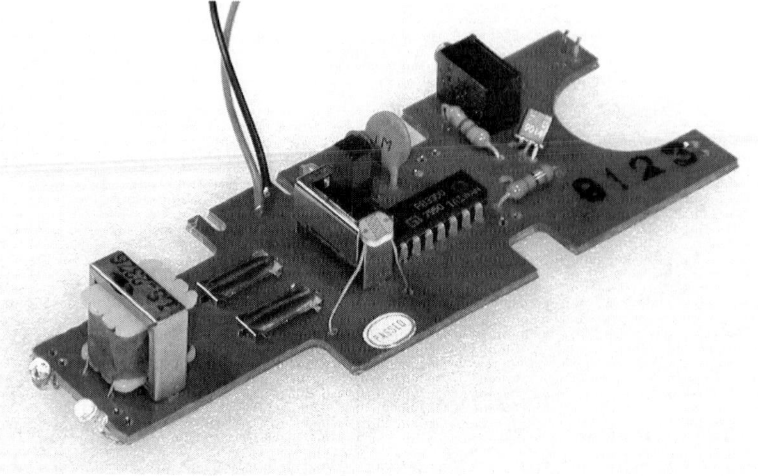

Figure 15.2
Photoresistor in place of clock resistor.

behaves when you pass your hand over the photoresistor or shine a flashlight on it. If you have more than one type of photoresistor swap them in and out, and listen to how different ones affect the circuit.

For maximum resistive range you must go from totally black (no light at all) to high illumination—carry your circuit into a closet and play it with a flashlight. More practically, you can mount the photoresistor at one end of an opaque tube (such as a drinking straw painted black, or a short piece of heat shrink tubing) to make it very directional in its light sensitivity: it will only respond to light aimed directly down the tube (see Figure 15.3). This is the core technology of certain carnival shooting galleries, where each "gun" fires a light towards similarly blinkered targets.

You can put the photoresistor in your mouth and make a very expressive controller that responds to changes in both the light level as you open and close your mouth, and in the conductivity of your saliva-laden tongue across the photoresistor's bare legs (a suggestively naughty extension of the licked-finger-on-circuit-board effect).

DON'T EVER *THINK* OF TRYING THIS WITH ANY CIRCUIT THAT IS EVEN BATTING ITS EYELASHES AT A WALL OUTLET!

Place a fan between a light source (such as a flashlight) and the photoresistor, or reflect light off a record turntable (put some delicately crumpled aluminum foil on the turntable instead of a record)—you should hear a vibrato effect or other wobbly modulation, which changes as you vary the speed of the fan or turntable.

If the toy has blinking lights or LEDs you can tape the photoresistor against one of the lights and the toy will modulate itself, producing possibly interesting patterns. Two toys with blinking lights and photoresistor-controlled clocks can modulate each

Figure 15.3 A shy photoresistor, hiding inside a section of heat shrink tubing.

other—curiously erotic electronics! The more toys, the greater your chances of creating artificial life.

A photoresistor can be a good compromise between the fluid, if somewhat unpredictable (and occasionally dangerous), effect of the finger on the circuit board and the more controllable but less expressive potentiometer. You can use it as a very responsive *performance* interface to interpret hand shadows or flashlight movement, or as an installation sensor, reacting to ambient light and the shadows cast by visitors. We'll look at more photoresistor applications in Chapter 18 and beyond.

As I mentioned earlier, although the zigzagged side is more sensitive to light than the backing, most photoresistors are made of translucent material, so that light striking the back will affect its resistance as well. It is important to cover the back if you want the greatest range. Besides burying the photoresistor in an opaque straw, you can seal off the back with black paint or electrical tape.

If you want to have both the gestural quality of the photoresistor and the controllability of the pot, you can combine the two: if you wire a pot in *series* with a photoresistor (see Figure 15.4), the pot will determine the *maximum* frequency of the clock in full light, and darkness will cause the speed to go *down* from that maximum. If you wire the pot in *parallel* with the photoresistor (see Figure 15.5), the pot will set the *minimum* frequency of the clock in full darkness, from which the speed will go *up* as light increases. (If this sounds confusing, just try it).

As I mentioned earlier, in total darkness, the photoresistor has a very large resistance—as high as 20 mOhm—much higher than any commonly available pot. If you want really low frequencies out of your toy a photoresistor is the way to go—moreover, multiple photoresistors can be strung together in series to drive a clock down into the glacially slow range.

ELECTRODES

Let's not forget the heady spontaneity of our youthful experiments with flesh-controlled circuitry back in Chapter 11. If you want to use your fingers to connect points on the board that are widely separated, or you just want a more formal playing surface, dimes or other silver-plated coins make excellent electrodes (copper tarnishes too quickly to

Figure 15.4
A potentiometer and photoresistor in series.

Figure 15.5
A potentiometer and photoresistor in parallel.

use in an urban or coastal environment—just take a look at the Statue of Liberty). Strip 1/4 inch of insulation off both ends of a few 5-inch pieces of wire. Solder one end of each wire to one of the "sensitive points" you've found on the circuit board, and solder the other end to a coin. Arrange the coins in a pattern that lets you bridge them easily with your fingers, but avoid direct shorts (see Figure 15.6). By the way, I was told as a child that it is illegal to solder or similarly deface US currency, so you might refrain from performing this one in the presence of the Secretary of the Treasury.

A nice way to combine the control certainty of a potentiometer with the gestural possibilities of finger-on-circuitry is to parallel the pot and a pair of electrodes, as we did with the photoresistor in Figure 15.5. When you solder your hookup wires to the lugs of the pot leave an extra inch of bare wire sticking up through the solder hole. When you go to mount the pot in the box that will hold the circuit (see Chapter 17), drill small holes to line up with the wire ends and two more about 1/2 inch away. Lead the bare wires up through the panel at the pot and then down again, so that they form two parallel strips (see Figure 15.7). You will have convenient electrode contacts immediately adjacent to the knob so you can slip your finger back and forth between precision adjustment and touchy-feely playing (you could also solder coins to these wires for a larger playing surface).

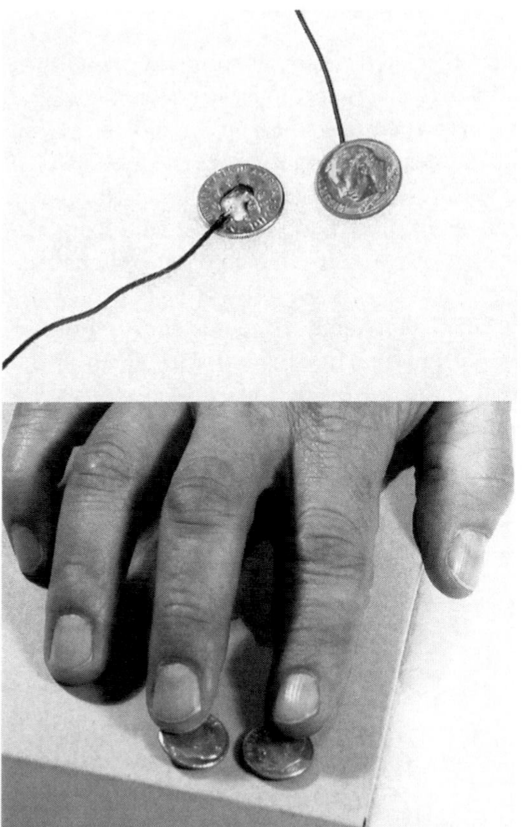

Figure 15.6 Coin electrodes and their performance.

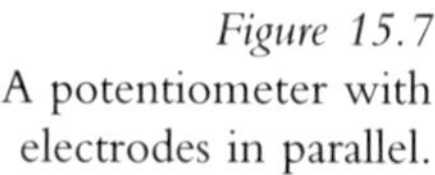

Figure 15.7
A potentiometer with electrodes in parallel.

CHEAP PRESSURE SENSORS

The squashy black "anti-static foam" in which integrated circuits are sometimes packaged has interesting electrical properties. Put a piece between two coin electrodes and measure the resistance with a multimeter as you squeeze them together—it gets lower as you apply more pressure (see Figure 15.8). This homemade pressure sensor can be used in place of a pot or photoresistor to make a pressure-sensitive controller for performance or installation (under chair legs to measure weight, for example). Anti-static foam can be bought in sheets from various online retailers, if you can't find an engineer's garbage pail from which to scrounge.

Vegetables and fruit also have resistive value. This value changes as they dry out or are squished. You can substitute small slices of produce for the anti-static foam in the above experiment, or poke bare wires directly into carrots or apples. (As some of you may remember from childhood science experiments, it is also possible to make a battery out of fruit or vegetables, but this lesson will wait until Chapter 29.)

Figure 15.8
Pressure sensor made from anti-static foam and two dimes.

ARTY

A pencil lead from a mechanical pencil makes an excellent, if delicate resistor (see Figure 15.9). Attach two clip leads to the clock resistor contact points in your toy. Clip one wire to one end of the pencil lead, and scrape the jaws of the other along the lead. Resistance is proportional to the distance separating the contact points, so the pitch of the toy should go down as you move the clip further from the end. (Most potentiometers are simply a neatly packaged strip of carbon with a moving wiper.)

The use of graphite as a resistor is not limited to pencil leads themselves. Draw two blots near the edge of a piece of paper, and clamp the jaw of a clip lead to the paper at each blot; clip the other ends of the leads to the clock resistor contact points in your toy (see Figure 15.10). Draw a line between the two blots and get the toy running— as you widen the pencil line linking the blots, or draw additional lines, the pitch should go up. Why? The wider the graphite path between the clips the lower the resistance (think fat pipes versus skinny pipes). Patrick McCarthy has made functional potentiometers using this technique (see Chapter 30 and his video on the DVD).

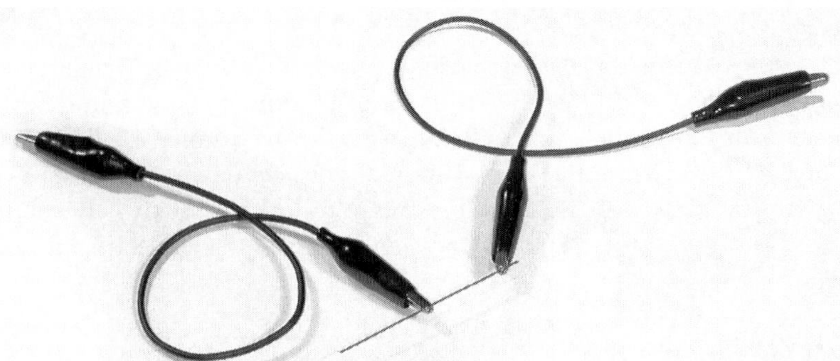

Figure 15.9
A pencil lead resistor.

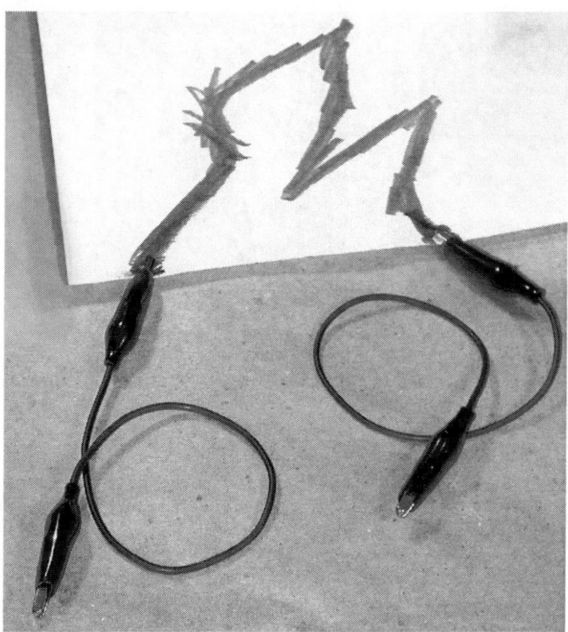

Figure 15.10
Drawing a resistor.

You can also use a strip of magnetic tape as an open-air resistor (old VHS tape works great): use two probes pressed against the surface, or an alligator clip at one end and one movable probe. As with the pencil line, resistance is proportional to distance. T. Escobedo's "Synthstick" is a glissando-based instrument, sort of a cross between a Theremin and a Stylophone, based on this principle.

ALMOST A SHORT CIRCUIT

Enough of clocks! There's more to life than pitch change. Sometimes a toy can be induced to make curious sounds if you make new connections between various locations on the circuit board. Take a resistor of about 100 Ohms and bend it into the shape of a croquet wicket. While listening to the toy, press one end of the resistor to a solder point on the solder side of the board; then touch the other end to various other points—if the circuit board is large you may need to use a clip lead to reach all over (see Figure 15.11 and audio track 13). You may (or may not) get some interesting-sounding circuit malfunctions. Disconnect *immediately* any connection that seems to cause heat, smoke, or flame.

Try different value resistors, but avoid shorting out the board with straight wire unless it's the only thing that works (see the 7th Rule of Hacking). If you find that the best sounds happen with the smallest value resistor, you can try plain wire, but do so gingerly and be prepared to remove the wire as soon as you feel or smell trouble. You can also try using pots or capacitors instead of resistors. You can go back to your radio and experiment with using resistors to jump between the hot spots you bridged with your damp finger.

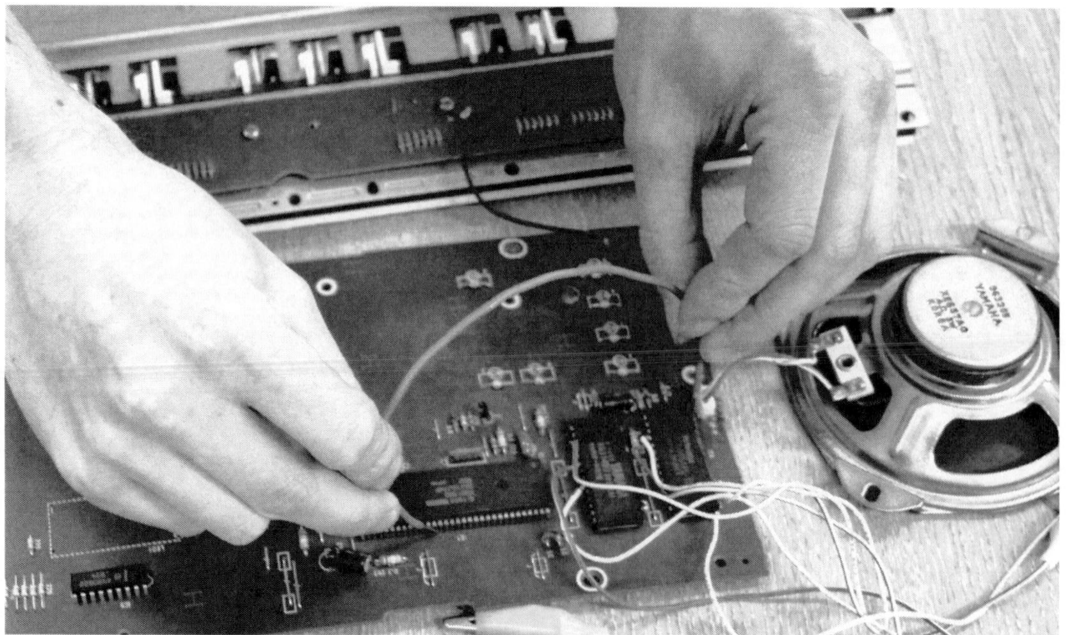

Figure 15.11 Phil Archer bridging the circuit board.

Once you find a useful connection you can solder the resistor permanently into place, or add a switch to connect and disconnect it as a performable change (see next chapter for switch information).

This technique is at the heart of Reed Ghazala's wonderful "Circuit Bending" philosophy of hardware hacking (see Art & Music 9 "Circuit Bending"), and is a very powerful and creative tool for extracting unusual sounds from almost any found circuit.

If you're feeling suicidal, you can follow Phil Archer's or hans koch's example and dribble water on the circuit board—watch their videos on the DVD, and watch out!

RUSTY

A corroded metal plate and a nail can serve as a quasi-random variable resistor, as we demonstrated with the jumping speaker in Chapter 5. Choose any two points on the circuit board that produce a change when they are connected to one another—this could be the clock resistor solder pads, or any of the hot spots we've just described bridging with resistors. Solder a short wire to each point, long enough to be grasped in the jaws of a clip lead. Clip the other end of one of the leads to a nail and the other lead to a sheet of rough or corroded metal (copper flashing, rusty baking sheet, file, etc.). Lightly scrape the nail across the metal and listen to the circuit twitter—the corrosion and intermittent hop, skip, and jump of the nail over the rough micro topography yields an ever-fluctuating resistance that can be steered (if not exactly controlled) by adjusting the pressure and speed of movement.

MOTOR-MOUTH

Noise, flashing lights, and frenetic activity—the essential attributes of any disaster are also the core components of a successful toy. After messing around with the sounds and lights, don't forget the motors that make tickled Elmo twitch and Billy Bass flap his tail. Some pretty sophisticated computer code goes into this electromagnetic choreography, and you can eavesdrop on it by placing a telephone tap coil on a motor and connecting the coil to your amplifier. There is often a beautiful rhythmic interplay between the toy's sounds, blinking lights, and movements, and a thorough hack can bring all this futurist polyphony to the ear. And don't forget to try another coil on the toy's speaker to make the basic sounds louder and more "hi-fi" (as we did with the radio in Chapter 11).

INTERCONNECTING TOYS

Once you've opened and hacked a few toys don't be afraid to experiment with interconnecting them. First connect a clip lead between the grounds ("–" end of the batteries) of both toys. Then use another clip lead to make random connections between any point on one toy and any on the other. Use a jumper between the clocks and you may get the toys to cross modulate each other; if you connect clock points between two circuits and remove the resistor from one, you can sometimes drive both in sync from a single clock.

ART & MUSIC 9

CIRCUIT BENDING

Traditionally, making functional electronic objects has necessitated a fair grasp of theory and a pretty clear idea of what you wanted to make *before* you picked up your soldering iron. David Tudor, Gordon Mumma, Composers Inside Electronics, and other musical designers began chipping away at these assumptions in the 1960s and 1970s. Being self-taught, they had only piecemeal knowledge of electronic theory and were less concerned about doing things "properly" than about making something that sounded cool. Immersed in a musical ethos that valued chance, they were highly receptive to accidental discoveries—in the pursuit of the "score within the circuit," they relished wandering down side paths, rather than race-walking toward a predetermined goal.

Then in the mid-1990s Qubais Reed Ghazala pushed serendipity back to the fore of electronic practice with his fervent advocacy of what he dubbed "Circuit Bending." Like Waisvisz (see Art & Music 7 "The Cracklebox," Chapter 11), as an adolescent in the late 1960s Ghazala encountered the sounds of accidental circuit interaction: an open amplifier left in his desk drawer shorted against some metal and began whistling. After some experimentation, Ghazala added switches so he could control the shorting, and Circuit Bending was born. He developed a series of techniques for modifying found circuitry—especially electronic toys, whose sonic sophistication grew in direct response to the boom of semiconductor technology in the 1980s—without the benefit of the manufacturer's schematics, or any engineering knowledge whatsoever. In 1992 he began publishing instructive articles in *Experimental Musical Instruments* (an influential journal for instrument builders) and acquired a cult following. In 1997 he launched his Web site and today a cursory Web search will reveal news groups, festivals, and workshops for Circuit Bending all over the world.

Circuit Bending is freestyle sound design with a postmodern twang—the perfect escape for artists bored by the powerful, but often stultifyingly rational, software tools that increasingly dominate music production, yet still hooked on the digitally inspired cut-and-paste aesthetic of scavenging, sampling, and reworking found materials. With its defiantly anti-theoretical stance and emphasis on modifying cheap consumer technology, bending has a natural egalitarian appeal (as well as some odd orthodoxies: looking at my instruments as I was setting up a demonstration at the "Bent 2004" Festival at The Tank Gallery in New York City, an audience member inquired, "Are they bent or hacked?" When I looked baffled he elaborated: "'Bent' means you have *no* idea what you are doing when you open up the circuit; 'hacked' means you have *some* idea"). But Bending's try-anything extreme experimentalism can produce wonderful results never anticipated by the original designers of the device being bent.

Some Benders specialize in particular adaptations: German musician Joke Nies has made a specialty of hacking an early digital instrument called the "Omnichord" (see Figure 15.12 and his video on the DVD); my ex-student Jon Satrom has based his VJ career on a specific V-Tech children's toy (see Art & Music 10 "Visual Music" in Chapter 24, Figure 24.6, and his video on the DVD). Texas Instrument's "Speak and Spell" has been a favorite from the day it was introduced in 1978, long before the term "Bending" came into use. Web sites abound with detailed instructions for specific cuts and jumpers on the boards of particular toys.

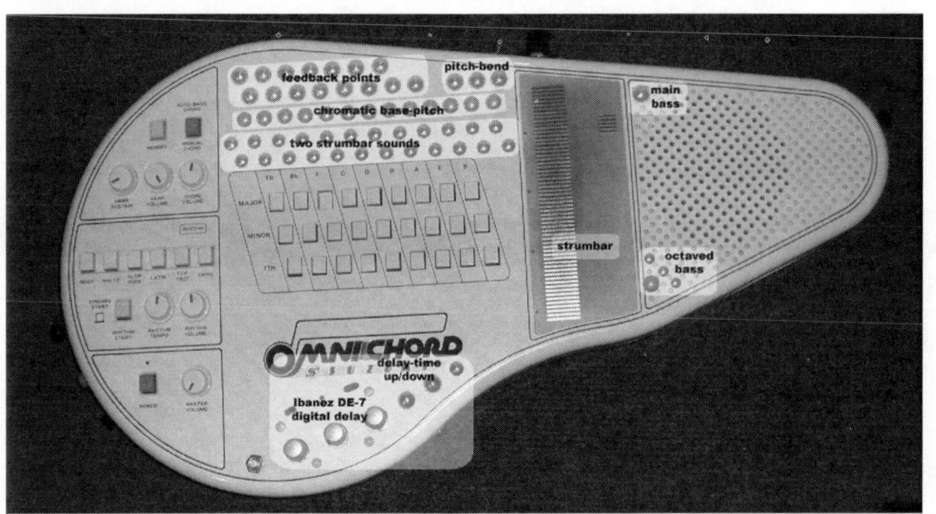

Figure 15.12 Bent Omnichord, Joker Nies.

Phil Archer (UK) and John Bowers (UK) are representative of the emerging generation of hackers, who effortlessly combine bending with Tudor-era contact mike technology and sophisticated computer programming. Archer did the "classic" bend to his Yamaha PSS-380 keyboard: exposing the circuit-board, placing the inverted instrument on the performer's lap, and making arbitrary connections between components on the board with a stripped piece of wire (see Figure 15.11 and audio track 13). "These connections," he writes, "induce tones, bursts of noise and corrupted 'auto-accompaniment' sequences from the device which are unpredictable in their details but generally 'steerable' overall with practice. The precision and control afforded by the standard keyboard interface is eschewed in favour of direct contact with the circuit, and the performer is continually forced to rethink and re-evaluate their relationship with the instrument in light of the sonic results." Most of his other instruments have a Frankenstein quality: a midget Hawaiian guitar whose single string is played by the sled mechanism from a CD player (see Figure 15.13); a set of small percussion instruments whacked and scraped by motors from a dot matrix printer; a music box mechanism activating "Bent" electronic keyboards and a keyboard played by dripping water (see Archer's video on the DVD).

John Bowers—in an ongoing struggle against his training as a computer scientist—"reinvented" what he has dubbed the "Victorian Synthesizer" (see Chapter 5 and audio track 4): it produces sounds with speakers animated directly by batteries, bereft of intervening electronic circuitry. Corroded metal, mercury-filled tilt-switches, and a handful of screws and washers complete instruments that could indeed have been built in the nineteenth century. His other "Infra-Instruments" combine similar electro-mechanical technology (mixing bowls filled with motors, magnets, contact mikes, and guitar pickups (see Figure 15.14); microphones embedded in a plank of wood; strings, stones, and guitar pickups strewn across a table with computers and rock effect boxes.

Figure 15.13 "CD Player Slide Guitar," Phil Archer.

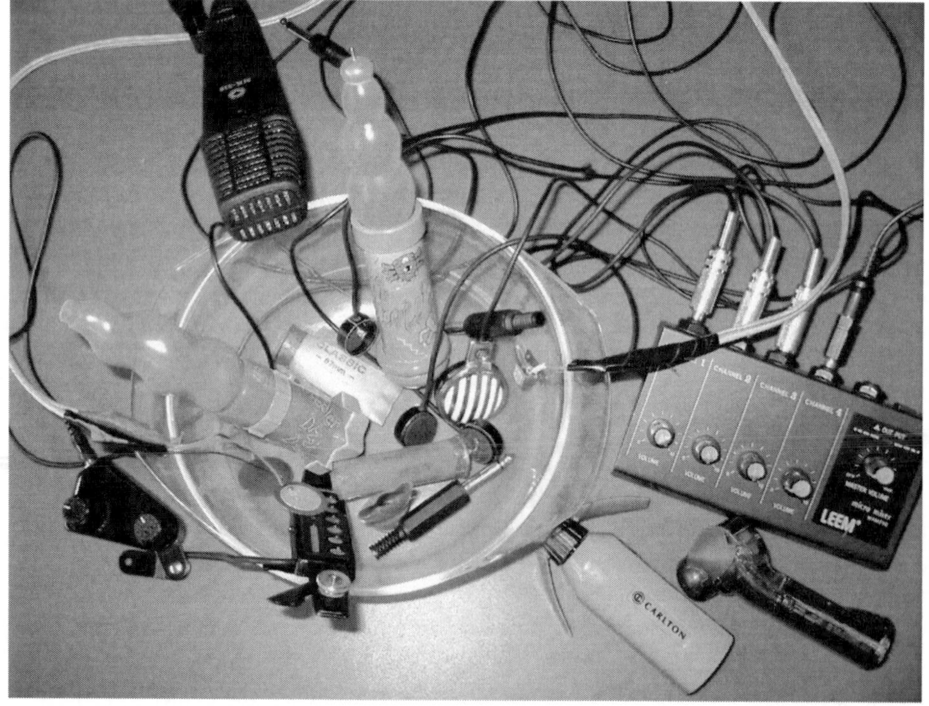

Figure 15.14 "Mixing Bowl," John Bowers.

Notable younger Benders include Knut Aufermann (Germany/UK), Xentos "Fray" Bentos (UK), David Novack (USA), Vic Rawlings (USA), Sarah Washington (UK), Chris Weaver (UK), and Dan Wilson (UK). Britain's particularly vibrant bending scene (including an "all bending ensemble," P. Sing Cho—see audio track 14) has roots in the prevalence of toys as affordable, alternative noisemakers among improvisers in the 1970s—most significantly Steve Beresford. As Sarah Washington says, echoing Tudor from four decades earlier, "I am an improvising musician . . . the choice of sounds is down to the circuit—whatever it comes up with is fine with me" (see Figure 15.15).

(For more information on the current state of Circuit Bending see Chapter 30.)

Figure 15.15 "Mao Tai," Sarah Washington.

Remember that you can also link circuits through blinking lights and photoresistors, as described earlier in this chapter.

Lest you get too distracted by the gizmo-factor of all these add-ons, don't forget the humble laying of hands. On larger circuit boards with multiple components (such as musical keyboards) a few damp fingers brushed against the circuit board can raise delightful havoc with the normal behavior of the toy—the Yamaha PS-140 is especially susceptible to fleshly corruption (see Figure 15.16). British bending iconoclast Dan Wilson lets worms crawl across his circuit boards (see Chapter 30 and his video on the DVD).

BEYOND TOYS

Most of the techniques described in this chapter can be used to extend the radio you opened up in Chapter 11 as well: electrodes can pull the radio's ticklish spots to the outside of a box, and make it easier to bridge multiple points with your fingers; pots, photoresistors, pressure pads, resistors, and rusty nails can be used to link these points as well. Toys and radios are cheap and plentiful, and thus an obvious flashpoint for hacking insurgency, but the same methods can be applied to almost any electronic circuit: CD players (see Figure 15.17), rock effect pedals ("stomp boxes") (see Figure 15.18), cassette players (see Figure 11.4), answering machines—you'll never know until you try them. Vic Rawlings (US) uses wire brushes, nails, fingers, and assorted metal junk to play a complicated array of effect pedals—see his video on the DVD and Figure 15.19. Neal Spowage (UK) has hacked metal detectors to make his "Electro Magnetic Wands" (see his video on the DVD).

Figure 15.16 Mike Challis (UK) playing his hacked Yamaha PS-140 by touching circuit board directly through a hole cut into the case.

Figure 15.17 "Sled Dog," hand-scratchable hacked CD player by Nicolas Collins.

Figure 15.18 Hacked guitar effect pedal by Chris Powers (US), with electrode contacts for toes.

Figure 15.19 Performance setup by Vic Rawlings, showing open circuits with wire ball.

You can buy dozens of different kinds of electronic kits—from strobe lights to electronic wind chimes—from online retailers (see Appendix A) and experiment with these kinds of modifications as you build them—hacking goes faster if you don't have to disassemble first. After savoring bespoke electronics you'll never accept off-the-rack again.

CHAPTER 16

Switches: How to Understand Different Switches, and Even Make Your Own

You will need:

- The electronic toy from the previous experiments.
- Some hookup wire.
- A Single-Pole Double-Throw switch (SPDT), momentary or toggle.
- A plank of wood, some short nails, and a large ball bearing.
- Soldering iron, solder, and hand tools.

This is possibly the most boring chapter in the book. Skip over it if you wish, but don't tear it out, because it may prove useful later.

Switches are useful for turning power on and off to a circuit to save battery life, for turning on and off specific sounds or functions, and for resetting a circuit if it freezes up. They are often described in catalogs, on Web sites or in packaging by arcane abbreviations. Here are the main distinguishing features.

MECHANICAL STYLE

A switch can be *momentary* pushbutton, like a door bell, that changes state (turns something on) when you press it, and returns to its default state (off) when you release it; or it can be a push-on/push-off switch that alternates but holds its state (like the bypass switch on a stomp-box). It can be a *toggle* switch with a lever, like a traditional light switch, that stays where you put it until you switch it back (usually). There is also the *rotary* switch, like the cycle selector on a clothes washer or the pickup selector switch on a Stratocaster, with which you choose between several positions, rather than just on and off. *Slide* switches, like the rotary switch, can select between two or more positions (see Figure 16.1). There are a few other oddball switches we'll discuss when they become relevant.

NUMBER OF THROWS

A switch is also described by the number of mutually exclusive connections it makes when moved or "thrown." A simple pushbutton that turns something on in one

Figure 16.1 Assorted switches.

position, but does nothing in the other (beyond turning off), is called a "Single Throw" (ST) switch (see Figure 16.2a). If the switch alternates between two possible connections, it is a "Double Throw" (DT) switch (see Figure 16.2b)—this could be a pushbutton, a toggle, or a slide switch. Rotary switches that can make several different connections are classified by the number of connections, i.e. a five-position switch would be abbreviated as "5T" (see Figure 16.2c).

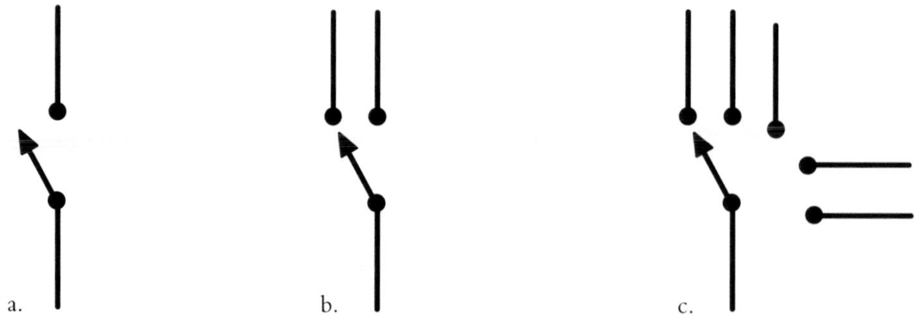

Figure 16.2
a. Schematic representation of a Single Throw (ST) switch.
b. Schematic representation of a Double Throw (DT) switch.
c. Schematic representation of a Five Throw (5T) switch.

NUMBER OF POLES

Sometimes a single lever or button can switch two or more separate circuits simultaneously (think of the huge double-bladed switches in *Frankenstein*, thrown by one ominous handle). Most pushbutton and toggle switches are either "Single Pole" (SP; see Figure 16.3a), meaning that they switch only one circuit, or "Double Pole" (DP; see Figure 16.3b), which switches two circuits. The dotted line on the schematic representation indicates that the two sections switch in tandem. Switches can have three, four, or more poles—they're just less commonly encountered.

TERMINAL DESIGNATIONS

In a Double Throw switch the solder terminal that is normally *off*, or unconnected, is designated "Normally Open" (NO). The one that is normally *on*, or closed, is the "Normally Closed" (NC). The terminal that is connected, by the movement of the button or toggle, alternately to the NO or NC terminals is the "Common" (C; see Figure 16.4).

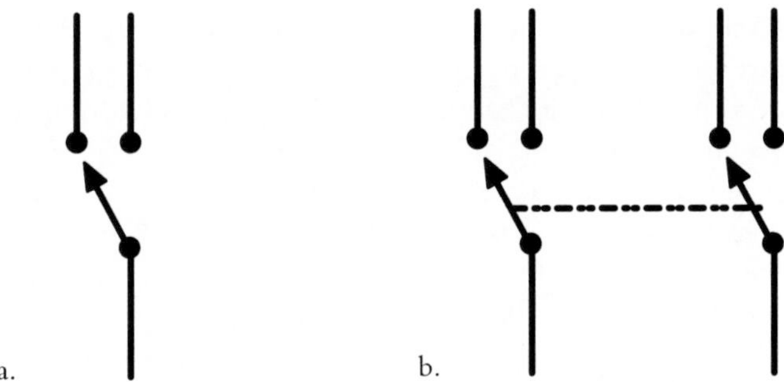

Figure 16.3
a. Schematic representation of a Single Pole (SP) switch.
b. Schematic representation of a Double Pole (DP) switch.

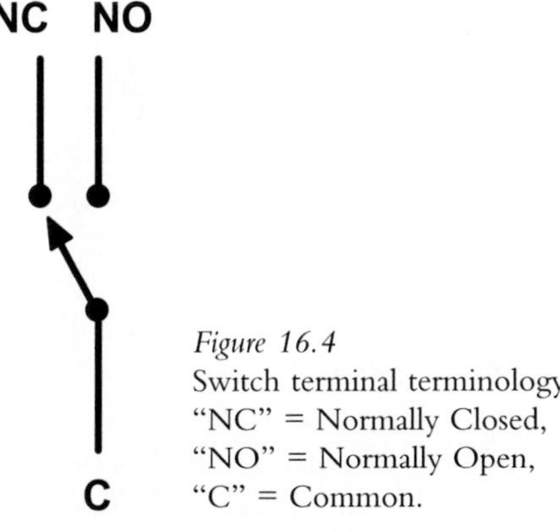

Figure 16.4
Switch terminal terminology:
"NC" = Normally Closed,
"NO" = Normally Open,
"C" = Common.

RESET SWITCH

You may have noticed that your toy occasionally freezes up, usually when the clock is run too high or too low, or you short out some part of the circuit. As per the 12th Rule of Hacking, momentarily removing the batteries will usually fix the problem. But this gets tiresome. The problem is that most modern toys do not have mechanical on/off switches—like televisions, they are usually only asleep when you turn them off. Current is still flowing through parts of the circuit, and the toy—like your computer—will not truly reset without *really* powering down and back on.

We can add a reset switch that lets you press a button or throw a toggle to disconnect the batteries temporarily, without the bother of actually removing them. You'll need a SPDT (Single Pole Double Throw) switch. It can be a *momentary* switch, if the toy already has a built-in power on/off switch; if you want to use the reset switch as a power switch as well, then get a toggle switch instead.

Cut one of the two wires connecting the batteries to the circuit board. Solder one end of the cut wire to the switch's Common terminal (C); solder the other end to the Normally Closed (NC) terminal. If the battery wire is very short you may want to extend one or both sections with some additional hookup wire. If the switch has more than three connectors, or they are unmarked, you should use a multimeter to figure out the switch logic.

With a momentary switch, the switch is normally *closed*, so the battery voltage flows into the C terminal and out through the NC terminal to the circuit; when you press the switch the C flips its connection to the NO terminal, breaking the connection to the NC terminal and disconnecting the batteries from your circuit. Next time your circuit crashes, return the clock speed pot to a middle setting, restore any other weird connections to their "safe" states, press the switch for a few seconds, and (hopefully) the circuit will "re-boot" when you release it—easier than removing the batteries, especially in front of a restless crowd at CBGBs or the Royal Albert Hall.

If you use a toggle switch the circuit will stay in the off state when switched, so it will function as a power on/off switch as well, saving battery life offstage.

Toggle or momentary switches can also be used to switch on and off the resistive jumpers you made in Chapter 15, or to switch between a fixed clock resistor and a variable one (or between two different variables resistors—such as a pot and a photoresistor).

HOMEMADE SWITCHES

Switches are useful, often essential, things. Unfortunately, they can also be the most expensive part of a hack: resistors, capacitors, wire, and many other electronic components you use in hacking typically cost fractions of a penny, but a switch can set you back several dollars. However, with a little mechanical ingenuity you can fabricate your own switches out of paperclips, springs, brass fasteners, and other scraps of metal—classic prison technology, like making a shiv from a bedspring. It's more of a mechanical problem than an electronic one: find two pieces of metal that conduct electricity, then figure out a practical way to move them in and out of contact with one another.

You can construct a nice multi-position tilt-switch by hammering a ring of brads into a piece of wood, soldering a wire between each nail and a point on the circuit

Figure 16.5
A homemade tilt switch.

board that needs switching, and dropping a big ball bearing into the corral (see Figure 16.5). You can use this switch to select different circuit jumpers (the almost-shorts you found in Chapter 15), or switch between the outputs of several toys to feed the input of your amp. Variations on this design can be made with loops of wire, strips of copper, or even blobs of mercury (once a common switch element, now banished behind the sign of a skull and crossbones—see Figure 16.6).

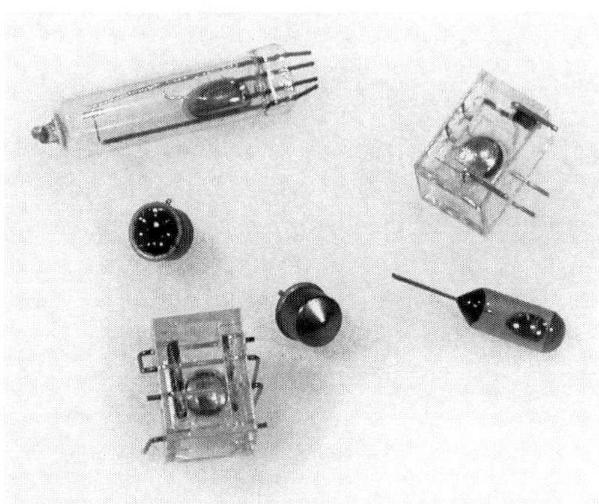

Figure 16.6
Some commercial tilt switches, using ball bearings and deadly blobs of mercury to connect contact points.

CHAPTER 17

Jack, Batt, and Pack: Powering and Packaging Your Hacked Toy

You will need:

- The electronic toy or radio from the previous experiments.
- A battery-powered mini-amplifier.
- Some hookup wire.
- One or more jacks for external audio connection.
- Some 1 kOhm resistors.
- A battery holder (if appropriate—see text).
- A box of some kind to house your circuit.
- Soldering iron, solder, and hand tools.

It's almost time to "close" your first hack, as they say in the O.R., but let's look at a few final modifications before Frank rises from the slab.

JACKS

Beyond retuning the clock and finding some musically viable almost-shorts, the most significant change you can make to a toy is replacing the little speaker with a big one, thereby confirming the Second Law of the Avant-Garde (introduced in Chapter 7) with increased loudness. A telephone tap coil resting on the little speaker is one approach, as discussed already. But if your circuit is *never* going to be played by saliva-drenched fingers it's safe to try a wired connection directly to an amplifier—a bit more preparation work than slapping the coil down, but ultimately cheaper and faster to patch together in performance.

By adding a jack to connect the circuit to an external amp and decent-size loudspeaker you not only make the sound much louder, which lets you hear more detail, but you will also hear low frequency components that aren't audible through the tiny, parent-friendly speakers inside most toys (and sometimes higher frequencies as well). It's easy to do:

1. Find the wires leading from the circuit board to the speaker.
2. De-solder them from the speaker terminals or cut them as close to the speaker as possible.

3. Solder these two wires to a female jack of your choice; usually it doesn't matter which wire goes to which terminal, but you must always have one wire going to the shield/sleeve connector and one to the hot/tip connector.
4. Plug it into a decent sound system and listen. Start at a low volume setting, since the output of a toy can be surprisingly loud. If there's lots of hum, reverse the hot and ground connection at the jack. If there's no sound at all, check your soldering. You may find that the raw sound is too much—too noisy or abrasive, too much extreme high or low—but that's where the equalization on a mixer, amp or graphic EQ "stomp box" can help you carve the sound you want out of the toy's raw material.

Let me repeat:

DO NOT ATTEMPT THIS MODIFICATION ON A CIRCUIT THAT WILL BE MAKING INTIMATE ELECTRICAL CONTACT WITH YOUR BODY (SUCH AS THE WET-FINGERS RADIO).

As long as you're adding one jack, why don't you see if there are any other interesting signals running around the circuit board yet unheard?

1. Solder a wire from the shield/ground connection on a jack to the place on the circuit board where the "–" terminal of the battery connects, or to the shield/ground terminal on the main output jack, if you've added one already. Use stranded wire if at all possible—less strain on the connections at either end, and easier to pack into the toy's case.
2. Solder another wire to the hot tab of the jack and strip and tin the other end.
3. Solder a 1 kOhm resistor to the tinned end of the wire.
4. Plug a cord between the jack and a battery-powered amplifier.
5. Turn the volume up just a little bit. Poke the free end of the resistor around the circuit board and listen to the different sounds (see Figure 17.1). Adjust the volume as needed. Sometimes you can find very odd noises that seem completely unrelated to the basic sound of the toy. Hold on to the body of the resistor, rather than the bare wire, to minimize hum.
6. When you find a place you like, solder down the free end of the resistor. Wrap the bare wire and resistor lead in electrical tape to prevent shorts (you can shorten the resistor leads prior to soldering to minimize the amount of bare wire running around your circuit, itching for a short circuit).
7. If you wish, add another jack and repeat the process, discovering and decanting hidden sounds. Or add a multi-position switch (such as a rotary switch or our homemade tilt-switch) to select among different circuit points to connect to a single jack.

If you get sound when one or the other of your jacks are connected to the amplifier/mixer, but not when you combine two or more, you have probably unintentionally crossed your grounds. De-solder one of the two incompatible jacks from its

Figure 17.1 Phil Archer hunting for interesting sounds.

wires (the speaker jack, if used, is the first choice) and swap the "hot" and "ground" connections.

CUT-OUT JACKS

You may have noticed that most audio devices with a built-in speaker and a jack for headphones will switch off the speaker automatically when the headphone plug is inserted. This is accomplished with what is known as a "cut-out jack," which basically combines the functions of an audio jack and a switch, and can be purchased at almost any retailer carrying electronic connectors. The jack's terminals will be designated "tip" or "hot," "sleeve" or "ground," and "normally closed" (NC—like the switches we discussed in the previous chapter). These designations might be printed on the jack's packaging (if you buy them at Radio Shack, for example), provided in the retailer's catalog or Web site, or you might have to decode them using your multimeter in the Ohms setting. You wire it as shown in Figure 17.2.

1. Solder a wire between the jack's ground lug and the circuit ground, in parallel to the speaker's ground connection (do not disconnect the speaker ground); or you can cut the ground wire in the middle and solder *both* ends of the cut to the ground lug on the jack (as shown in Figure 17.2).

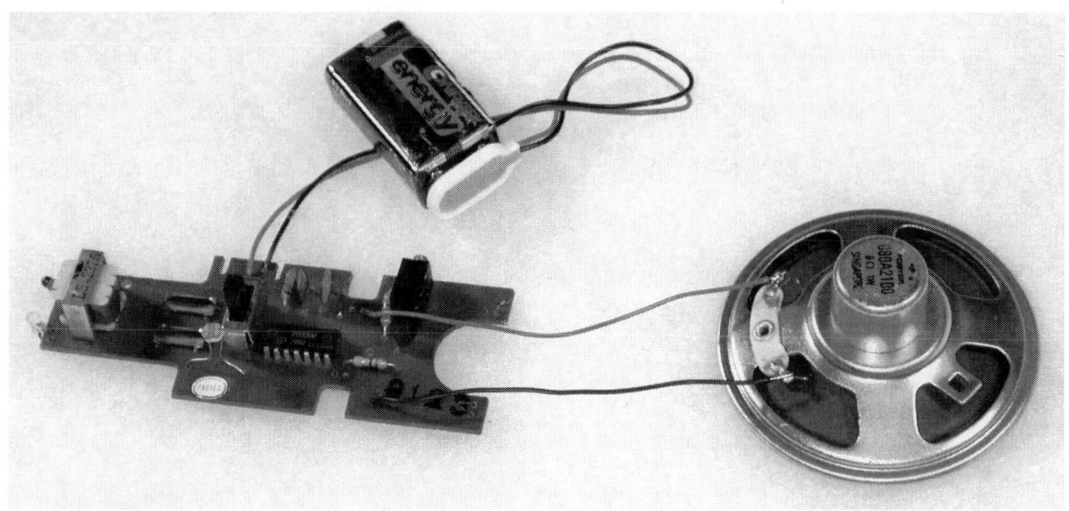

Figure 17.2
Wiring a cut-out jack.

2. De-solder the wire from the speaker's hot terminal and solder it instead to the jack's tip/hot.
3. Solder a new wire from the speaker's hot terminal to the NC terminal on the jack.

Turn on the circuit. You should hear the sound through the built-in speaker as usual. Now plug a cord from the cut-out jack to your amplifier—the speaker should shut off while the signal passes to your amp through the patch cord. If not, check your wiring logic and soldering joints—it's easy to make a mistake.

VERY IMPORTANT: BE VERY CAREFUL CONNECTING A HACKED CIRCUIT TO A MIXER OR AMPLIFIER THAT GETS ITS POWER FROM THE WALL. IT'S BEST TO TEST FOR ANY POSSIBLE ELECTROCUTION HAZARD BY GINGERLY TAPPING THE CIRCUIT BOARD, JACKS, AND POTS WITH A DRY FINGER AND FEELING FOR ANY BUZZ OR TINGLE. ALTERNATIVELY, LET A SQUIRREL RUN ACROSS THE BOARD.

Figure 17.3
Homemade stereo isolation box made with output transformers removed from an API mixer, Nicolas Collins.

And always remember the 14th Rule of Hacking:

Rule #14: Kick me off if I stick (Zummo's rule).

Always have a buddy nearby when there is a risk of electrocution, and chant Peter Zummo's mantra before you power up.

An excellent insurance against electrocution is to insert a DI box between your circuit and the AC-powered world—one of these can be purchased from almost any retailer selling electric guitars, microphones, keyboards, etc. Or, hacking at a somewhat more advanced level, you can wire what is known as an "audio isolation transformer" between the circuit's output and the rest of your sound system (see Figure 17.3). But if you are unsure of your hacks or the power grid, JUST STAY AWAY FROM ANY OUTLETS! Use a telephone tap on the speaker instead, if you want to get loud.

BATTERY SUBSTITUTION

Almost all toys use batteries that put out either 9 volts or 1.5 volts. Most 9-volt batteries look the same: bricks with two connectors that resemble android navels (both innies and outies). However, 1.5-volt batteries come in all sorts of packages: cylindrical ones, like D cells (the biggest kind, in the flashlights of Southern sheriffs), C cells (a little smaller), AA cells ("penlight" flashlight batteries), and AAA cells (even thinner and a bit shorter, like some metric mismatch of an AA battery), and button cells (the watch and camera batteries that are infernally small, come in a zillion different sizes and shapes, and are way too expensive and hard to find). A few button cells put out 3 volts (inside the casing they stack 2 tiny batteries), and there are some oddball batteries (like the one that looks sort of like an AAA but puts out 12 volts, often used in remote controls for garage doors), but the vast majority of batteries follow the above rules.

Nine-volt batteries are usually used singly, but 1.5 volt are often combined to add up voltage to power a circuit—commonly one will find them in sets of two, three, or four. The larger (and heavier) the battery the more *current* it provides, which means it lasts longer and can power a larger circuit, so:

Rule #15: You can always substitute a larger 1.5-volt battery for a smaller one, just make sure you use the same number of batteries, in the same configuration.

This means you can replace those little button battery cells with the same number of AA cells and run the circuit much longer and much cheaper, and afterwards you can find replacement batteries almost anywhere. All you need to do is:

1. Disconnect the existing battery holder, noting which wire connects to the "+" end of the battery stack, and which connects to the "–" end (look at the labeling on the batteries or holder, or measure the voltage with a meter).
2. Get a battery holder for larger batteries of your choice.
3. Connect it to the circuit, observing the proper polarity of the wires you de-soldered earlier.

Some low-current 6-volt circuits (i.e., using four AA, AAA, or button cell batteries) will run on a 9-volt battery, and might even react to the additional juice with extra perkiness, but others will succumb to cardiac arrest. Unfortunately there's really no way to know until you try it, so proceed with caution (and a duplicate circuit, if at all possible) and disconnect the new battery if you smell smoke or feel a component on the circuit board getting hot.

As you accumulate circuits you will be tempted to minimize your energy costs by using a single battery (or set of batteries) to run several devices. This, unfortunately, is not always a good idea: sharing often induces noise, weird interference, and interaction, especially if you try to connect more than one of the circuits to the same amplifier.

Rule #16: It's always safer to use separate batteries for separate circuits.

How big a battery to upgrade to has as much to do with fitting them inside the device as any electrical consideration, which brings us to our next topic.

PACKAGING

As your first hack nears completion you'll want to think about how to package it. You have a few basic style choices:

Stealth: Keep the original packaging, with added knobs, switches, and jacks, as needed (see Figure 17.4 and 17.5).

Camp: Go for other recycled housing, such as a cigar box, a cereal box, or a human skull (see Figure 17.6 and 27.5 in Chapter 27). David Tudor favored plastic soap boxes. Since the rise of the DVD, plastic VHS boxes fill the dumpsters outside rental shops, and they make perfect homes for circuit boards (see Figure 26.8). Cigar boxes are great because they can usually be had for free, and you can open them easily to change batteries or touch the circuit (see Figure 11.2 in Chapter 11). If the clock speed is controlled by photoresistors inside the box, the pitch will glide up and down as you open and close the box, like a cubist trombone. Unfortunately, the

Figure 17.4 Stealth packaging: bent keyboard by Alex Inglizian.

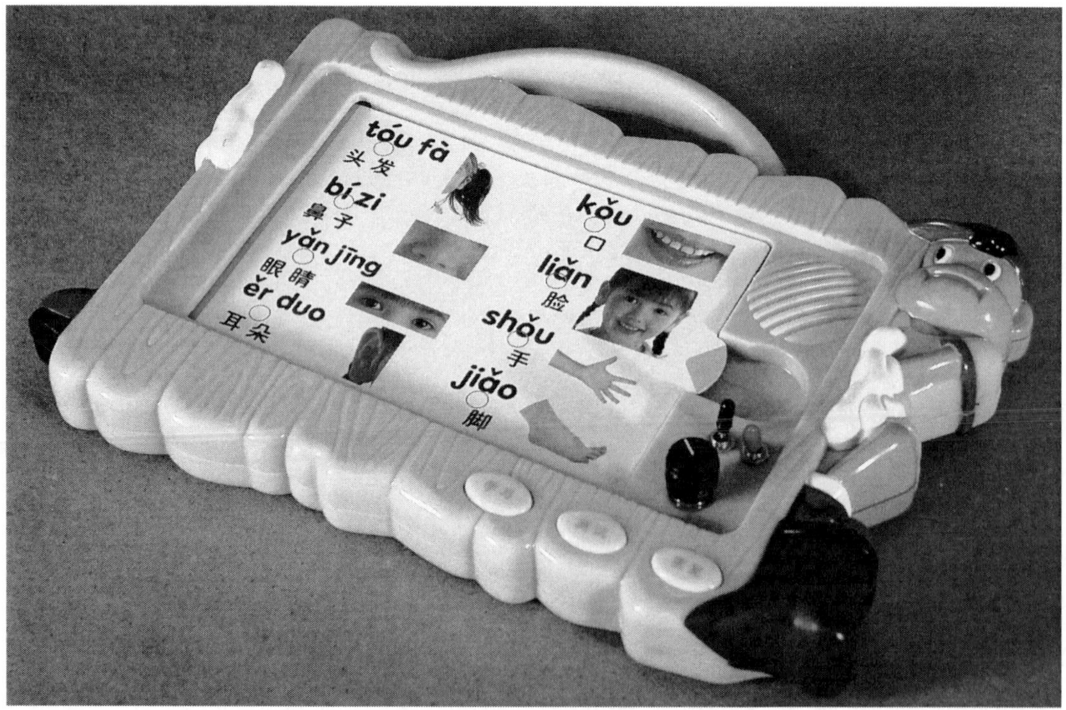

Figure 17.5 Stealth packaging: bent Chinese language trainer by Nicolas Collins.

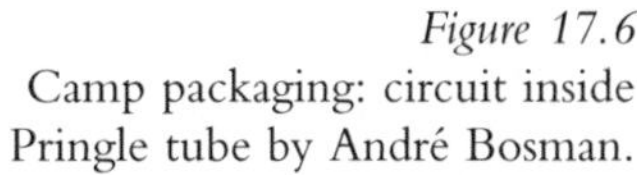
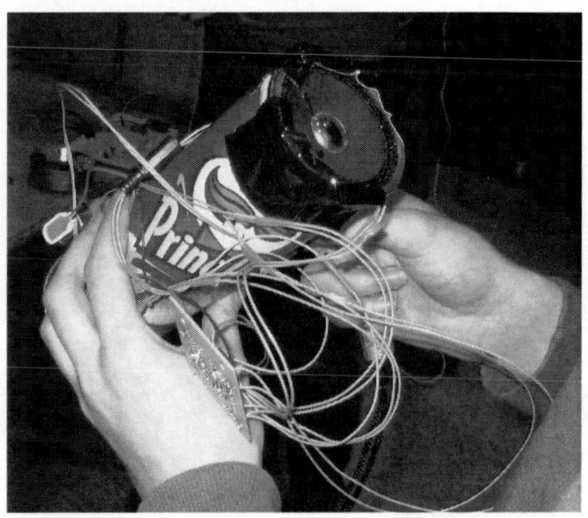

Figure 17.6
Camp packaging: circuit inside Pringle tube by André Bosman.

wood of a cigar box is sometimes a little too thick for some jacks and pots to mount easily—you may need to countersink the mounting holes in order to secure any nuts that screw down; alternatively, you can pass shielded cable through holes in the wood and solder plugs or jacks directly to the ends, and end up with a cigar box octopus. Altoid tins are good for small circuits (they hold a 9-volt battery neatly), but since they are made of metal you must make sure the circuit does not short out against the case—the discrete application of electrical tape to the bottom of the circuit board or the inside of the tin should do the trick (a sheet of thin cardboard works too); one can also buy plastic "standoffs" for this purpose. You can see Altoid tins in action throughout this book, the legacy of my son's long-standing obsession with minty breath.

Sandwich: Two slabs of acrylic plastic or thin wood with a circuit board in between (see Figure 17.7). Fast to make, and clear plexi lets you see what's going on inside.

Traditional: One of those plastic or metal boxes from Radio Shack, or numerous online retailers, that make your product look "professional" (or boringly geeky, depending on your perspective; see Figure 17.8). Remember that a bare circuit board will short out if placed in a metal box unless it is isolated from the metal, as mentioned for the Altoid tin, above.

The decision is partly topological (how do I fit in the all new jacks, pots, and switches?), partly practical (what's the easiest material to drill?), but largely aesthetic (what looks coolest?).

Now that your first hack is resting comfortably in a beautiful box, the time has come to show it off. Get out of the shop, pick up the phone and book that gig! Or at least upload something to YouTube.

Figure 17.7
Sandwich packaging: David Behrman's "Kim 1" microcomputer (1977).

Figure 17.8
Traditional boring packaging: small amplifier by Nicolas Collins.

PART IV

Building

CHAPTER 18

The World's Simplest Circuit: Six Oscillators on a Chip, Guaranteed to Work

You will need:

- A plastic prototyping board ("breadboard").
- 1 CMOS Hex Schmitt Trigger Integrated Circuit (74C14, CD4584, or CD40106).
- Assorted resistors, capacitors, pots, and photoresistors.
- Some small signal diodes, such as 1N914.
- Some solid hookup wire, 22–24 gauge.
- A plug to match your amp.
- A 9-volt battery and connector.
- A battery-powered mini-amplifier.
- Hand tools.

In the contrarian spirit of hacking, the first circuit we build from scratch is based on the *misuse* of an integrated circuit (IC) never intended for making sound. The "Hex Schmitt Trigger" is a CMOS digital logic chip consisting of six identical "inverters." An inverter takes a logical input, 1 or 0, and puts out its opposite (so 1 becomes 0, 0 becomes 1)—it is one of the fundamental Boolean building blocks that goes into the design of the computers we've come to depend on for so much of our work and play. This particular implementation of an inverter is useful to us because it runs for a long time on a 9-volt battery, it is very cheap, and it contains a circuit element known as a "Schmitt Trigger" whose fine points you don't need to understand at this point but, trust me, transforms the chip from a simple digital no-man (as opposed to a yes-man) into a versatile sound generator.

The Hex Schmitt Trigger may be labeled with the numbers 4584, 40106, or 74C14. There may be prefixes, suffixes, or additional number strings that you can ignore, but chips with a different "innerfix" may not work: if labeled 74HC14 or 74AC14 it will not run on a 9-volt battery, and so is less suitable for this project. If you purchase it online, sight-unseen, make sure it is specified as having "dual in-line" (DIP) packaging, and is not a "surface mount device" (SMD), as the latter format is infernally small and difficult for prototyping (this packaging requirement applies to all the chips we will use in this book).

Figure 18.1 shows the 74C14's internal configuration and external connections. This is the information you need in order to hook the chip up to a battery and the additional components needed to metamorphose a mute bug to a buzzing oscillator.

We will build our circuit on a "solderless prototyping board," commonly referred to as a breadboard. On it you can assemble and rearrange components quickly, without solder. It consists of a plastic block with lots of little holes, beneath which are springy channels of metal arranged in a matrix (see Figure 18.2). These strips, called "buses," run in one or two long horizontal strips along the top and bottom edges of the block, and in numerous shorter vertical strips that extend above and below a central groove.

There are some variations on the designs shown here: some breadboards are longer, or have only one horizontal bus at the top and bottom instead of two, or consist of multiple plastic modules on a metal plate. But they all employ the same underlying bus and matrix system.

The holes are of an appropriate diameter for the leads of most electronic components (resistors, capacitors, integrated circuits, etc.) and hookup wire. Thanks to the metal channels underneath, any wire or component lead stuck down into a hole is connected

Figure 18.1 74C14 pinout.

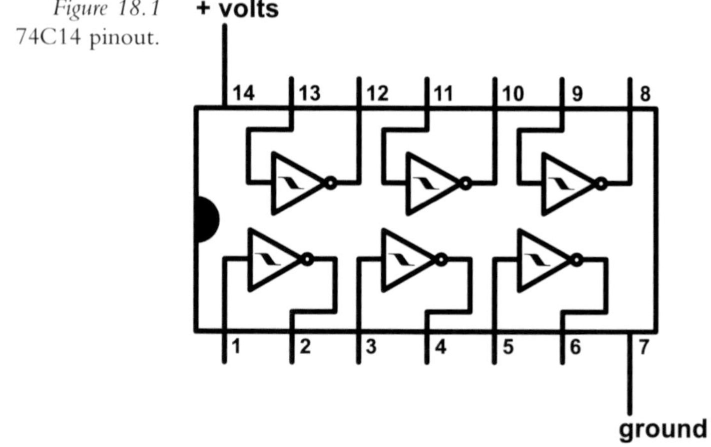

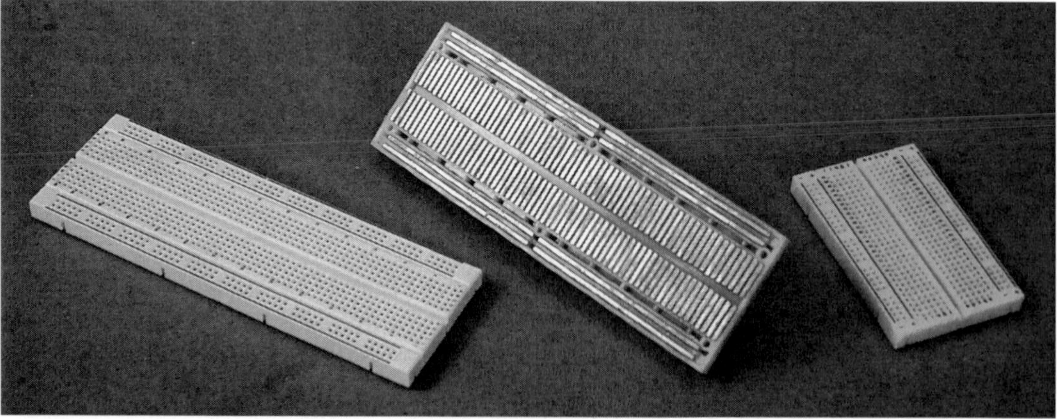

Figure 18.2 Some solderless breadboards, showing bus strips on underside of board (center).

electrically to anything else stuck into another hole in the same row or column of the matrix—as if you were clipping them together with test leads, but much tidier. Circuits are built up by inserting components into the holes on the board and linking rows and columns with short strips of wire (often called "jumpers").

YOUR FIRST BEEP

Place the breadboard on the table so the trough-like central groove runs horizontally, from left to right. Strip, twist, and tin 1/2 inch from the ends of each lead of a 9-volt-battery hookup clip, so that the stranded wire can fit neatly into a hole on the breadboard. Push the end of the red wire into one end of a horizontal bus on the upper edge of the breadboard and the black wire into a horizontal bus along the lower edge. Anything inserted in the upper bus will now be connected to +9 volts, while anything plugged into the lower one will be connected to ground (0 volts).

If the breadboard is longer than the one shown here, the upper and lower horizontal buses may be "broken" in the middle, rather than extending the full length of the board. If so, you will notice a slightly larger gap in the pattern of 5 holes, gap, 5 holes, etc. You can surmount this difficulty by jumping the gap with a small piece of wire, as shown in Figure 18.3.

Press a Hex Schmitt Trigger IC into the breadboard, taking care not to bend over any pins. Be sure that the notch and/or small dot is at the left side—these markings are there to help you orient the otherwise disorientingly symmetrical chip. Strip the insulation off either end of a short strip of solid wire and use it to connect between any free hole in the vertical matrix column containing pin 14 and any in the +9-volt horizontal bus. Use another piece to connect pin 7's column to the ground (0 volt) bus (see Figure 18.4). If there are two horizontal buses at the top and bottom be careful to link to the active one, and not its parallel twin (a very easy mistake to make).

Connect a 0.1 uf capacitor between pin 1 and the ground bus: push one lead of the capacitor into a hole in the same vertical bus into which pin 1 has been inserted, and the other end in to a hole in the horizontal ground bus—you can use any of the holes in these buses, but with components with closely-spaced leads (such as capacitors) it's easier to use holes that are close together (see Figure 18.5).

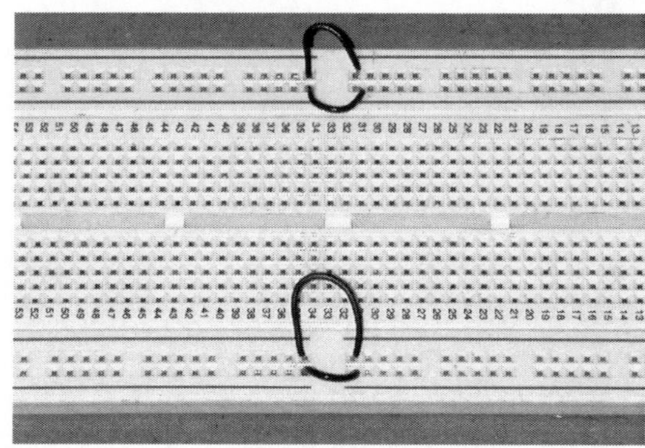

Figure 18.3
Using jumpers to join split power buses.

Figure 18.4 74C14 in place, with power connections.

Figure 18.5
Capacitor added.

Bend the leads on a resistor of about 100 kOhm so it resembles a difficult croquet wicket. Push one leg into the bus connected to pin 1 of the chip, which should already have one leg of the capacitor attached. Insert the other leg into the bus connected to pin 2 (see Figure 18.6).

Connect the tip/hot of a jack or plug to the pin 2 vertical bus, and the sleeve of the jack/plug to any point along the horizontal ground bus. You can solder some light-gauge solid insulated wire directly to the jack or plug, and strip and insert the other ends into the buses (see Figure 18.7); or you can clip test leads to short pieces of bare wire inserted in the appropriate buses, and clip the other ends of the leads to the input of your amplifier, via an open audio plug.

Snap the battery into its hookup clip. Connect the jack/plug to your amplifier, turn it on and listen (keep the volume low—this circuit is *loud*!) You should hear a strident, steady pitch—a square wave.

If not:

- Check your connections. It's very easy to be off by one hole to the left or right when you insert component leads and jumper wires.

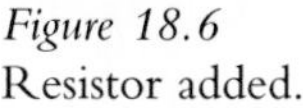
Figure 18.6
Resistor added.

Figure 18.7
Jack added.

- If there are double power buses at the top and bottom of the board make sure you've connected the chip and other components to the buses you are using (and bridge the gap if the buses are split, as shown in Figure 18.3).
- If the chip is HOT disconnect the battery immediately and start checking your wiring—in particular make sure you haven't put the chip in the board backwards, or reversed the battery connections. Hooking the voltage up backwards is one of the only things that can destroy this plucky little silicon warrior—if the chip was cooking for several minutes before you disconnected the battery you may need to start over with a fresh chip, after a hasty burial for the fallen hero.
- Make sure the component values are correct—too small a capacitor or resistor will cause the circuit to oscillate at a frequency too high for you to hear (you might notice the dog complaining, though); very large values in turn produce sub-audio frequencies, the slow tick-tock of a metronome, that may not be so audible through a tiny speaker.
- Check that none of the chip's pins have been folded, yogi-like, *under* the chip instead of inserted into a hole.
- If the circuit oscillates, but erratically, and is sensitive to placing fingers on chip pins and component leads, then you may have forgotten to connect the + or - power to the chip, or connected them to the wrong pins, or left one leg of the capacitor or resistor unconnected. Sometimes the circuit *almost* works without a direct power hookup, by sucking voltage through other connections you have made (spooky!)
- Confirm that the chip has none of the mysterious "innerfixes" I mentioned at the head of the chapter, since those chips only run reliably on 5 volts, and will behave very erratically with a 9-volt battery.
- If nothing else works, strip everything off the breadboard and start over, in a different location, with different components. This is one of the simplest circuit designs in the world, and usually works first time, every time. But with every connection comes the possibility of failure, and in rare circumstances it can take a second try to get things humming.
- Check all your connections again . . . and again . . . and again. . . .

YOUR FIRST SCHEMATIC

This is as good a time as any to start getting familiar with *schematic* representation, which conveys a circuit design independently of the physical arrangement of its components on a breadboard (see Figure 18.8).

The big triangle represents one inverter—any of the six in the 74C14 package; the flat left side is the input, which could be pin 1, 3, 5, etc., while the pointed right side is the output (2, 4, 6, etc.). The zigzag line is the resistor, between input and output. The two vaguely parallel lines (one straight, one curved) symbolize the capacitor, connected between the inverter's input and the ground bus. Ground is represented by the weird runic arrangement of three lines. The signal output appears as a single line with an arrowhead at the end. This is the signal that would go to the tip of a plug; the ground half of the connection is implied, rather than drawn in (see Rule 10). Likewise the power connections to the chip (+9 volts to pin 14, ground/–9 volts to pin 7) are implicit in this drawing.

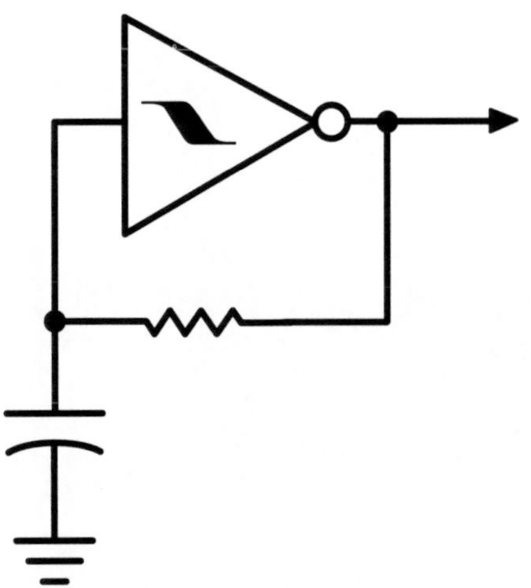

Figure 18.8
Schematic representation of our oscillator.

The translation from this symbolic schematic to objects on a breadboard may not seem obvious at first, but once you get more familiar with the language of electronics you'll see that the schematic is a useful way to represent the way a circuit *functions*, rather than just the way it goes together. As I introduce new circuits I'll try to present both a picture of a typical physical arrangement on a breadboard and a schematic representation. It's worth familiarizing yourself with the language of schematics, since it opens up a huge world of circuit possibilities—in books, in technical magazines, and on the web.

VARIATIONS

You are justified in taking great satisfaction in producing your first electronic tone, but after a while you may wish for a change of pitch. Try substituting different resistors and capacitors and listen to the effect. Take a pot or photoresistor and use it instead of a fixed value resistor—the wiggly resistor symbol in the schematic above can be taken to mean any form of resistor, including variable ones such as pots or photoresistors. We can specify that the resistor is a potentiometer if we use the symbol shown in Figure 18.9—"A" and "C" are the ears of the pot, while "B" is the nose (as demonstrated in Figure 13.6 in Chapter 13.) A pot gives you direct, repeatable control of frequency.

A photoresistor-controlled version of the oscillator is shown in Figure 18.10. A photoresistor turns this simple oscillator circuit into a wonderful, Theremin-style instrument controlled by light and shadow (see Figure 18.11 and audio track 12). You can wiggle your hands between the circuit and any ambient light; or you can go at it with a flashlight and fan or other forms of interruption (as described in Chapter 15). What you lose in the accuracy of control you gain in performability.

The capacitor determines the *range* through which the variable resistor will sweep the pitch. Larger values (greater than 2.2 uf or so) lower the frequency from the range of audible pitches to that of rhythm—you'll hear the oscillation as a tick-tock instead of a buzz. With a very small capacitor (less than 0.001 uf) the circuit will emit ultrasonic tones that only dogs and bats can hear.

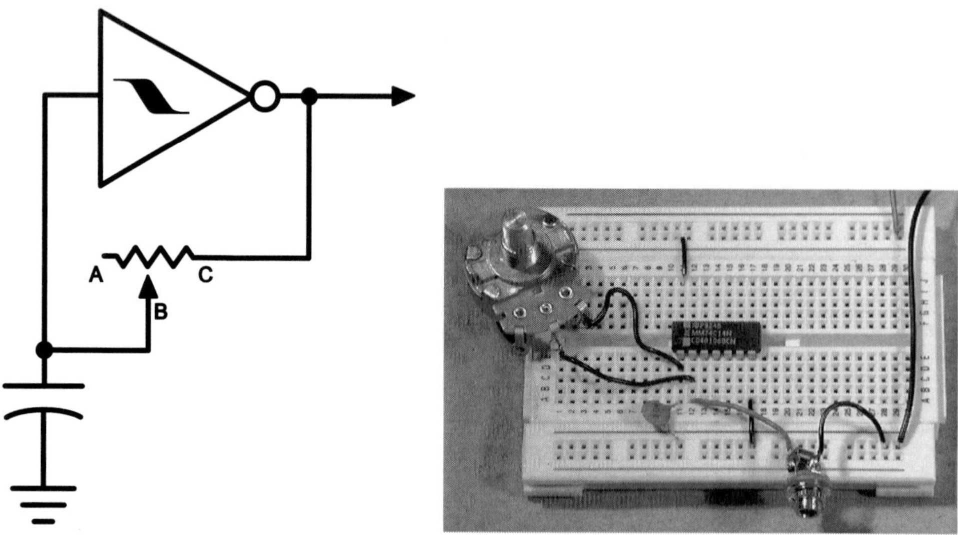

Figure 18.9 Potentiometer-controlled oscillator: schematic and photo.

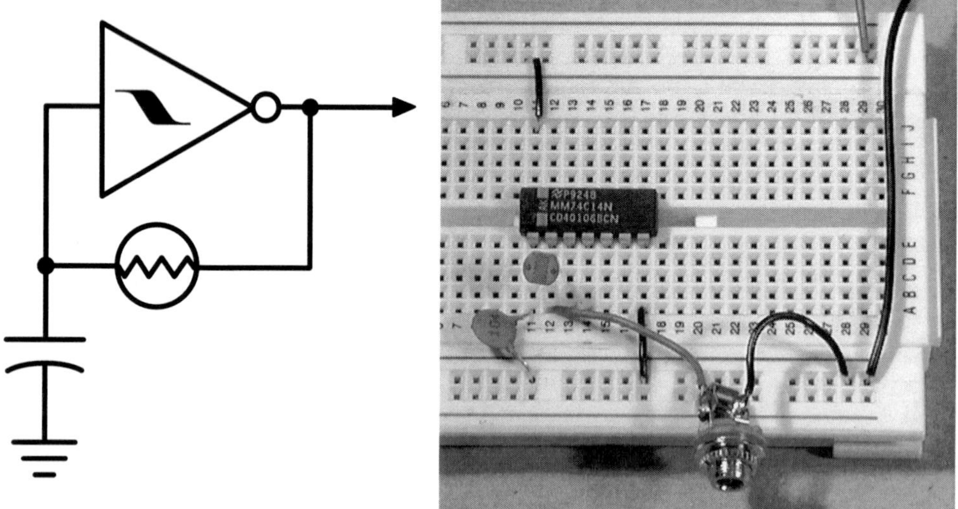

Figure 18.10 Photoresistor-controlled oscillator: schematic and photo.

Capacitors 1 uf or larger are usually found in a style called *electrolytic*, and take the shape of little plastic cylinders. Electrolytic capacitors have polarity, like a battery: one leg will be shorter and labeled "–" with a string of minus symbols or a black stripe along one side of the cylinder. When using electrolytic capacitors, be sure to observe the correct polarity and hook the "–" side to the ground bus (in a schematic the "–" side of the capacitor is indicated by the curved line). The circuit will work if the capacitor is connected backwards, but it may not be as stable or go as low as you expect. Electrolytic capacitors usually have their value printed quite clearly.

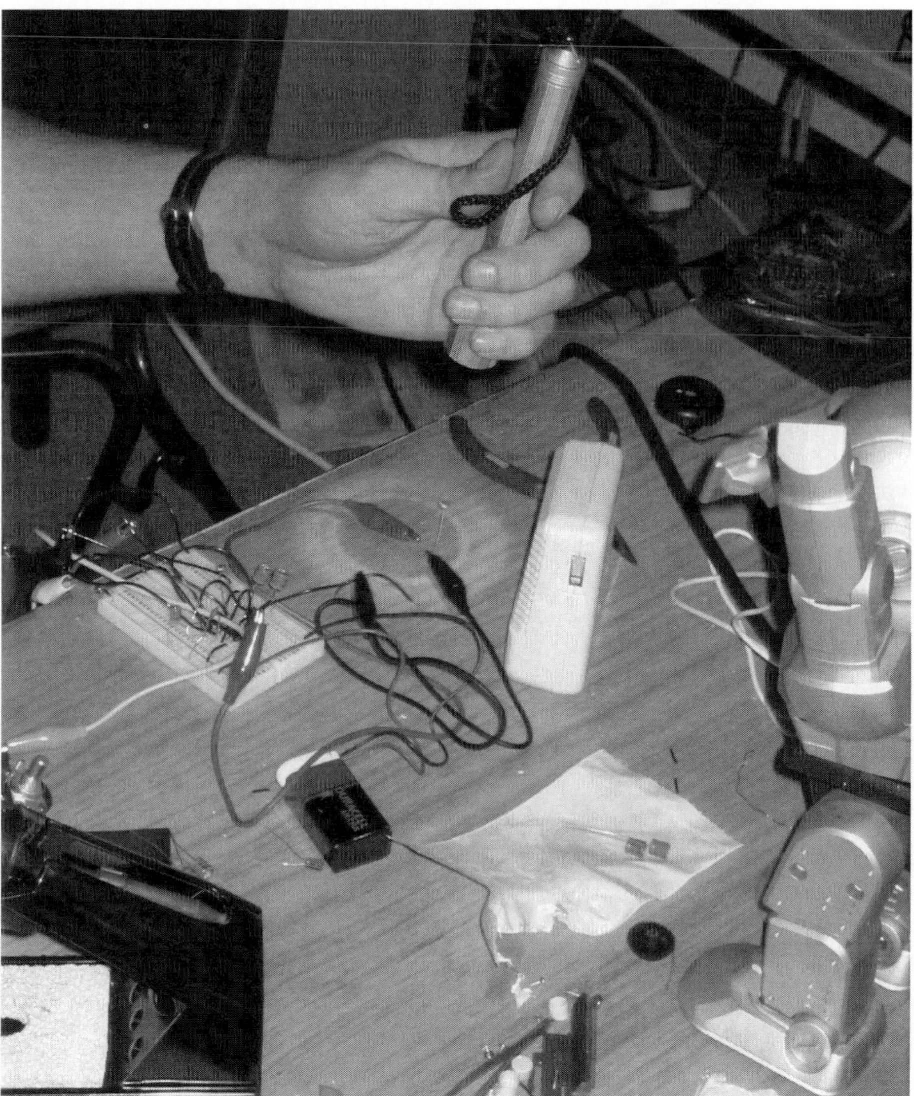

Figure 18.11 Playing a photoresistor-controlled oscillator with a flashlight.

The smaller value, flat capacitors often employ an arcane code of "most significant digit + multiplying factor" (similar to the resistor color code): 104 = 0.1 uf, 103 = 0.01 uf, 202 = 0.002 uf, etc. The easiest way to familiarize yourself with this argot is to buy a few capacitors of known value and look for correspondences between the numbers printed on the capacitor and its actual value, as specified in the packaging or ordering information. Or you can download a capacitor identification table from the Web.

When you use a pot you will notice that at one extreme of the rotation the pitch will go too high to be heard—often regardless of the size of the capacitor when the resistance approaches 0 Ohms. This typically does not cause the circuit to hang up and crash, as did the toys when the clock was pushed too fast, but the chip drains more current at very high frequencies, which shortens battery life. More significantly, this ultrasonic dead zone limits the useful range of the pot, forcing you to make all your musical decisions within fewer degrees of rotation (like playing on a 1/4-scale Suzuki violin if you have big hands). Therefore, if you plan to use a potentiometer you may want to put a fixed resistor of modest value in series with one leg of the pot (see Figure 18.12) to set a maximum pitch that is within the range of hearing—as we demonstrated in Chapter 14 when we limited the upper frequency of a toy clock (see Figure 14.1). Start with something around 10 kOhm, then experiment with various larger and smaller values to find the one that provides the best balance of resolution and range.

Similarly, if you wire a pot in series with a photoresistor, you can use the pot to set the upper pitch limit of a light-control instrument, combining the accuracy of the pot and expressiveness of the photoresistor, as we did with the toy clock in Chapter 15 (see Figure 15.4).

Speaking of which, now would be an excellent time to go back to Chapter 15 and review our collection of alternative resistors, any of which can be used with our new circuit. In fact, many of the subtle differences in response and range are more apparent when the resistor is coupled to a simple oscillator than to a toy that produces more complex sound patterns. To recap (and throw in a few new ones as well), the most promising things to try would be:

- Potentiometer with frequency-limiting resistor in series.
- Photoresistor.

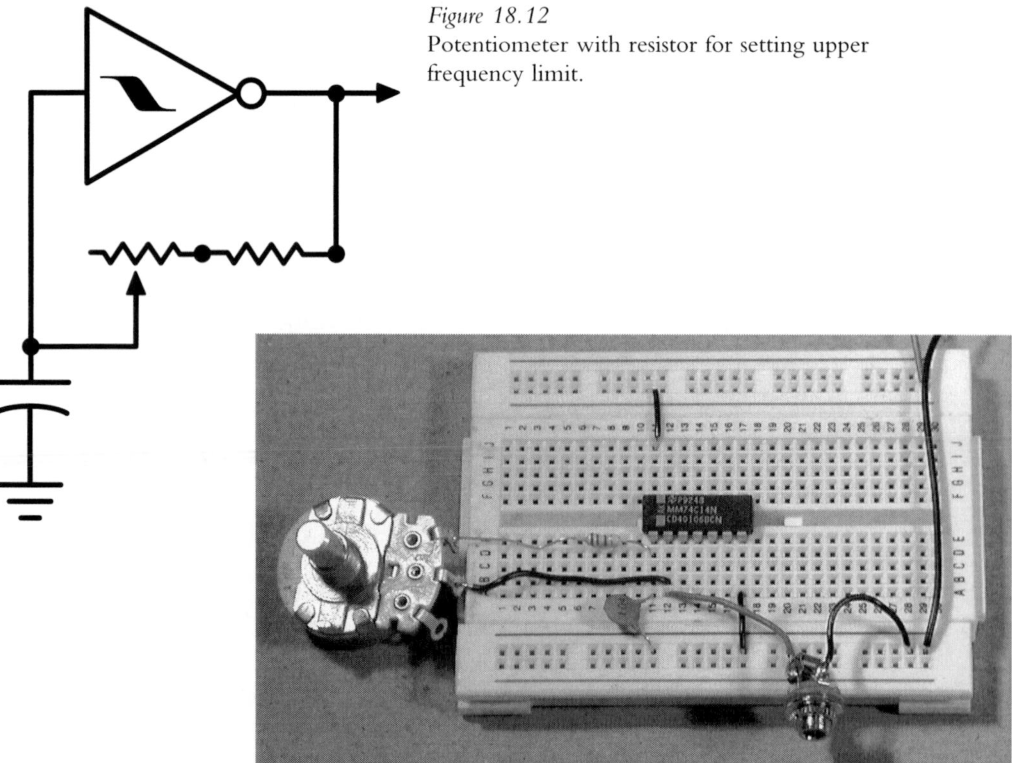

Figure 18.12
Potentiometer with resistor for setting upper frequency limit.

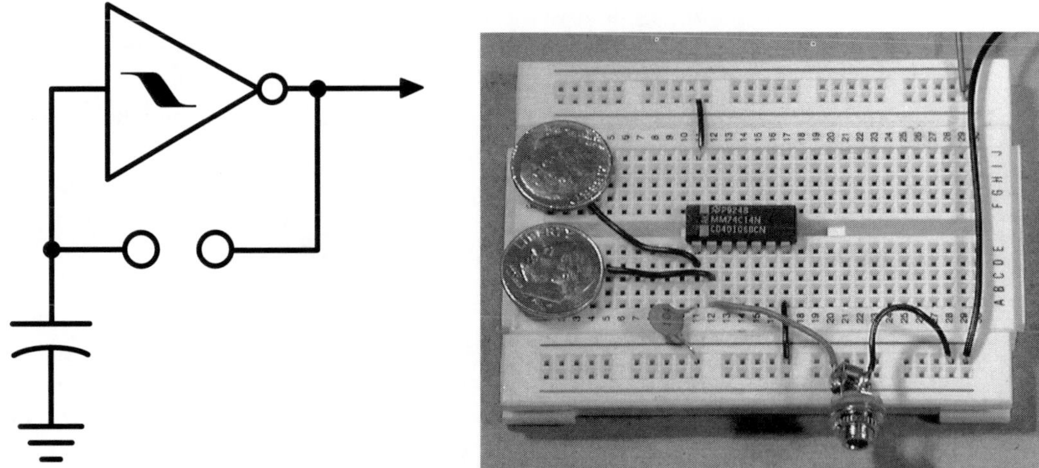

Figure 18.13 An electrode-controlled oscillator.

- Photoresistor in mouth.
- Potentiometer with photoresistor in series (see Figure 15.4 in Chapter 15).
- Electrodes (coins or bare wire) (see Figures 18.13 and 15.6). Good party fun can be had with "social electrodes": gather a group of friends and ask them to hold hands in a line; give one electrode to the person at either end; listen to the group sweat (you'll need to use a rather small capacitor in the oscillators to get an audio pitch rather than rhythm, since the resistance of several bodies in series is very big). See the video of Lauren Carter's and Joe Gimm's "Picnic" on the DVD. Dan Wilcox harnesses huge pieces of metal for electrodes—see his video on the DVD.
- Potentiometer with electrodes in parallel (see Figure 15.7).
- Pressure-sensitive foam rubber (anti-static foam) (see Figure 15.8).
- Corroded metal.
- Pencil lead (see Figure 15.9).
- Graphite drawing on paper (see Figure 15.10).
- Magnetic tape.
- Vegetables or fruit (for nice examples of this see Grégoire Lauvin's video on the DVD).
- Another circuit. Use clip leads to connect from the input and output of the inverter to any two points on another circuit board in lieu of a resistor. Try dead cell phones, fax machines, computer motherboards, radios, TVs, DVD players, disk drives, answering machines—anything. Just make sure the appliance is switched off and NOT PLUGGED INTO THE WALL!!

WHY? (IF YOU CARE)

Our circuit oscillates because of the principle of argumentation. Each inverter stage, represented by the big triangle in the schematic, puts out the opposite of whatever signal appears at the input: if a binary "1" is applied to the input, then a "0" appears as the output. In a circuit running on a 9-volt battery, that theoretical 1 is represented by a 9-volt signal, while logical 0 equals actual 0 volts. The 0 at the output flows through the resistor back to the input. When the 0 appears back at the input it now causes the output to switch to 1 (9 volts), which flows back to the input and the whole process begins again, causing the circuit to flip back and forth between two states, generating a square wave that alternates between 0 and 9 volts. The speed of the flip-flopping is the pitch we hear, and it depends on the values of the resistor and capacitor: just like in our earlier clock experiments, the smaller the values the higher the pitch.

It's like the Monty Python argument sketch, transferred to a bar: I disagree with everything you say, so our output keeps flipping between yes and no according to how fast each of us can reply. The resistor and capacitor act like beer—the more you add the slower the argument goes, ergo the lower the pitch. Having brushed you off earlier, I will now confide that the appropriately Germanic-sounding Schmitt Trigger part of the inverter prevents indecisiveness in the argument: the inverter snaps completely from one state to the other, from 0 to 1 and back, and never vacillates in between or proffers a "maybe" at the output.

Since I've already started down the slippery slope of liquid analogies, I can offer one more parable for those of you who want a deeper understanding of the interaction of the capacitor and resistor. Pretend for a moment that the inverter processes water rather than voltage (see Figure 18.14). Water flows from the output back to the input through a pipe, which is our resistor. If the pipe is skinny (high resistance, left side of figure) it takes longer for the water to flow back than if it is fat (low resistance, right figure). At the input to the inverter is a basin, which represents the capacitor. Halfway up is our "logic line": if the water is below that line we have a 0; when it rises above we have 1. If the basin is a wide washtub it takes longer to reach the line (big capacitor, left figure) than if it is a narrow modern vase (small capacitor, right figure), since the pipe outflow must top up a greater volume of water. Got it? Now commence the rinse cycle.

Remember how I warned you earlier that the output of this circuit is very loud? And just above I mentioned that its square wave swings from 0 to 9 volts? This is more than 10 times the signal put out by a CD player or iPod, and more than 100 times

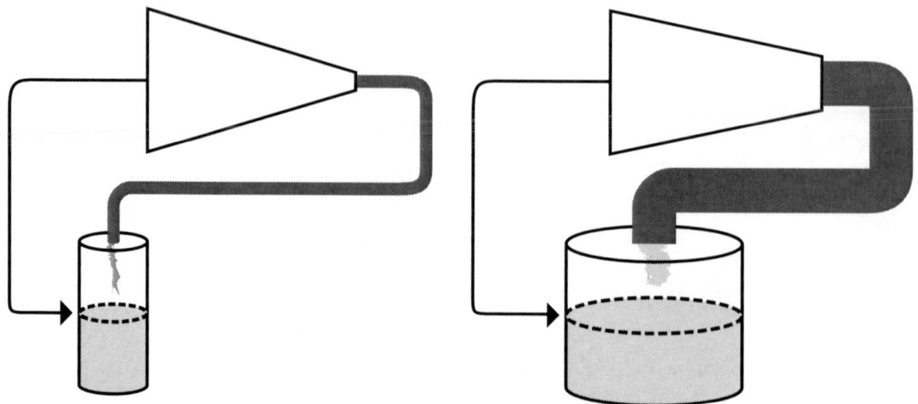

Figure 18.14 Pipe and bucket model of inverter signal flow.

greater than the output of a guitar or microphone. If you've been using a nice little high-gain amplifier, like the Radio Shack mini-amp or one of those shrunken Marshalls, you may notice that the oscillator saturates and overdrives the input with the volume control barely open a crack. Soon we'll learn how to cut down the oscillator's level to a more workable range; but in the meantime, if you have access to a simple pair of amplified speakers (such as are used with computers or MP3 players) this would be a good time to change over to this from a mini-amp with a high-gain input stage.

POLYPHONY

As you might be able to intuit from the pinout in Figure 18.1 or any dormant knowledge of classical Greek, the *Hex* inverter has six identical sections. You can make an additional oscillator with any of these sections, just duplicate the connections we made for our first oscillator with another set of components, attached to another set of pins: capacitor between any input and ground; resistor, pot, photoresistor, etc. between that input and its appropriate output; connect the output via a jack to the amplifier.

Remember when working on the "top" side of the chip (pins 8–14) that the capacitor must go between the chip and *ground*, not to the + supply that mirrors the ground bus on the upper side of the board. If there are two parallel horizontal buses at the top and bottom of the board, you can connect a wire jumper between the lower ground bus and the upper bus that is not being used for +9 volts; the second upper bus is now an additional ground bus, into which you can insert the legs of the capacitors used on the upper half of the chip (see Figure 18.15). Be careful to connect to the new ground bus, not the 9-volt bus, and don't accidentally link the ground and power buses together with a wire, or you will soon have a very hot, very dead battery on your hands.

Alternatively, you can push the grounded legs of the capacitors into any free *vertical* bus on either side of the chip, and jumper that vertical bus to the lower ground bus, creating a secondary, ground "satellite-bus" (see Figure 18.16). Finally, if the breadboard only has one horizontal bus at the top and one at the bottom, you can use *both* for ground (link them as in Figure 18.15), and connect the +9 volt (red) wire of the battery holder directly to pin 14, rather than via a long bus.

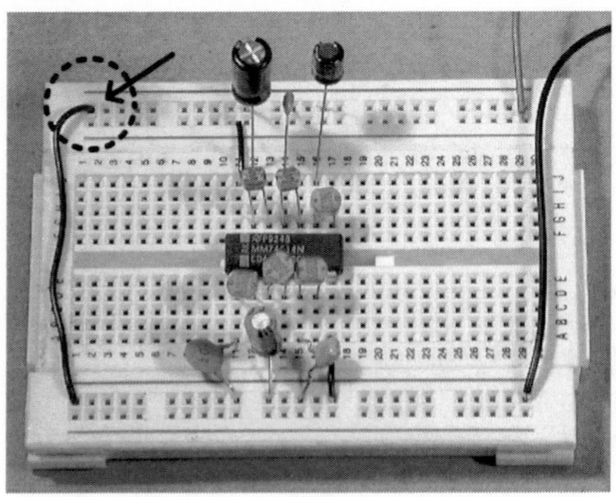

Figure 18.15
Six-voice oscillator, with capacitors connected to secondary ground bus parallel to + voltage bus along upper edge of breadboard.

Figure 18.16
Six-voice oscillator, with capacitors connected to satellite mini ground bus.

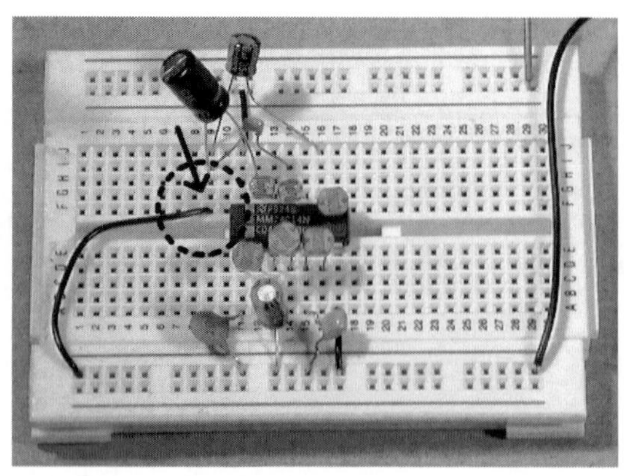

You can use a different-size capacitor for each oscillator, so each of the six covers a different range, from low BPM to ultrasonic pitches. Or you can add a switch to select among different capacitors for each oscillator—a rotary switch gives you clear control, while a multi-position tilt-switch makes a nice wobbly interface (see Chapter 16 for switch information).

A joystick is an expressive device for controlling pairs of oscillators (see Figure 18.17). You can salvage one from an unneeded game controller (perhaps the one you will eviscerate in Chapter 28). A proper analog joystick consists of two potentiometers controlled by the *X–Y* movement of a shaft (some games just use four switches activated by the four quadrants of movement). If you de-solder the joystick from its original circuit you will notice the familiar three terminals on each of the small pots. Connect the nose and one ear of each pot to each oscillator. You may need to try various connections before you arrive at the most satisfying interaction between the two oscillators, but when you get there you'll be rewarded by square waves careening off each other like radio-controlled mosquitoes.

So far I've assumed that each oscillator on your breadboard will be connected to a separate output jack, each of which is in turn connected to a separate amplifier or channel of an external mixer (remember that every oscillator output needs a ground connection to its output jack, as well as the signal from the output pin). In the interest of efficiency and economy, however, you may want to mix all your voices down to one (mono) or two (stereo) outputs. Sadly, you can't just jumper all the outputs together with bare wire, since this confounds the logical decisions the inverters must make in order to oscillate: it sends an unpredictable mix of 1s and 0s to each input, causing an indecision that even Herr Schmitt cannot overcome—and when an inverter can't decide what to do it does nothing, which is very quiet indeed.

Figure 18.17
Joysticks.

Instead, to mix more than one oscillator to a single jack, connect each output to the jack through a resistor of about 10 kOhm (see Figure 18.18)—anything from about 3 kOhm to 1 mOhm will work, but use the same value for every oscillator if you want them all the same loudness; otherwise those that pass through the smaller resistors will be louder than those mixing though the larger ones (see Chapter 27 for more information about "proper" mixing).

You can add a tilt-switch to select different oscillator outputs as you tip and wobble the circuit, instead of mixing them all together all the time.

The resistors form a simple linear mixer, through which you can hear each oscillator distinctly—like a recording mixer or PA mixer with the knobs glued in place. If you mix the outputs together using a component called a "diode" instead, the individual oscillators will interact and distort in an archetypically "electronic music" way—they

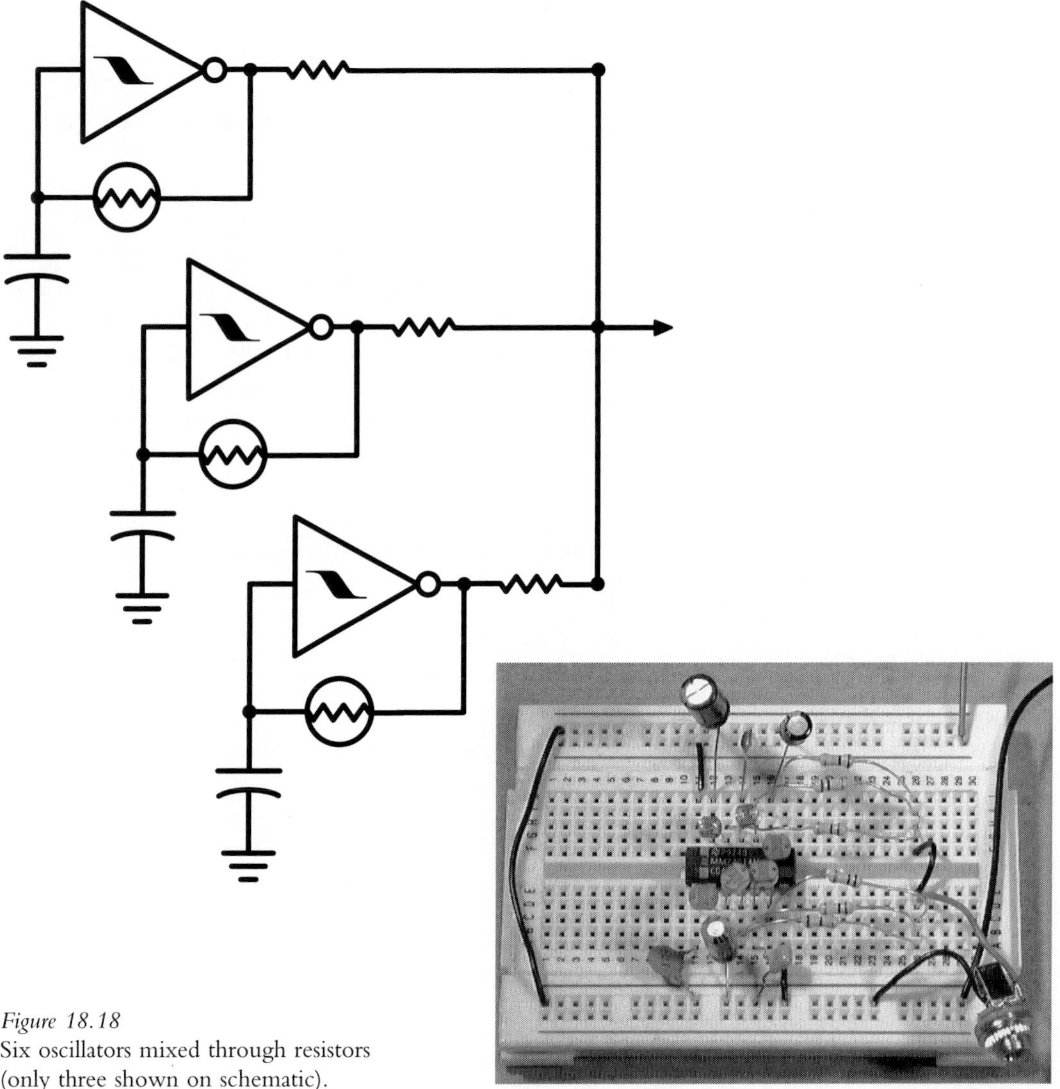

Figure 18.18
Six oscillators mixed through resistors (only three shown on schematic).

produce a "ring modulation"-type sound, in which sum and difference frequencies are exaggerated and the individual source pitches obscured. Pull out the resistors we used above and substitute diodes as shown in Figures 18.19 and 18.20.

If some of the oscillators are running at low frequencies (in the metronome range) while others are in the audio range, the low ones may appear to gate the high ones on and off (see Chapter 20 for more advice on cross-modulating oscillators). Alternatively, if all the oscillators are tuned to ultrasonic frequencies (use very small capacitors, less than 0.001 uf), the diode mixing casts difference-tone "shadows" down into the audible range—very spooky sounds indeed.

Diodes are odd little devices that only allow signals to pass in one direction. Current flows toward the leg indicated by the single stripe that appears at one end of the component. This mixing scheme works best when *all* of them face the same direction, with all the stripes at the "jack side" of the circuit. Exactly why they make such a cool-sounding mixer for oscillators is a somewhat complicated story, and pretty irrelevant to appreciating their sonic contribution, so I won't try to explain it. Which brings us neatly to Rule #17:

Rule #17: If it sounds good and doesn't smoke, don't worry if you don't understand it.

If you sum three or more oscillators together with diodes you might notice that the signal gets oddly noisy or drops in level; if you don't like these artifacts (which depend largely on the nature of the input stage of amplifier or mixer you plug into) you can mix three oscillators to one jack and three to another for a nice stereo sound field, and/or connect a 10 kOhm resistor between the output point (where all the diodes tie together) and ground—this is indicated as "R load" in the schematic. This latter fix helps make the behavior of the circuit more predictable as you move between different sound systems (for better or for worse).

Six square-wave oscillators make a wonderful din. With six photoresistors or electrode pairs, control is somewhat unpredictable but nonetheless intuitive and playable. There's a glorious tradition of music made with masses of homemade oscillators, from David Behrman (see audio track 11) through Voice Crack and beyond. Indulge in it.

CAVEAT

There is a complement of sorts to Rule #17 that is worth noting before we get much further in the construction of our own circuits:

Rule #18: Start simple and confirm that the circuit still works after every addition you make.

Don't assume after you get one oscillator buzzing that you can smote the remaining five in one blow. When you find yourself gazing down at a rat's nest of wires that makes neither sound nor sense, strip your board back to one oscillator and start over, one voice at a time, listening as you go. Make sure you can hear each new voice clearly before building the next.

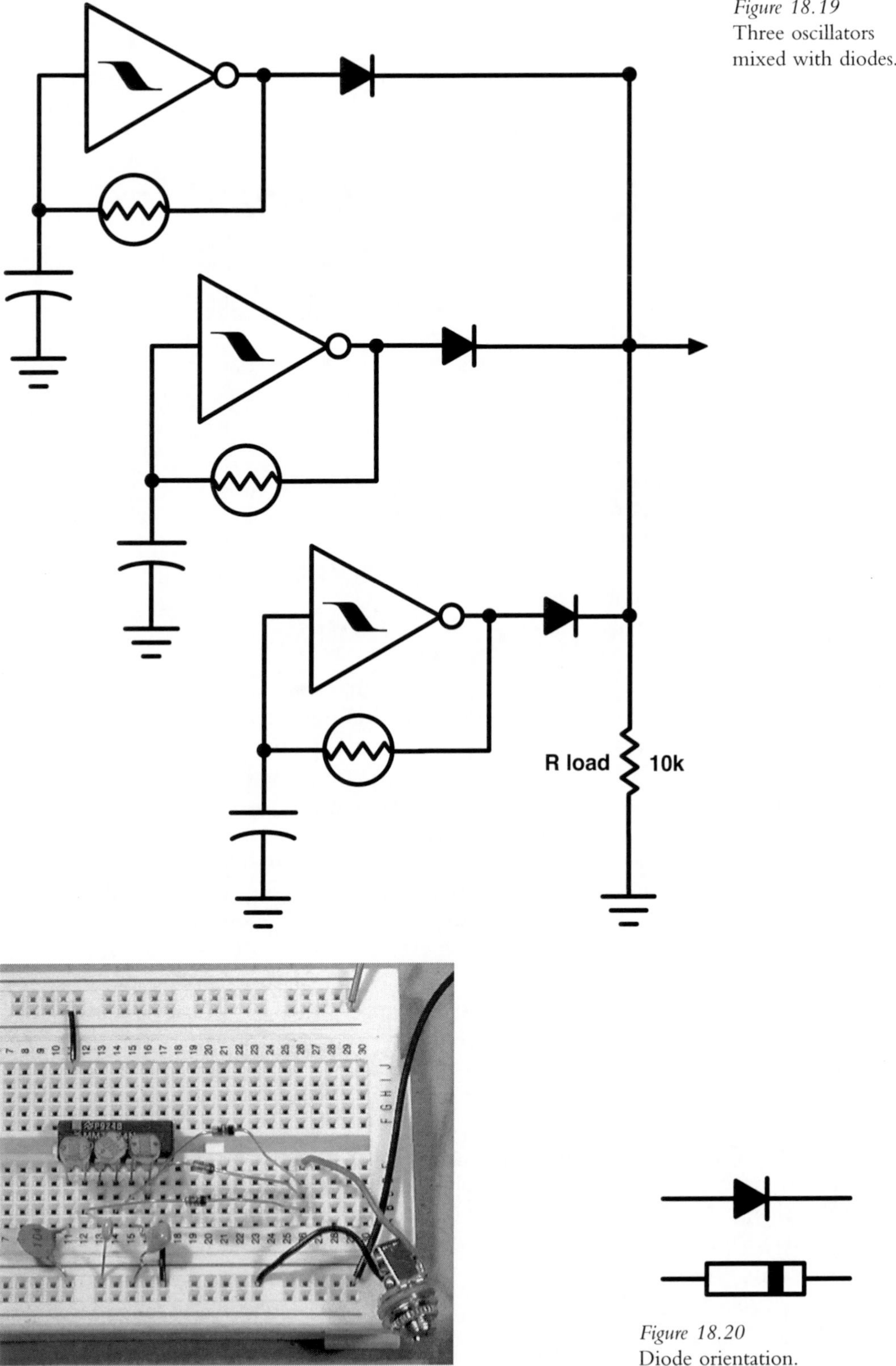

Figure 18.19
Three oscillators mixed with diodes.

Figure 18.20
Diode orientation.

CHAPTER 19

From Breadboard to Circuit Board: How to Solder Up Your First Circuit

You will need:

- Your breadboarded circuit from the previous chapter.
- A full duplicate set of parts used in the circuit.
- A prototyping circuit board.
- 14-pin IC socket.
- Solid and stranded hookup wire.
- Hand tools and soldering iron.

The breadboard is great for trying out designs: components can be easily swapped, and mistakes undone without burning your fingers. But it's not very stable if you want to take your music on the road. At some point you may wish to solidify the circuit. This means soldering the components down onto a generic "Printed Circuit Board" (PCB) specifically intended for prototyping and developing new designs (as opposed to the board you find inside a toy, radio, or mixer, which was designed for one particular circuit). You don't *have* to turn on your iron now—you can skip over this chapter and continue breadboarding more circuit variations before settling on one to solder. But if you're happy with something you've done, and want to preserve it before clearing off the breadboard, read on.

Prototyping circuit boards come in various styles, with patterns of individual copper pads that can be linked together with components and bits of wire any way you wish (see Figure 19.1). Some designs, such as Radio Shack part #276–170, or part #32–3008 from Unicorn Electronics (see Appendix B), mimic a typical breadboard almost exactly, and make it much easier to transfer your circuit from breadboard to soldered board (see Figure 19.2).

If the board you've chosen is larger than needed for your one-chip circuit, you may want to cut it down before you start soldering on parts—scribe along the dotted line with a sharp knife, then snap the board over the edge of a table. You can use the vacant half later. Alternatively, you can build your first circuit at one end, and add designs as you develop them.

Once you've got your circuit board cut down to size, pick up a 14-pin IC socket. The socket is essentially a tiny version of the breadboard. Since chips can be easily damaged by the excess heat of sluggish soldering, and are very difficult to de-solder if

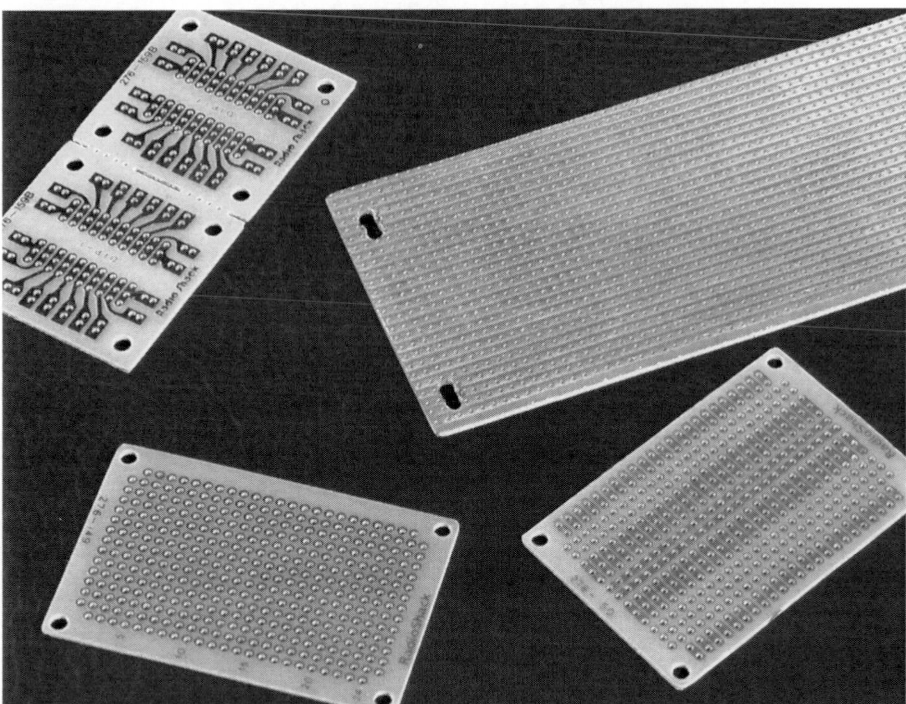

Figure 19.1 Assorted printed circuit boards, showing solder side.

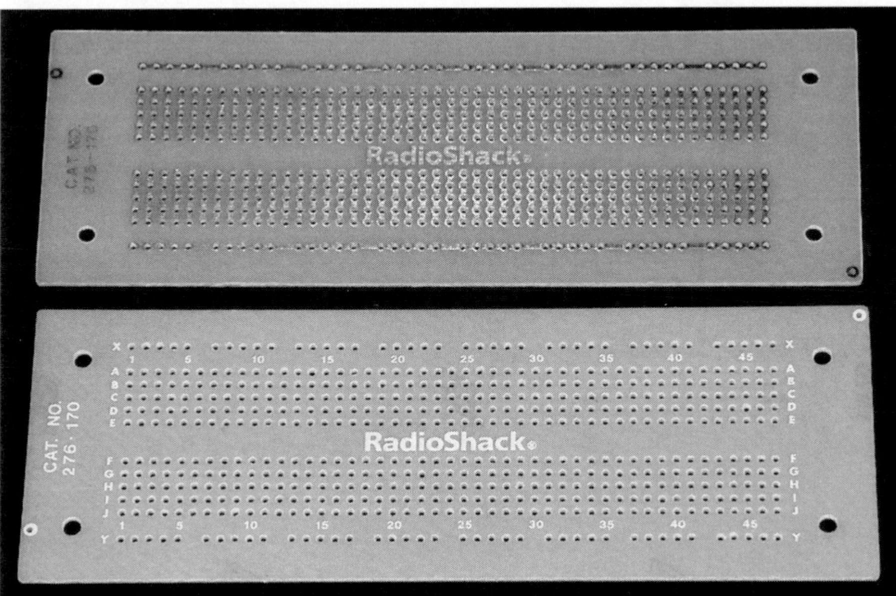

Figure 19.2 Printed circuit board mimicking breadboard, showing solder side (top) and component side (bottom).

they have to be removed, we will solder in a socket instead, and wait to insert the chip itself until after we have finished all our hot work. If for some reason the chip doesn't work (or fails later) we can remove and replace it as easily as on our breadboard.

Place the socket on the side of the board that does *not* have the copper paths and pads (this is the *component* side—the other side is the *solder* side). Push the pins gently through the holes, ensuring that, if the traces form a matrix of rows and columns like on the breadboard, the socket is similarly positioned, with rows fanning out from each pin, not shorting them all together. Make sure all the pins go through fully, and none are bent over on top of the board. Solder the socket pins carefully: avoid letting blobs of solder short together adjacent pins or copper traces (see Figure 19.3).

Now collect a full identical *duplicate* set of resistors, capacitors, and other components that you used on your breadboard. Don't be cheap:

Rule # 19: Always leave your original breadboard design intact and functional until you can prove that the soldered-up version works.

This makes it much easier to debug any mistakes, by comparing the working version on the breadboard with the miscreant on the circuit board (see Figure 19.5).

Bending the leads as necessary, place the parts, one by one, on the component side of the board. Insert the leads into the appropriate holes, solder them into place on the solder side of the board, and clip off the excess wire before going on to the next part. Follow *exactly* their placement on your breadboard. You can press resistors, capacitors, etc. snugly down against the circuit board, rather than leaving them waving in the air.

Use thin, insulated solid wire to interconnect components: strip insulation off the ends (as you did to make the jumpers for the breadboard) or use bare wire if there is no danger of it shorting against another wire or component lead; link circuit traces by running the wire along the component side of the board, passing it through holes in the appropriate strips, and soldering it to the pads on the solder side (see Figure 19.4). Make sure the uninsulated bits of wire or component legs do not short against each

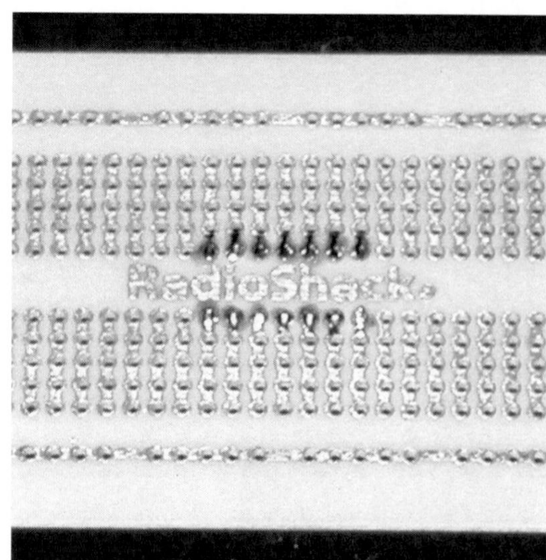

Figure 19.3
Socket soldered to PC board, solder-side view.

Figure 19.4
Completed circuit.

other or adjacent components or solder pads. Use solid wire for the on-board jumpers (since it's easier to work with) but stranded wire is better for attaching the pots and jacks because the wire can flex easily without breaking when you mount the circuit in a case. And don't forget that every jack needs signal connection and a ground wire (Rule #10). Depending on how you plan to package your circuit, you may want to extend the leads of any photoresistors with stranded wire, so that you can mount them in holes through the box (wrap some electrical tap around the bare legs to keep them from shorting).

If you use a circuit board that mimics a breadboard closely (such as the Radio Shack or Unicorn board), it will have horizontal buses at the top and bottom of the board that you can wire just like on the breadboard. If you are working with non-clones, however, you may have to create your own satellite mini-buses for grounds as we explained in the previous chapter (Figure 18.16). If you made a multi-channel oscillator and the circuit board only has one horizontal bus along each edge you may need to use *both* buses as grounds for the capacitors, in which case you should connect the +9 volts directly to pin 14 (see Figure 18.5).

As mentioned earlier, this circuit has a very low current drain, and will run a long time on a single 9-volt battery. But unless your uncle works for Union Carbide you'll still want to turn the thing off if you're not using it for a while. You can always just disconnect the battery from the clip when you put the circuit to bed. But if you prefer a real on/off switch:

1. Pick up a SPST or SPDT toggle switch.
2. Solder the "NO" terminal to the +9-volt line (red wire) of your battery clip.
3. Solder a wire from the "C" terminal of the switch to the +9-volt bus or directly to pin 14.
4. Finish by soldering the ground wire (black) from the battery clip directly to the ground bus on the board.

After all soldering is finished, carefully insert a (duplicate) chip in the socket, and check to make sure its orientation is correct (i.e. pin 14 goes to +9 volts, pin 7 to ground, not vice versa). Look to see if you made any unintentional "solder bridges" between traces. (After transferring your first design from breadboard to circuit board you will see how important it is to have a good soldering iron with a very fine tip.) Compare your connections against the breadboard one more time before connecting the battery and turning on the circuit. If the battery or chip gets hot when the circuit is on, shut it off immediately and check again for mistakes.

If your soldered circuit doesn't work, compare the placement of every part and wire between the breadboarded version and the soldered one—look sharp, since it's easy to miss a connection that's off by one hole. Check for wires you pressed though the board but forgot to solder, as well as blobby grey cold solder joints. Make sure there are both signal and ground connections for the audio output. If it makes sound, but is much quieter than the breadboarded version, or just acts weirder than it should, check to make sure you remembered to hook up the battery's "+" and "–" connections to pins 14 and 7 of the chip. Make sure that none of the copper traces have torn from repeated soldering and de-soldering, or the strain of wires

If this is starting to sound familiar, yes, it's a recapitulation of the de-bugging process outlined in the previous chapter—turn back there for more advice if needed.

A tip for keeping this (and any other) circuit running cleanly is to solder a 0.1 uf capacitor between the "+" pin (14) and the "–" pin (7), keeping it as close to the pins as possible (not at the other end of the board and linked by wires). This "decoupling" capacitor helps filter noise that can spread through power supply connections.

A circuit board that mirrors your breadboard exactly makes the transfer process much easier (see Figure 19.5). If you can't get such a board you must make adaptations carefully, checking your connections as you go (sometimes it helps to draw out your layout before starting to solder, as nerdy as this sounds). Once you have transferred a few designs to clone boards, and have gotten comfortable with the "topology of circuitry," you can advance to various other patterns of circuit board that give you the freedom to rearrange your design between the breadboard and the soldered version, sometimes shrinking it down to fit in a smaller box. And, as we mentioned earlier, larger boards can always be cut into smaller sections for simple one-chip circuits, such as our first oscillator.

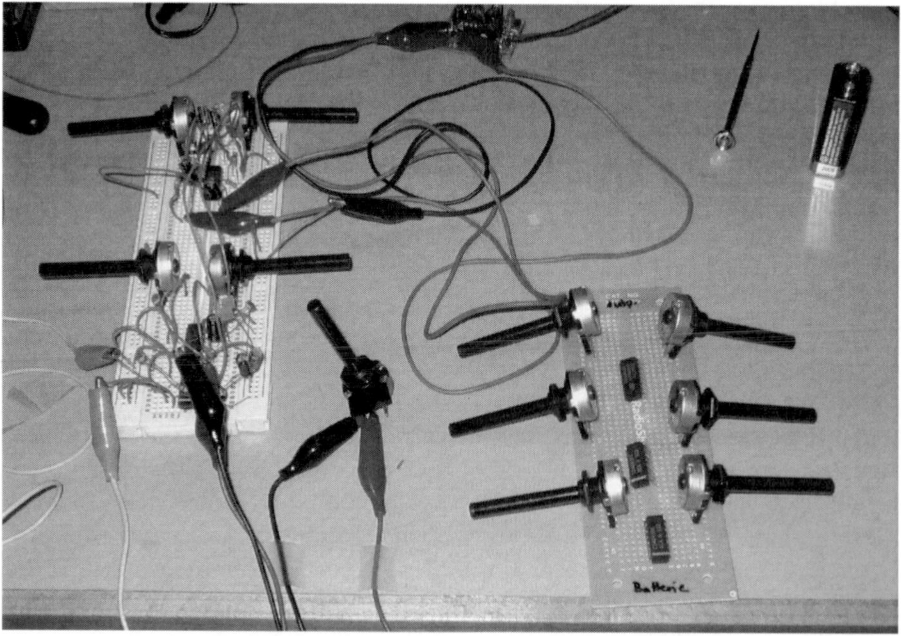

Figure 19.5 Breadboarded and soldered circuits, side by side.

When your circuits start to get complicated you may find that regular hookup wire is too thick and messy on the board. To lighten up, and move from spaghetti to capellini, buy yourself a roll of what is called "wire wrapping wire" (see Appendices A and B for sources). At 30 gauge, it's real thin, stays in place when snaked around the board, and comes in nice, bright, child-friendly colors that will cheer you through the ordeal of soldering.

Once you've confirmed that everything works you can move on to finding a box and drilling a mess of holes (see Chapter 17 and Figure 19.6). That's the fun part (but remember to insulate the circuit board from any metal surfaces!).

A PROVISO

There is one complication to Rule #19: some electronic components—capacitors in particular—have sufficiently wide tolerances that two parts with identical marked values may cause the breadboarded and soldered versions of your circuit to sound slightly different. If this is unacceptable you might have to transfer the original parts from the breadboard to the circuit board as you solder, rather than working with a duplicate set. In this case debugging is greatly helped by taking a photograph of the working breadboard before you start the transfer (is your cell phone handy?), or—in a pinch—making a detailed drawing.

Figure 19.6 A boxed photoresistor-controlled Hex Oscillator; note photoresistors flush with circuit board, and tiny on-off switch soldered directly to the board.

CHAPTER 20

Getting Messy: Oscillators That Modulate Each Other, Feedback Loops, Theremins, Tone Controls, Instability, Clocks for Toys, Crickets

You will need:

- A breadboard.
- One CMOS Quad NAND Gate Schmitt Trigger Integrated Circuit (CD4093).
- A 74C14 Hex Schmitt Trigger.
- Assorted resistors, capacitors, pots, and photoresistors.
- Some solid hookup wire.
- A jack to match your amp.
- A 9-volt battery and connector.
- An amplifier.
- Hand tools.
- A toy.
- A matchbox or balsa airplane.

The 74C14 presents a great introduction to making musical circuitry: with just a handful of components and a few minutes of time you can throw together reliable oscillators whose pitch can be easily swept over a wide range. It also opens the door to the slightly warped world of making sound with digital logic, rather than traditional analog circuitry. With another chip from the same CMOS family we can implement some more-advanced control functions commonly associated with classic analog synthesis.

GATED OSCILLATOR

The Schmitt Trigger circuit element that turns each inverter in the 74C14 into a snappy oscillator is also found in other CMOS chips. Most useful is the CD4093 Quad NAND Gate (see Figure 20.1)

This chip contains four identical NAND gates. There are two gates on each side of the chip. Each gate has two inputs (on the flat side) and one output (on the rounded

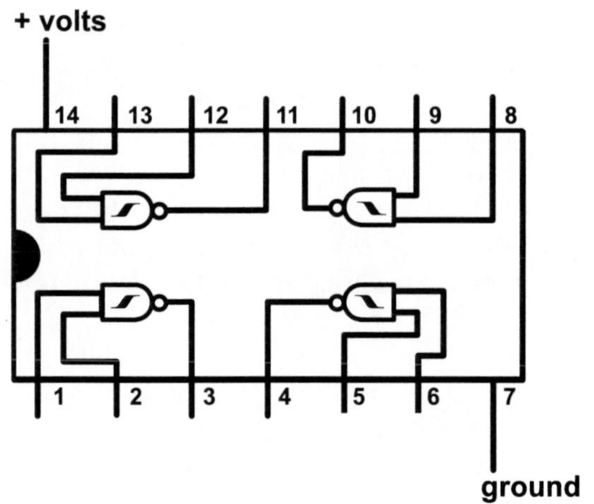

Figure 20.1
CD4093 Quad NAND Gate pinout.

end). Unlike the spawning salmon of the 74C14, these gates are arranged in mirror symmetry, like rutting elks: the outputs of each gate face each other, rather than the same direction. Note that this chip has the same power connections as the 74C14 Hex inverter chip we used in the previous two chapters: + voltage to pin 14, ground to pin 7. All of which brings us to an important new rule:

Rule #20: All chips may look alike on the outside without being the same on the inside—read the fine print!

A NAND gate is a variant of the basic binary function of an AND gate, whose "truth table" follows:

INPUT A	INPUT B	OUTPUT
0	0	0
1	0	0
0	1	0
1	1	1

You can see that the output only goes "true"/1 when both inputs are true/1—democracy in action: we go to the zoo because it's the place *both* kids agree would be fun.

A NAND (NOT + AND) gate adds an inverter stage after the AND logic to flip the output to the opposite state, like this:

INPUT A	INPUT B	OUTPUT
0	0	1
1	0	1
0	1	1
1	1	0

Democracy is replaced by contrarian despotism: dad *avoids* turning off the highway to visit Mammoth Caves specifically because both kids have been whining to see it for 200 miles.

This added inverter stage introduces the principle of knee-jerk denial (discussed in Chapter 18) that transforms this logic circuit into a "gateable oscillator."

Hook up the circuit shown in Figure 20.2 with +9 volts connected to the upper bus and ground to the lower one. Note that the basic design is similar to our earlier oscillator: a capacitor between an input and ground; a feedback resistor from the output back to the input. But where each stage in the Hex inverter package had just one input, each NAND gate has *two* inputs. Because of the combinatorial logic of the NAND gate (explained above), the second input of the gate can be used as a control input to turn the oscillator on and off: the output of the circuit will only change state (i.e. oscillate) when the control input is held "high" (+9 volts). If you connect the second input of the gate to ground (0) the circuit stops oscillating and the output remains in a "1" state (see Figure 20.3); silent.

Try both configurations. Plug a bit of wire into the breadboard near the gate input (pin 1 in Figure 20.2 and Figure 20.3) and alternate connecting the other end to the +

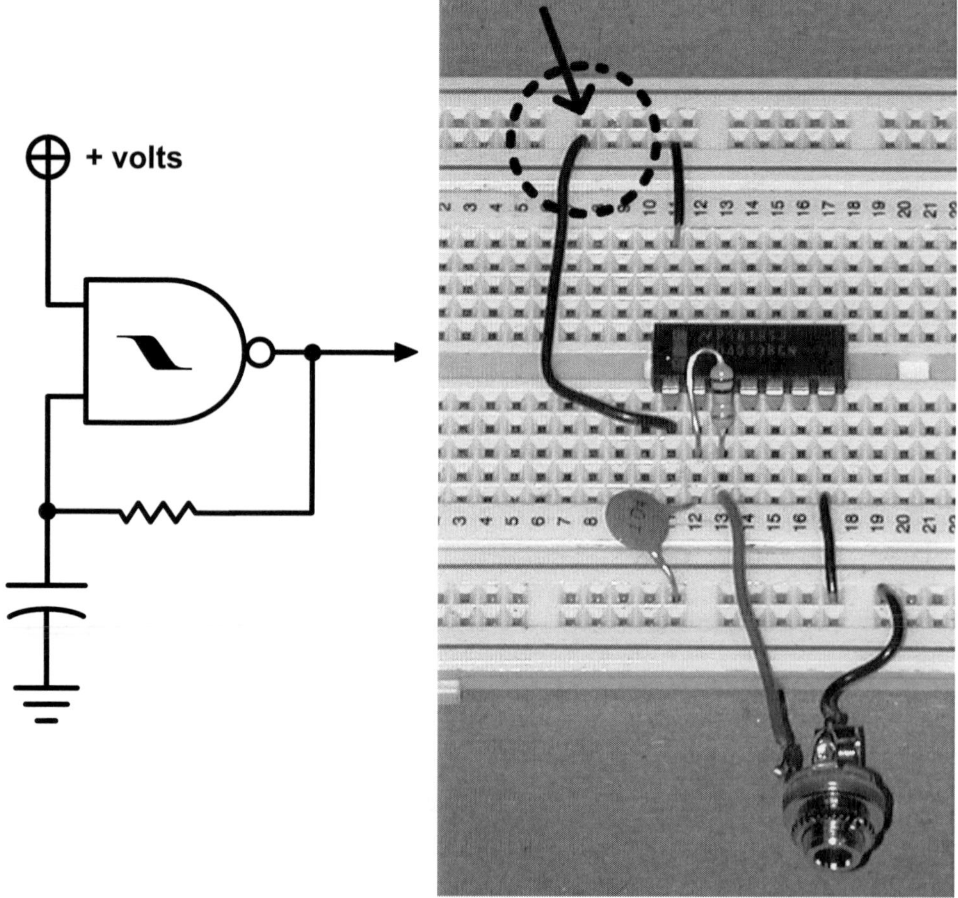

Figure 20.2 Basic NAND Gate oscillator, enabled.

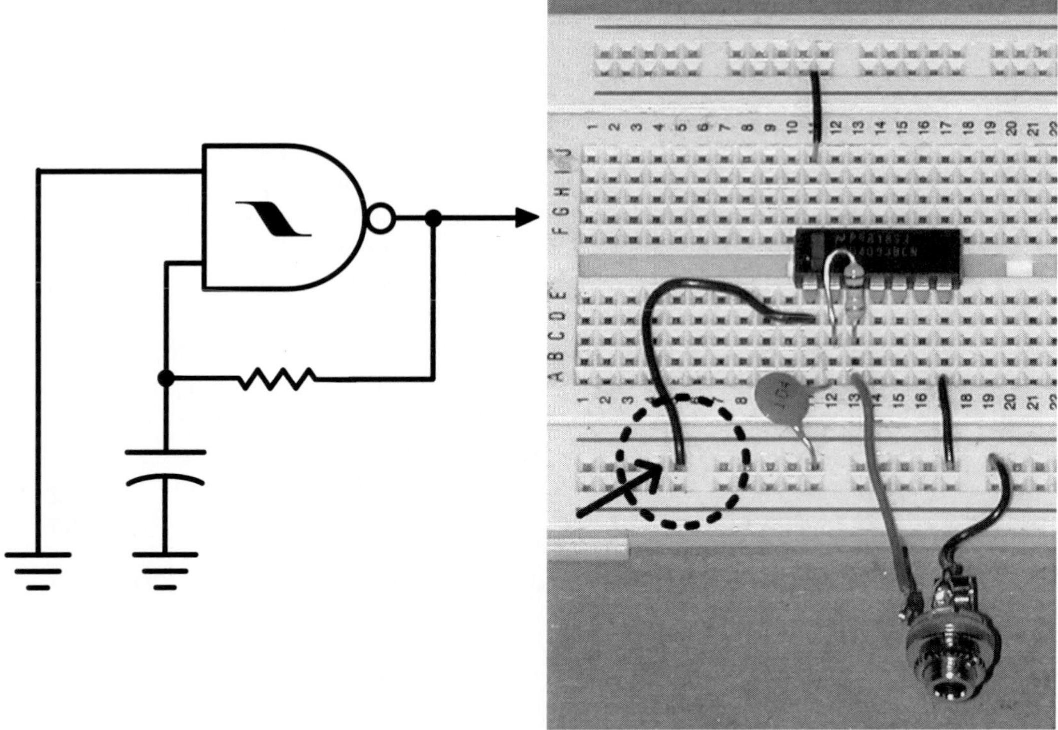

Figure 20.3 Basic NAND Gate oscillator, disabled.

and ground buses. The two inputs to each gate are identical—it doesn't matter which you use for the "control input" (the one with the wire moved between + and ground) and which for the "feedback input" (the one with the capacitor and feedback resistor), as long as you don't mix up the two and connect the capacitor to one and the feedback resistor to the other. And you can use any of the four NAND gates on the chip—they function identically—the photographs show just one of four possible hookups.

Note that even though a control input (pin 1 in the figures) might be connected to +9 volts we still need to connect +9 volts to pin 14 and ground (– voltage) to pin 7. The connections to the supply voltages have two distinct functions in our new circuit: through pins 14 and 7 they provide *power* to the chip, needed to run its internal operations—this is the "gas"; but + and – voltage also have *logical* value, and are evaluated as part of the (admittedly simple) mathematical calculations that the circuit performs in order to oscillate—that's the connection to pin 1. As with our previous circuit with the 74C14, sometimes this chip will make sound *without* proper power connections, but it will be "coasting," and probably will not perform reliably or predictably. Which brings us to Rule #21:

Rule #21: All chips expect "+" and "–" power connections to their designated power supply pins, even if these voltages are also connected to other pins for other reasons—withhold them at your own risk (or entertainment).

MODULATION

The oscillator oscillates when the control input is connected to +9 volts; it turns off when the control input is connected to ground. Big deal, you say, we can do this by simply connecting and disconnecting the battery. But, because an oscillator's output consists of a square wave swinging between +9 volts and ground, we can also use the output of one oscillator to switch another oscillator on and off. Breadboard the circuit shown in Figure 20.4. Use a large capacitor (2.2–10 uf) and 1 megOhm pot for oscillator 1 (shown here using pins 1, 2, and 3), and a 0.1 uf capacitor and pot or photoresistor (as here) for oscillator 2 (using pins 4, 5, and 6).

The control input on oscillator 1 (pin 1) is tied directly to +9 volts, so it runs all the time, as we demonstrated earlier in Figure 20.2. But the control input of oscillator 2 (pin 6) is connected to the *output* of oscillator 1 (pin 3), which gates oscillator 2 on and off as it swings between 9 volts and ground. If oscillator 1 (the control oscillator) has a large capacitor and runs slowly (like a metronome), you can hear oscillator 2 (the modulated oscillator) switch on and off at a regular tempo. As we tune oscillator 1 higher and higher, the obvious on/off function transforms into a kind of frequency modulation that is heard as a change in the timbre rather than tempo.

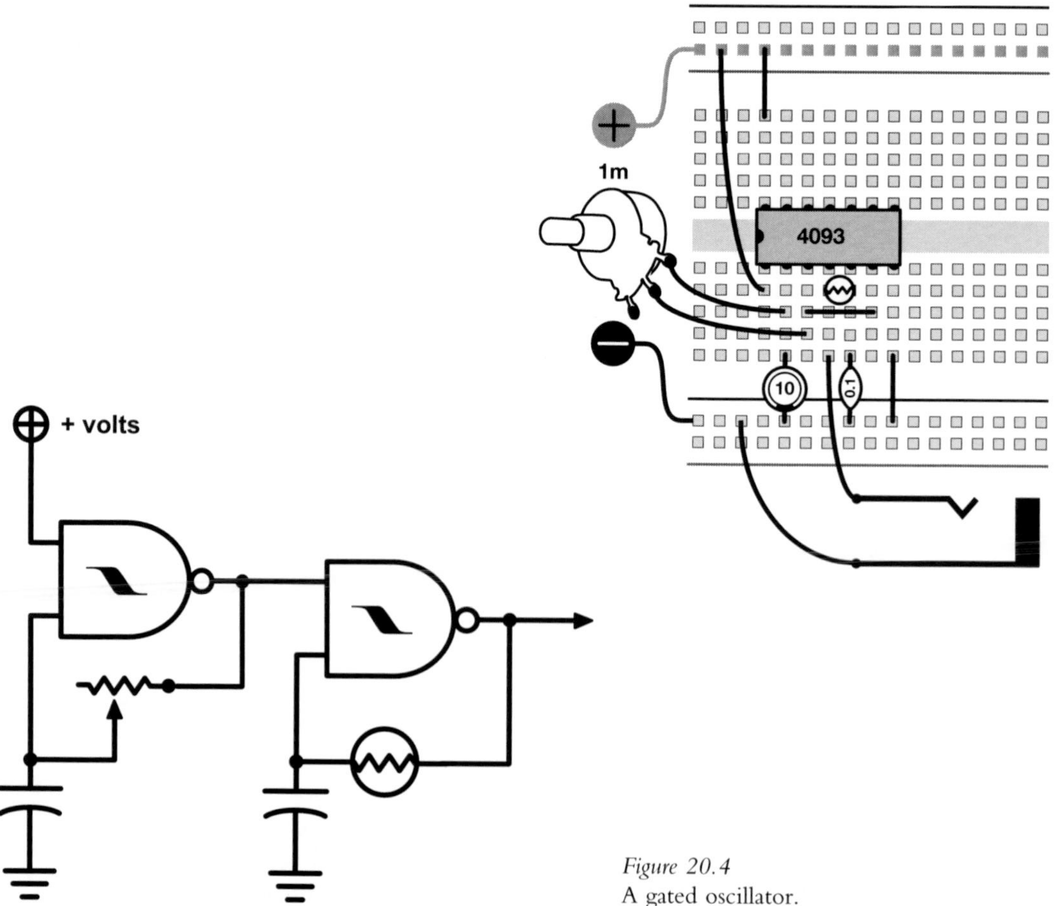

Figure 20.4
A gated oscillator.

Very cool flangey/wah-wah noises occur when the control oscillator and the modulated oscillator are both in the audio range and are very close in frequency—try using 0.1 uf capacitors and identical photoresistors for both stages. Now substitute 10 uf capacitors and 1 megOhm pots for both stages: careful tuning of the pots results in interesting polyrhythms.

You can cascade three or four oscillators (see Figure 20.5) to create tone clusters or rhythmic patterns, depending on capacitor sizes. The timbres/rhythms get more complex with each stage you add, but with diminishing returns: you can add a second 4093 chip to add four more stages, but it might only yield a minor change in the timbre. You'll have to configure parallel or satellite ground buses for the gates on the top half of the chip, as we did for the multi-voice oscillators in Chapter 18 (see Figure 18.15 and 18.16).

Experiment with different value capacitors and pots for the different stages. You can use photoresistors, electrodes, or any of the other alternative resistors discussed in Chapters 15 and 18.

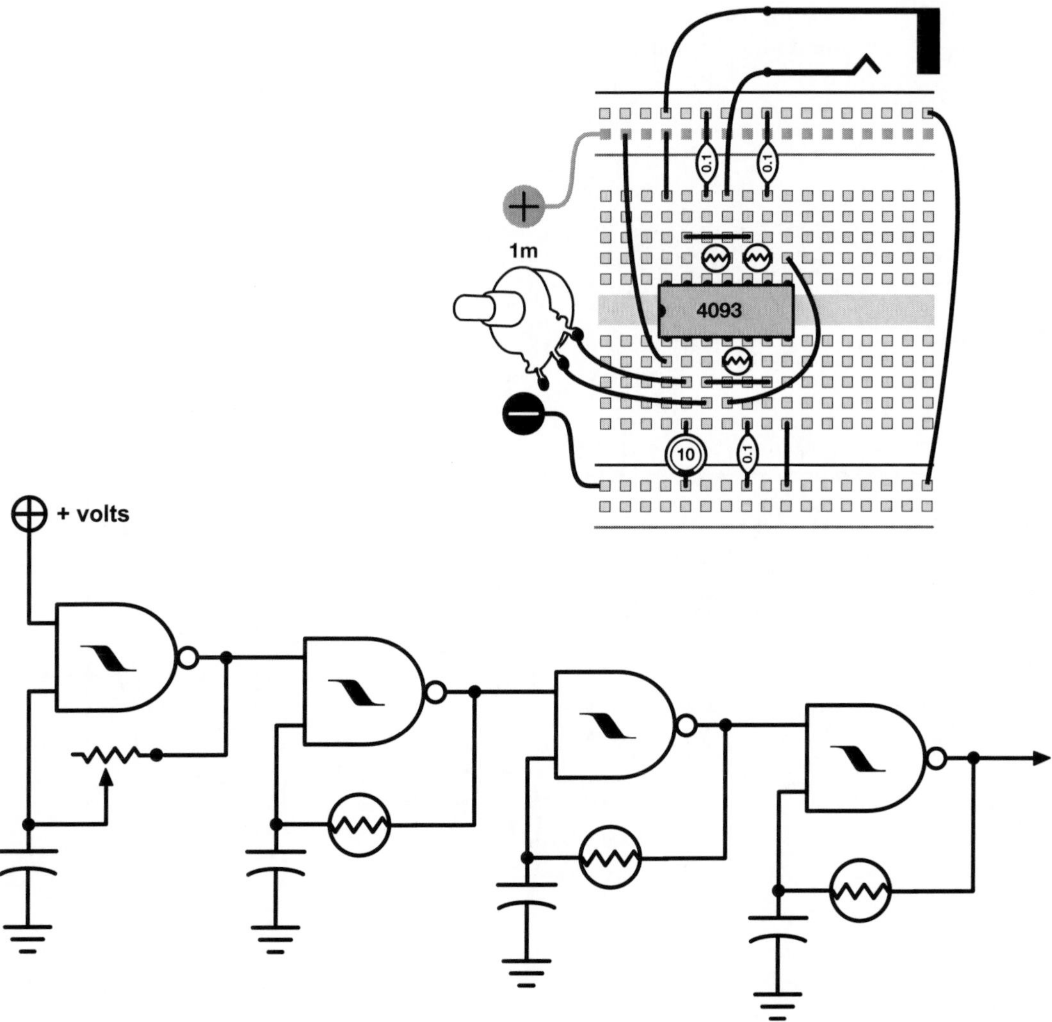

Figure 20.5 Four cascaded gated oscillators.

You can ratchet up the complexity of the sound by adding feedback from the output of the last stage to the control input of the first (see Figure 20.6). This works best if you observe the following guidelines:

- You must "bootstrap" the control input (pin 1 in the previous figure) by connecting a large resistor (around 100 kOhm is good) to +9 volts—without this the oscillator might never start running.
- The feedback only works around an *even* number of cascaded stages—i.e. after 2 NAND gates, 4 gates, 6 gates, etc.

Otherwise this should be a pretty intuitive process: just insert a jumper between the output of the last stage and the input of the first, and vary the resistors and capacitors.

This is as good a time as any to reiterate why I've chosen the CMOS family of integrated circuits (of which the 74C14 and the CD4093 are members) for our experiments:

- They consume very little current and can run on a wide range of voltages, which makes them ideal for battery operation.
- They are rather difficult to blow up.
- They were never intended as for audio applications, which adds pleasurable frisson to our experiments.

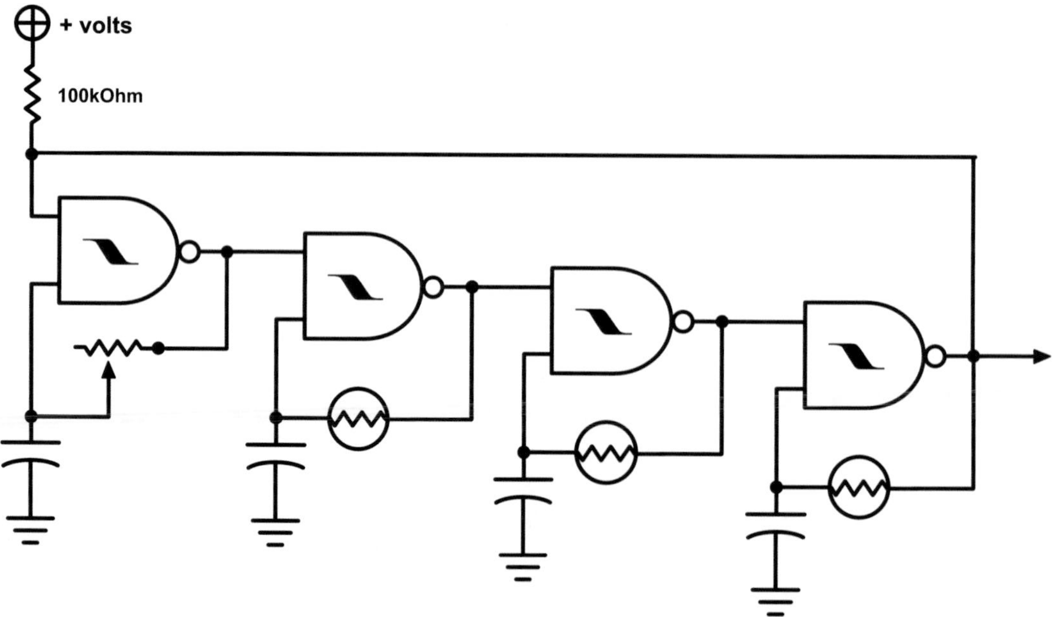

Figure 20.6 Four NAND gates with feedback connection from output of last stage to control input of first (note pullup resistor).

VOLUME CONTROL

Because these oscillator circuits put out such a hot signal (both the 74C14 and 4093 designs) they may overload the input to your amplifier or mixer, causing distortion (not always a bad thing—see Chapter 22) or limiting the useful range of your faders. If you want to drop the level down to a volume that matches your other line-level equipment (like a CD player), try adding the circuit shown in Figure 20.7 to the output of each oscillator, or, if you are mixing multiple signals, after the summing resistors.

This deceptively simple add-on to any oscillator forms the basis for several more dynamic circuits.

For example, if you want to fade the volume up and down, rather than just drop it down to normal line level, you can substitute a pot for the resistors, as shown in Figure 20.8 (the two fixed resistors in Figure 20.7 can be thought of as a pot frozen in one setting). Any pot whose value is 10 kOhm or greater will do. If you can find a pot designated as having an "audio taper" it will make the fade feel smoother (see Chapter 26 for a discussion of pot characteristics, and information on making simple mixers).

You may notice with this volume control, or when feeding your oscillator directly into some mixers or amplifiers, that at one extreme of the pot the pitch changes slightly. This is a drawback of the very simple design of the oscillator, and a function of the loading effect of the circuit that follows it. You can usually fix this problem by putting a reasonably large fixed resistor (say 10 kOhm) between the oscillator output and the ear of the pot, rather than connecting the pot directly to the inverter. Your maximum loudness will be decreased slightly, but the pitch should be more stable.

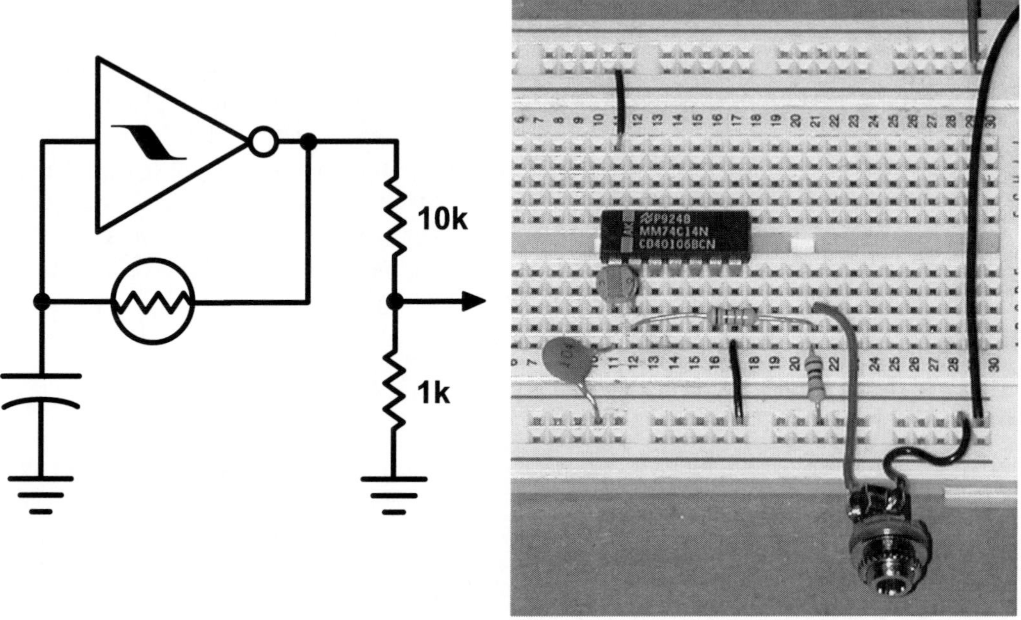

Figure 20.7
A volume dropping circuit.

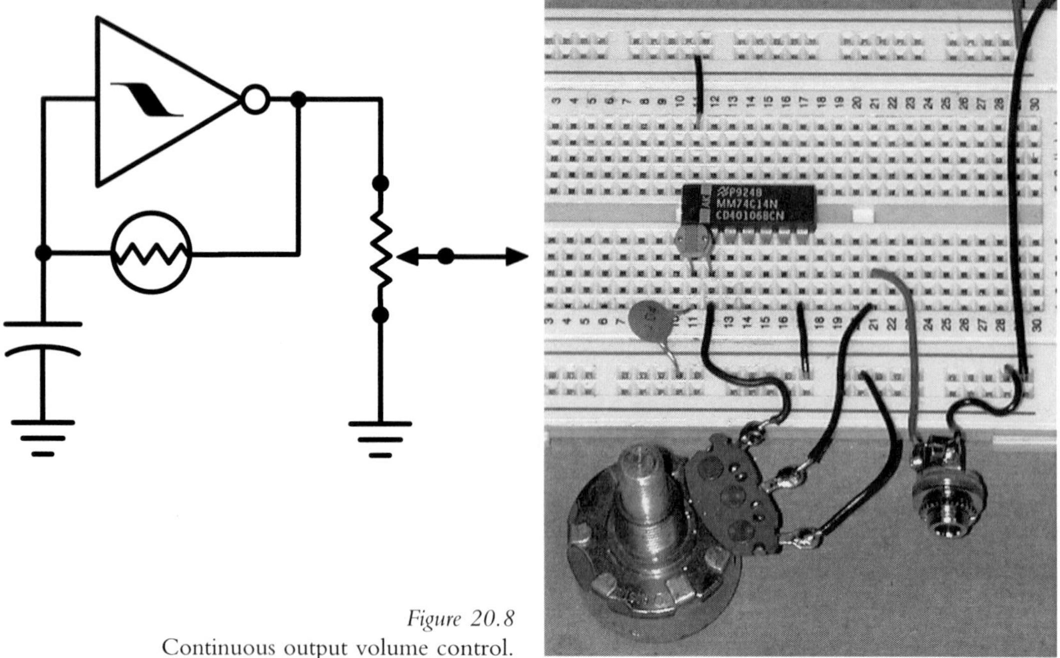

Figure 20.8
Continuous output volume control.

A simple Theremin-like instrument can be made using photoresistors to control both the pitch and the output level. As you can tell from Figure 20.9, all we've done is replace the 10 kOhm resistor in the volume drop circuit in Figure 20.7 with a photoresistor, and upped the value of the 1 kOhm fixed resistor to 10 kOhm. This resistor determines the dynamic range—you may want to experiment with various values between 1 kOhm to 10 kOhm. When less light strikes the output photoresistor (i.e. when you move your hand closer) the volume gets quieter. This simple circuit will never mute the sound completely, but it can provide expressive control of dynamics. The physical arrangement of the two photoresistors is critical to the effectiveness of this circuit: try gluing the two down, about a hand-span apart, on the top surface of an opaque plastic or wooden box (like a cigar box) holding the circuit; make sure you have a direct, stable source of light from above (see Figure 20.10).

FILTERING

These oscillator circuits aren't just loud: they are very, very, very bright—as befits the squarish waves they generate. (I confess that the signal isn't a perfect square wave: if you were to look at the waveform on an oscilloscope you'd notice that it does not have the requisite perfect 50 percent duty cycle, but—thanks to Herr Schmitt—is slightly asymmetrical, which is what gives these circuits their distinctive sonic charm. However, it's close enough that we can use the term "square" to save a few keystrokes.) You can always throw in an external circuit to tweak the timbre: a 10-band graphic EQ footpedal is an excellent accessory for any hacker; you can also use the equalization on your amplifier

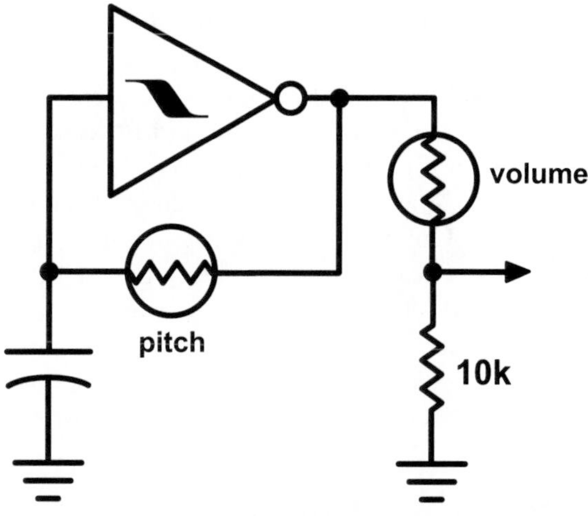

Figure 20.9
Theremin-esque circuit, with photoresistors for control of pitch and volume.

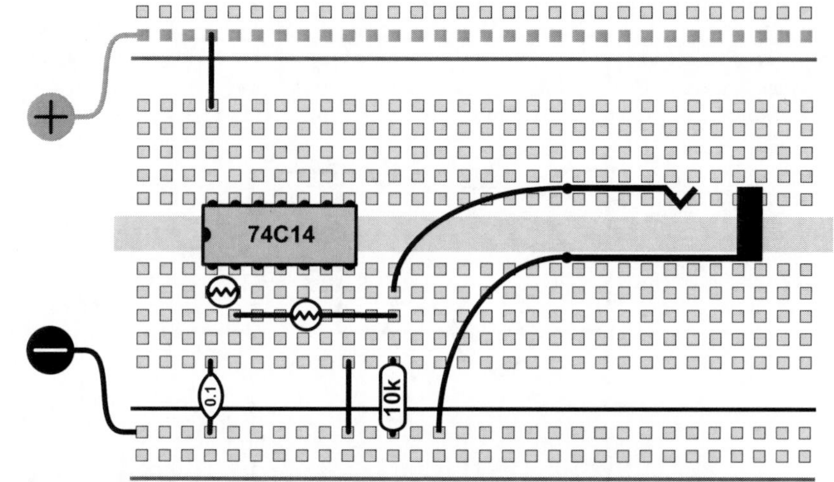

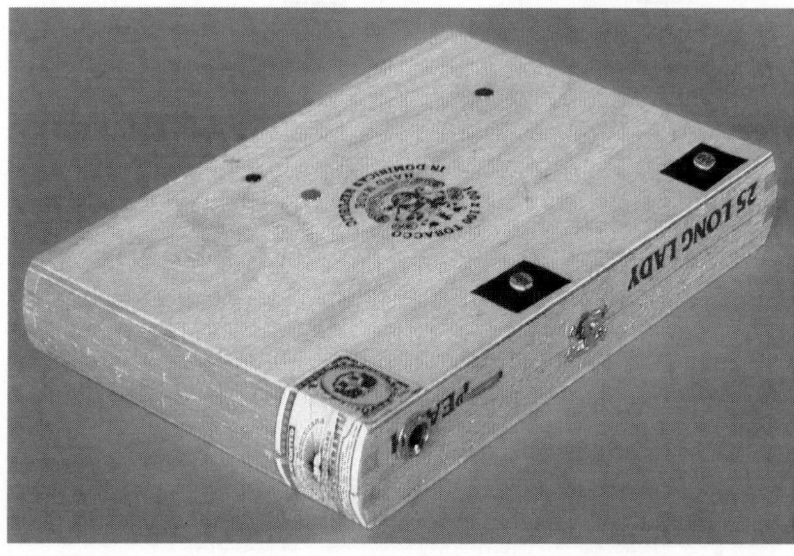

Figure 20.10
Cigar box pseudo-theremin.

or mixer. But with the addition of a few more common parts you can modify the timbre before the signal leaves your circuit board.

The waveform exiting an inverter or NAND gate may be pretty square, but after feeding back through a resistor and capacitor to the input it sounds (and looks, on a scope) astonishingly much like another synthesizer classic, the triangle wave. The overtones of a triangle wave are considerably quieter than those of its rectilinear brother, and consequently it has a much mellower tone. Unfortunately, connecting an output jack directly to this input point interferes with the delicate balance that keeps the oscillator swinging, and often mutes the circuit or causes the pitch to shift. So we must isolate the circuit from the outside world with a big resistor, such as the 100 kOhm shown in Figure 20.11. Moreover, since the triangle wave "floats" on a DC bias equal to one-half the supply voltage, we add a capacitor (here 10 uf, but the value is not very critical)

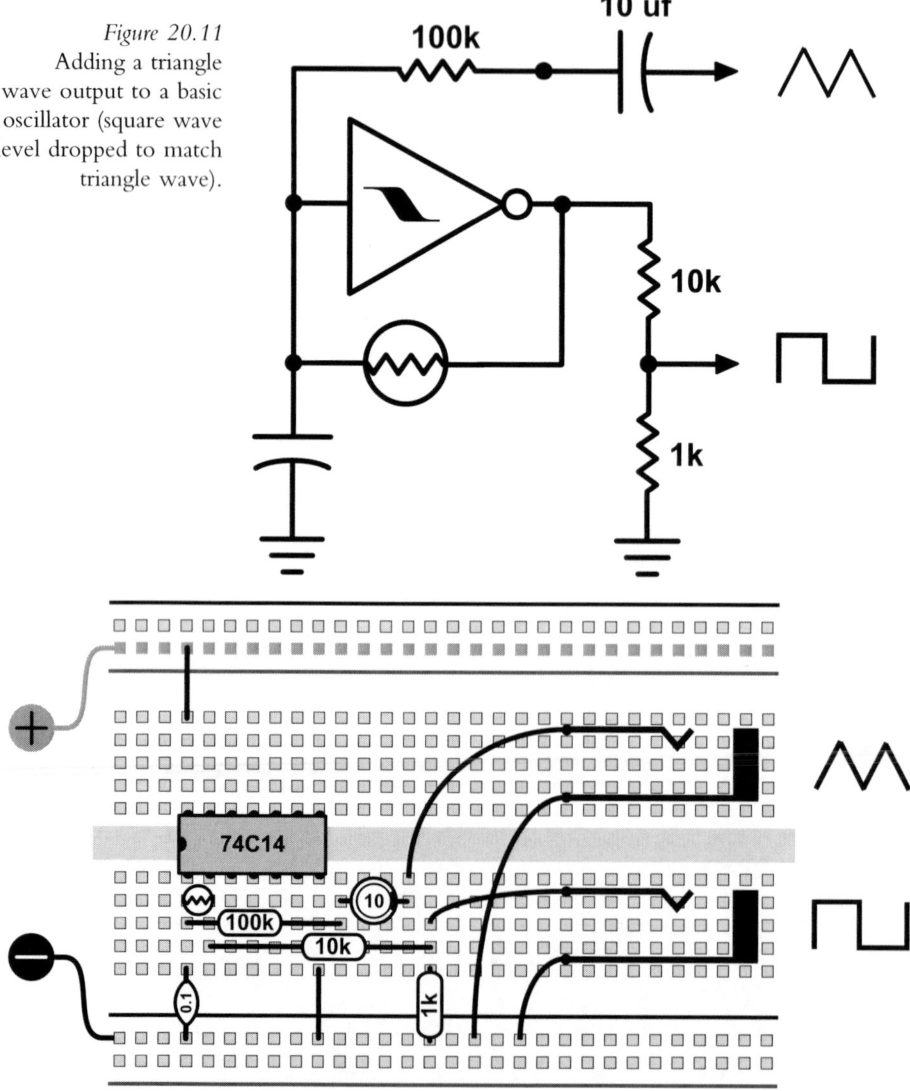

Figure 20.11 Adding a triangle wave output to a basic oscillator (square wave level dropped to match triangle wave).

to block the voltage from reaching your amp or mixer (like we did to block the electret power supply from reaching the mike preamp way back in Chapter 10).

We can use the square wave and triangle wave outputs separately and simultaneously, but when you hook up this circuit you'll notice that the triangle wave is considerable quieter than the square wave if the latter is taken directly from the output of the inverter: the combined effect of the isolation resistor and the inherent behavior of the chip's input stage drops the triangle wave level close to the typical "line level" we approximated on the square wave in the circuit in Figure 20.7. So I've added a similar dropping circuit here to balance the loudness of the two waveforms—you can eliminate it if you don't mind having the square wave be much louder than the triangle.

You may notice that the load of the circuit that follows the triangle wave output may affect the frequency slightly—a drawback of the extreme simplicity of this design. If this bothers you, skip ahead to Chapter 22 and build a little preamp circuit to buffer the oscillator output.

Another solution to the excessive brightness of the square wave is to *filter* it, instead of taking the triangle wave output. The circuit in Figure 20.12 is a crude fixed frequency "low pass" filter: it rolls off all the overtones above its cut-off frequency, which is

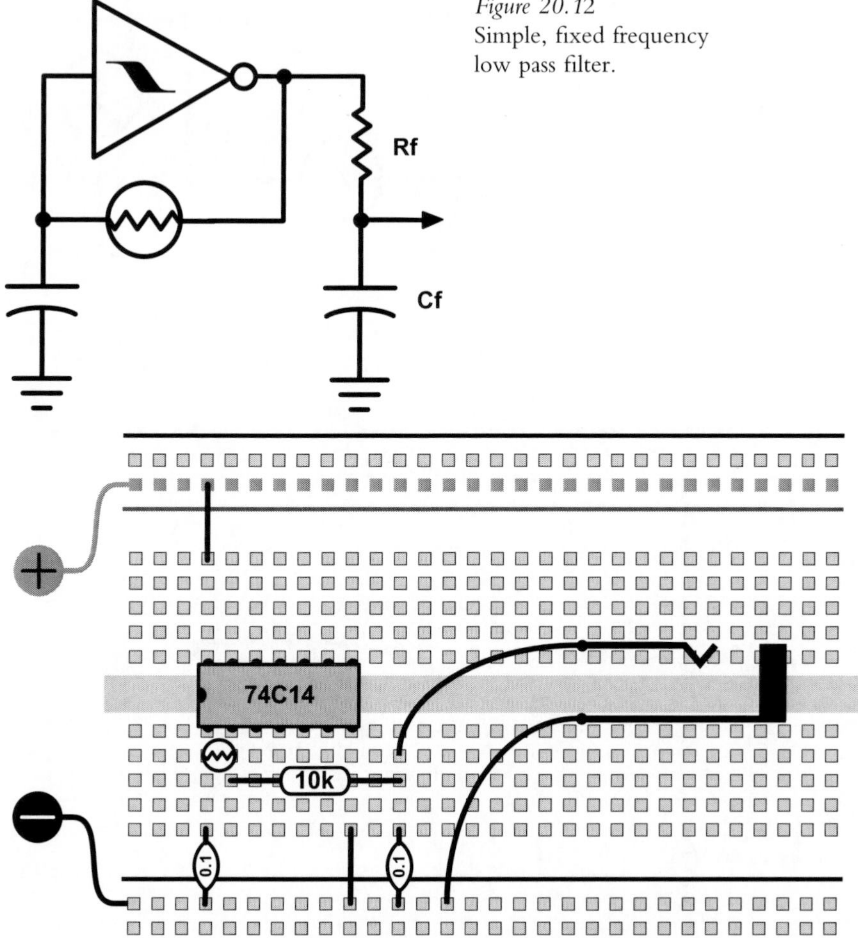

Figure 20.12
Simple, fixed frequency low pass filter.

determined mostly by the values of Cf and Rf. Good starting values are 0.1 uf for Cf and 10 kOhm for Rf. Substituting a smaller capacitor makes the circuit brighter, while a larger one will roll off more highs. You can also change the value of Rf: as with the capacitor, a smaller value raises the cut-off frequency, while a larger one lowers it. Which brings us to our next circuit.

The fancier version of a low pass filter shown in Figure 20.13 was suggested by one of my cleverer hacking students, Massimo Gennara. We substitute a photoresistor for the fixed resistor Rf of the previous circuit, and now we can use light to control the filter cut-off frequency. Start by setting Cf equal to the capacitor value used to set the range of the oscillator (i.e. 0.1 uf), and experiment with smaller values for a brighter tone, larger ones for a darker one. If both photoresistors receive the same amount of light (if they are immediately next to each other, for example) the filter will "track" the frequency of the oscillator—i.e. the filter will go up when the oscillator sweeps up, and down when the oscillator goes lower. If the photoresistors are placed far enough apart,

Figure 20.13
Low pass filter with photoresistor-controlled cut-off frequency.

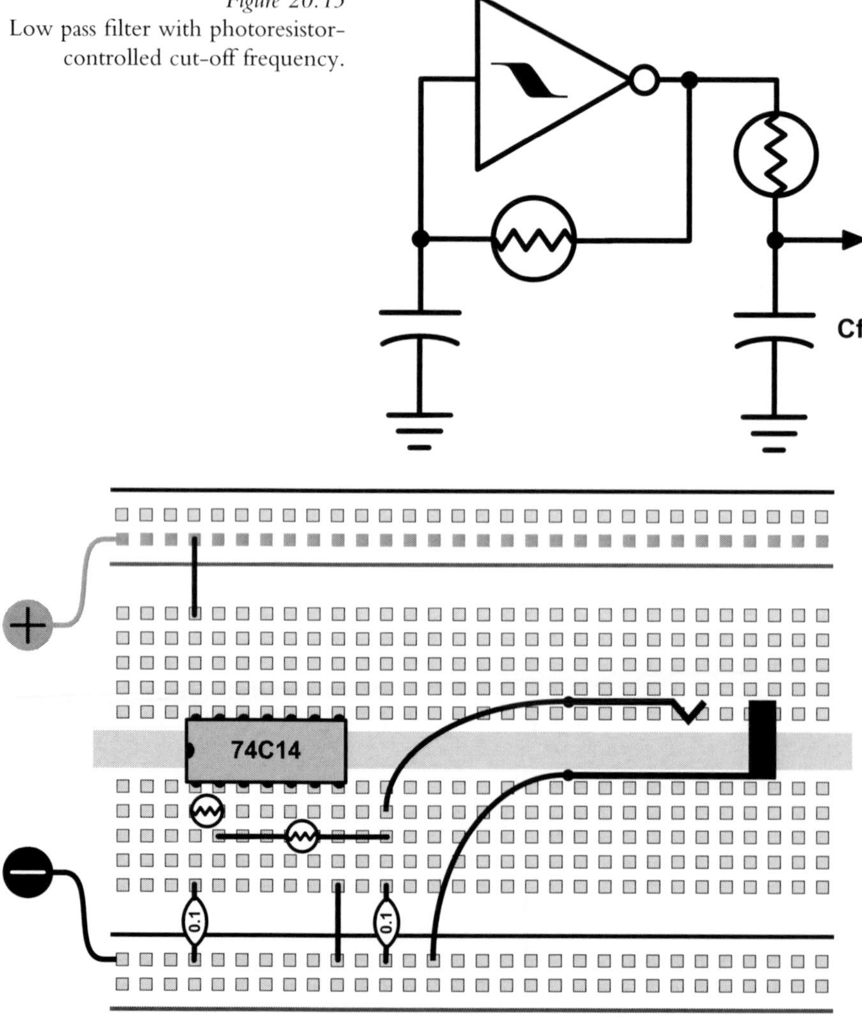

on the other hand, you can play the pitch and filter separately. So, as with our pseudo-Theremin, you need to think about the physical layout of your instrument. You can substitute a potentiometer for the photoresistor that controls the filter in this circuit, if you want more direct control—a big one, like the 1 mOhm we've been using to tune the oscillators and toy clocks, is a good starting value.

Before any "real" engineers start accusing me of leading my hackers astray, let me explain that filters and equalizers typically are dauntingly complex circuits. I confess that I have simplified them greatly for inclusion here. You might notice that altering the cut-off frequency can affect volume as well. Moreover, the behavior of the filter can be influenced by the "load" of the circuit that follows it (i.e. your amplifier or mixer). But they sound pretty good nonetheless. Trust me, a mongrel EQ with a slight limp can prove as true a companion as a purebred Moog filter (although I recently came across a Web site displaying a very similar passive filter design drawn by the hand of Robert Moog himself—see References).

I've shown the volume control and filter circuits as applied to the 74C14 oscillator, but these methods can be added to the output of the 4093 circuits as well. The Theremin and variable low pass circuits are especially effective on cascaded 4093 oscillators.

RANDOM WALKS

Once you get the hang of these basic circuits and the way they are *supposed* to work, don't be afraid to experiment with alternate configurations, even if you have no idea what you're doing (see Rule 17). It is almost impossible to destroy the chip by making "wrong" connections between its various pins (unless you connect the battery backwards); at worst, certain configurations will be mute.

Try random connections and component substitutions. Create multiple feedback paths by linking the outputs of some oscillators to the inputs of others—you can do this with plain wire, resistors, capacitors, pots, photoresistors, etc. Add multiple electrode touch contacts to set up feedback paths through your skin. You may need to add pull-up resistors to the inputs of the circuits to start them oscillating, as we did with the 4093 in Figure 20.6: connect a resistor in the 100 kOhm–1 mOhm range between the input pin and +9 volts where needed. Arranged in matrices, sometimes the oscillators produce unstable, complicated patterns of pitch and rhythm not displeasing to the ear (or brain). When you hear something good, stop and make very careful notes of what is wired to what, because you may never find it again. And probably no one will ever be able to explain why it sounds the way it does.

An excellent route to instability and surprise is "voltage starving" your circuit. Instead of hooking the +9-volt output of the battery directly to the circuit, connect it through a low value pot, as shown in Figure 20.14. As you increase the resistance of the pot by turning it, you diminish the voltage reaching the chip. At one extreme of the rotation (fully CCW as shown in the figure) the circuit gets full voltage and should operate "normally." At the other extreme (fully CW) there should be so little voltage reaching the chip that it falls completely silent. In between the two limits you should find a range of settings at which the circuit "almost works"—it's just a question of whether any of these states produce sounds that interest you. It's important to find the value of pot that gives you a decent range of behavior. Start with 1 kOhm pot. If this has no effect no

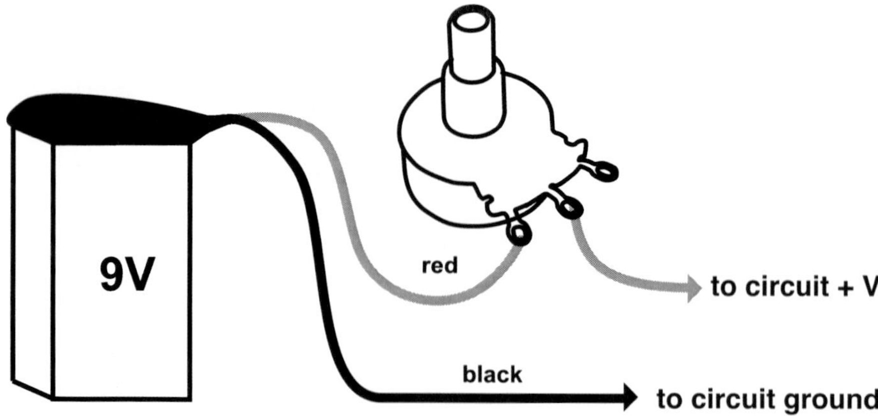

Figure 20.14 "Voltage Starve" circuit.

matter what the setting of the pot, then go up to 5 kOhm; if, on the other hand, the 1 kOhm pot is too sensitive, try 500 Ohm.

Voltage starving is especially effective on more complex circuits, such as the cascaded 4093 chains, with feedback, since different elements of the circuit often collapse at different voltages. A lot of otherwise predictable toys also benefit from a restricted voltage diet—it might be time to retrofit some of the instruments you made in the second part of this book.

CLOCKS FOR TOYS

The oscillator circuits we've made have a number of applications beyond just plugging into an amplifier and droning away. For example, you can use the output of a high frequency oscillator as a substitute for the clock circuit in a toy. This is especially handy if you want to control the speed of a toy that uses a crystal for a clock instead of the directly hackable resistor/capacitor circuit.

Breadboard an oscillator (using the 74C14 or 4093 design) that runs too high to hear—use a capacitor between 100 pf and 0.01 uf and a big pot to adjust the speed. But do not hook up a battery to the circuit. Instead connect it to the toy's batteries: use clip leads to link the ground of the toy (where the battery's negative terminal is soldered to the board) to the ground bus on your breadboard, and the positive voltage of the toy (where the battery's positive terminal joins the board) to the positive bus of your circuit. If the toy uses a resistor/capacitor pair for its clock, remove the timing resistor (as we did in Chapter 13) and solder short pieces of wire to each of the now vacant pads. If the toy uses a crystal this part will usually take the form of a tiny metal can, or a shiny plastic capacitor-like thing, sometimes with three legs in a line instead of two; de-solder the crystal and solder short wire jumpers into the holes left behind.

Use a test lead to connect the output of your oscillator to the free end of one of the clock wires you soldered onto the toy's circuit board. Does the toy run? You may need to disconnect and reconnect the batteries to re-boot the toy, and adjust the oscillator speed with the pot. If it doesn't work, try connecting to a different clock point; you might have to connect one or more of the unused clock pads to ground. If no configuration works try a different value for your oscillator's capacitor. Or, if you get frustrated, another toy.

If and when you find a toy and a hookup configuration that does work you can proceed with applying all of our oscillator variations (weird timing resistors, cascaded 4093s, etc.) to modifying the toy's performance—the irregular waveform of cascaded 4093s can have a distinctively odd effect on a toy's behavior. Moreover, you can connect one oscillator to several toys, so they all track the pitch of your new master clock.

POWER STRUGGLES WITH CRICKETS

As I alluded earlier, this family of integrated circuits consumes only tiny amounts of power, which makes them well suited to battery power. Power consumption is directly proportional to the frequency of the oscillator: the chip only uses power when it changes state from low to high or back again. The lower the frequency, the less power consumed in a given period of time—a metronome built with this chip can run for a year or more on one battery. A high-pitch audio squealer will go through batteries faster. Disconnecting the battery when you leave the breadboard for the day is a good idea in any case.

The miserly power consumption of these chips manifests itself in their output signal: although the output is very "hot" (swinging 9 volts peak-to-peak, compared to 0.7 volts of a typical CD player output), it carries very little current. This means you have to plug the circuit into a power amplifier in order to drive a speaker—if you wire the oscillator output directly to an ordinary speaker you will probably hear nothing, since it does not produce enough current to move the coil. Piezo disks, on the other hand, require very little current to function as (admittedly low fidelity) speakers, and these CMOS chips can drive one directly, without need for an additional amplifier. Simply connect the oscillator's ground bus to the disk's ground (metal perimeter), and connect its output to the hot wire coming from the disk's center. Don't use the transformer from Chapter 8 in this configuration—the CMOS chips can't put out enough current to drive the 8 Ohm load of the primary either. Hook up the battery and sweep the oscillator through its range.

Low frequency pulses yield a pleasing clicking sound, while audio pitches buzz like insects and frogs on a sultry summer evening. If it's too quiet for your taste, clamp the piezo disk to a cookie sheet or Styrofoam cup, or place it on a wooden matchbox with a stone on top, or glue it to the wing of a balsa glider. Dutch sound artist Felix Hess has made beautiful large-scale installations with multiple small circuits pinging piezo disks (see Art & Music 5 "Drivers," Chapter 8). Laurie Anderson embedded a piezo in molded bamboo fiber to create a portable sound generator for her "Walk" installation (see Chapter 24, Figure 24.16). If you leave the piezo disk inside the plastic lollipop packaging this resonator will also increase the loudness of the chirping, at the cost of emphasizing the one frequency resonant to the plastic chamber.

By now your ringing ears (and domestic partner) are probably begging you to take a break from oscillators. So let's move on to something rather different . . . .

CHAPTER 21

On/Off (More Fun with Photoresistors): Gating, Ducking, Tremolo, and Panning

You will need:

- Two sound sources to process, such as MP3 players, CD players, radios, electric guitar, etc.
- Some photoresistors.
- A flashlight.
- A 74C14 or 4093.
- A breadboard.
- Some LEDs (Light Emitting Diodes).
- Some heat shrink tubing (optional).
- Assorted resistors, capacitors.
- Some solid hookup wire.
- Some plugs and jacks.
- Clip leads and Y-cords.
- A 9-volt battery and connector.
- Two amplifiers.
- A toy with switches.
- Hand tools.
- Plastic electrical tape.

As we have seen in our earlier experiments with toy clocks and simple oscillators, the photoresistor changes resistance in response to changes in light level. We've harnessed this change in resistance to control the speed of a toy's clock and the pitch, volume, and filtering of an oscillator. The photoresistor can also be used as a generic form of gate or volume control to pass, block, fade, or pan any audio source (such as the output of an MP3 player, a hacked toy, or a microphone), and can be used as a substitute for certain switches in toys, keyboards, and other circuits. In this chapter we'll take a look at some of these applications.

FLASHLIGHTS

Assemble the simple circuit shown in Figure 22.1. Using clip leads, connect one leg of a photoresistor to the "hot" or tip of any audio signal (such as the output of an MP3

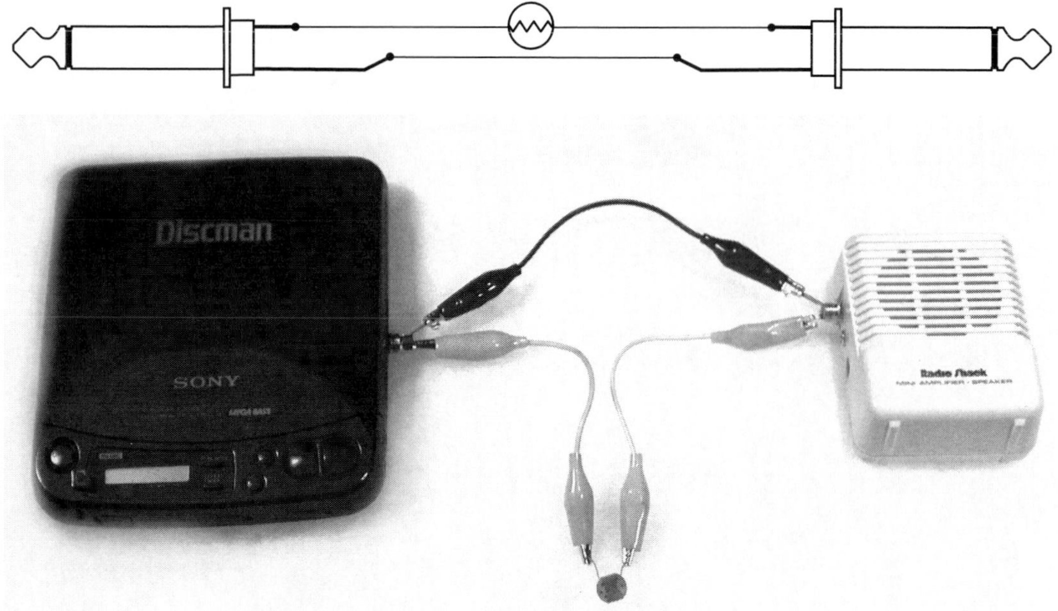

Figure 21.1 A basic photoresistor gate circuit.

player or CD player), and connect the other leg to the hot/tip of your amplifier input. Connect the shield/ground of your audio source to the shield/ground of the amplifier input (remember Rule 10: every audio connection needs signal and ground). Turn on the amp, play the MP3/CD, and confirm that audio passes through.

Now take the whole rat's nest into a dark place, like a closet, or turn off your lights and draw the blinds. The sound should get quieter. Switch on your flashlight and pass the beam across the photoresistor—the sound should get louder when the cell is lit, quieter when the cell is dark. This super-simple circuit won't shut off the sound completely, but you should hear a significant volume difference between light and dark. The back-side of a photoresistor is usually translucent, so total darkness and quieter level can only be achieved if you fully darken the cell: enclose it in your hand for example; if you cover the back with black electrical tape you can control it fully by adjusting the light falling on the front. Make sure you don't let the legs short against one another or the signal will pass through unattenuated regardless of light level.

You can increase the dynamic range of this circuit (the difference in loudness between "on" and "off") by adding a resistor of about 10 kOhm between the *output* side of the photoresistor and ground, as shown in Figure 21.2. Without getting into unnecessary technical detail, the resistor "clamps" the output to ground when the circuit is off, minimizing bleed-through of the diminished input signal, and thereby increasing the depth of the muting when the circuit is in its "off" state. Note that, until we add the clamping resistor, this circuit works in both directions: either jack can be used as an input or output. Once the clamp is connected between a jack's hot and shield we fix that jack as the output, and the other as input.

This circuit is an absurdly simple but nonetheless very effective light-controlled audio gate. You can use it with any audio signal, but it is more effective with nominally

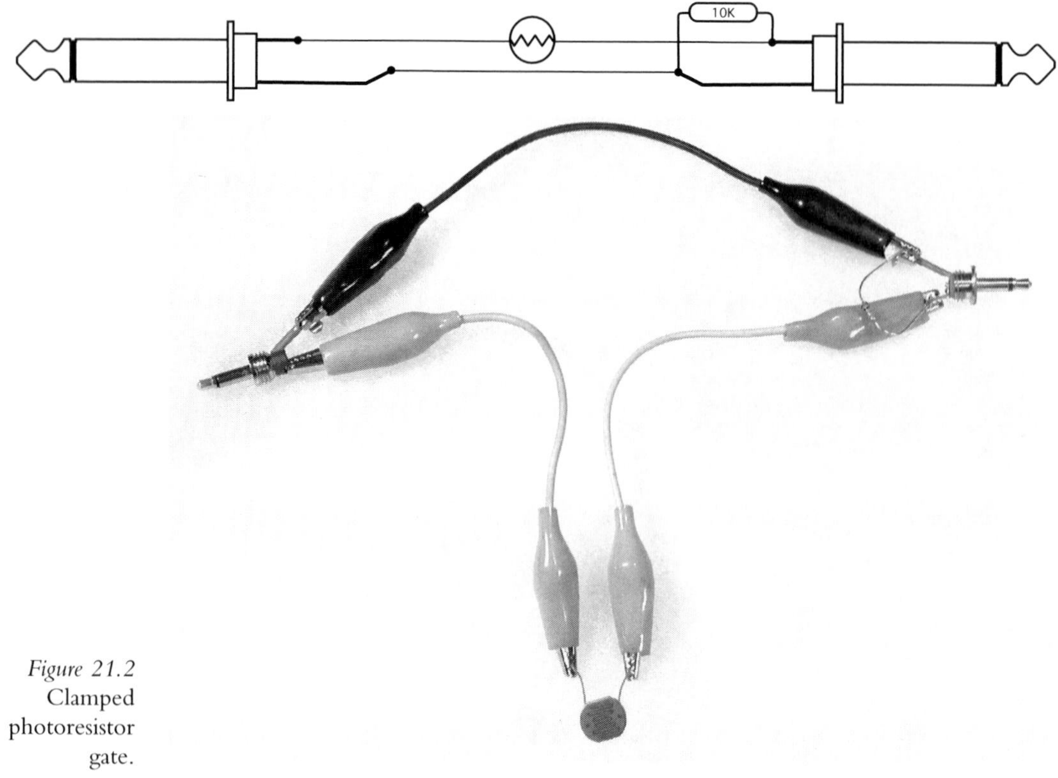

Figure 21.2 Clamped photoresistor gate.

line-level signals, such as the CD/MP3, rather than the very high signals generated by your oscillators—not because of any flaw in our new circuit, but because the proportionally minimal leak-through of such a hot signal is still high enough in absolute terms that it can swamp the high gain input stage in an amplifier like the Radio Shack mini-amp. This is a "passive" signal processor, which means that it needs no batteries (although the flashlight does). The only requirement is darkness, which makes it well-suited for stage use, camping trips, or persons with difficulty paying the electrical bill.

PING PONG, PONG PING

By adding a second photoresistor and some more connectors and audio devices we can expand our gate into a light-controlled panner or mixer. Hook up the configuration shown in Figure 21.3.

Connect an audio signal through two clip leads to two photoresistors. Clip the free leg of one photoresistor to the hot/tip of a plug connected to one amplifier, and connect the free leg of the second photoresistor to the hot/tip of a plug connected to a second amplifier. Link together the grounds on all three connectors with two more clip leads, as we did in the previous experiments. Seek the cover of darkness once again, play your source material, turn on the amplifiers, and flick the flashlight beam back and forth across the two photoresistors: the sound should pan between the two speakers, following the movement of the light across the two sensors.

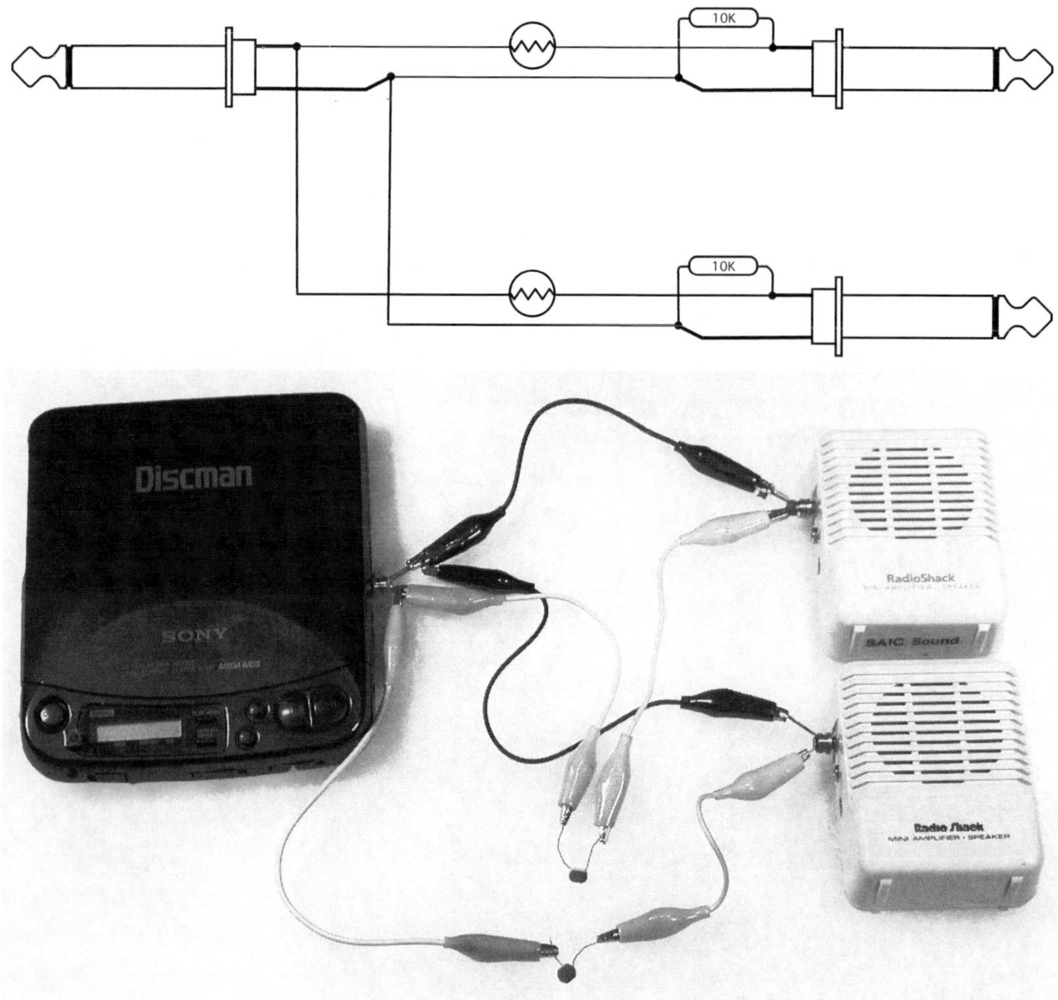

Figure 21.3 Light-controlled panner.

Now rewire the circuit slightly, as shown in Figure 21.4. This time connect one audio source through a clip lead to one leg of one photoresistor. Connect a second source through a second clip lead to one leg of a second photoresistor. Connect the free legs of both photoresistors through clip leads to the input of one amplifier, and link all the grounds together. Turn off the lights. Now when you pass a flashlight across the two photoresistors you should be able to cross-fade and mix between the two sources, like cutting between turntables.

You can solder up this circuit with two input jacks and two output jacks as shown in Figure 21.5, so it can be easily re-configured to be a panner, a mixer or a two-channel gate. To make a panner, use an audio Y-cord (available at Radio Shack and elsewhere) to connect a single audio source to both input jacks, and patch the outputs to two amplifiers. For a mixer, connect a different source to each input, and use a Y-cord to mix the outputs of both photoresistors to a single amplifier input. For two independent

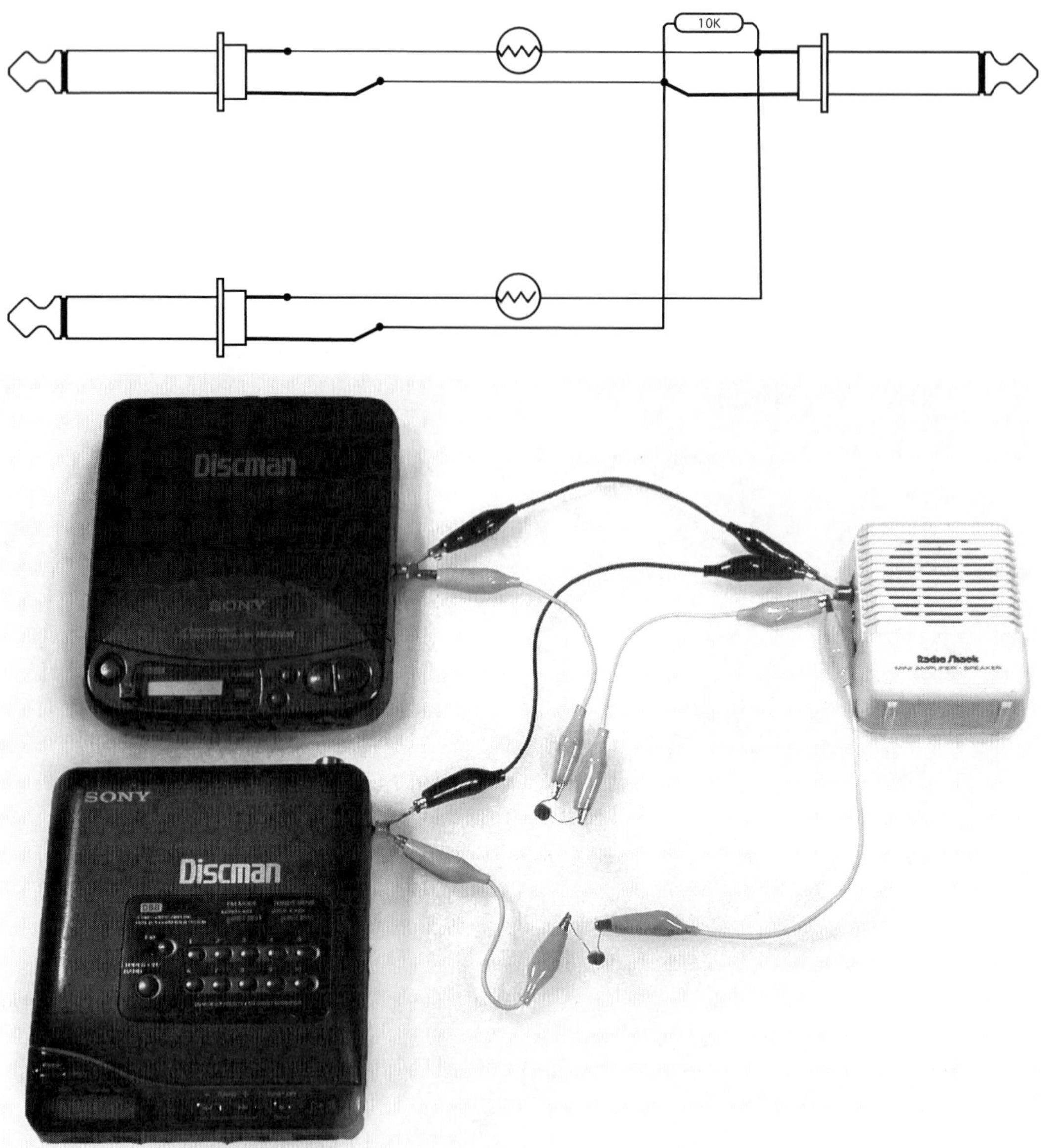

Figure 21.4 Light-controlled mixer.

channels of gating hook two audio sources to two amplifiers with no interconnections (like the two CD players and amps shown in the photograph).

This basic panner/mixer circuit can be expanded with more photoresistors, inputs and/or outputs to make four-channel panners, multi-channel mixers, etc. Note that male plugs are shown in all these drawings, since they make it easy to distinguish the signal/tip from ground/shield, but of course you can wire female jacks instead—as you would if you built the parts into a box of some sort.

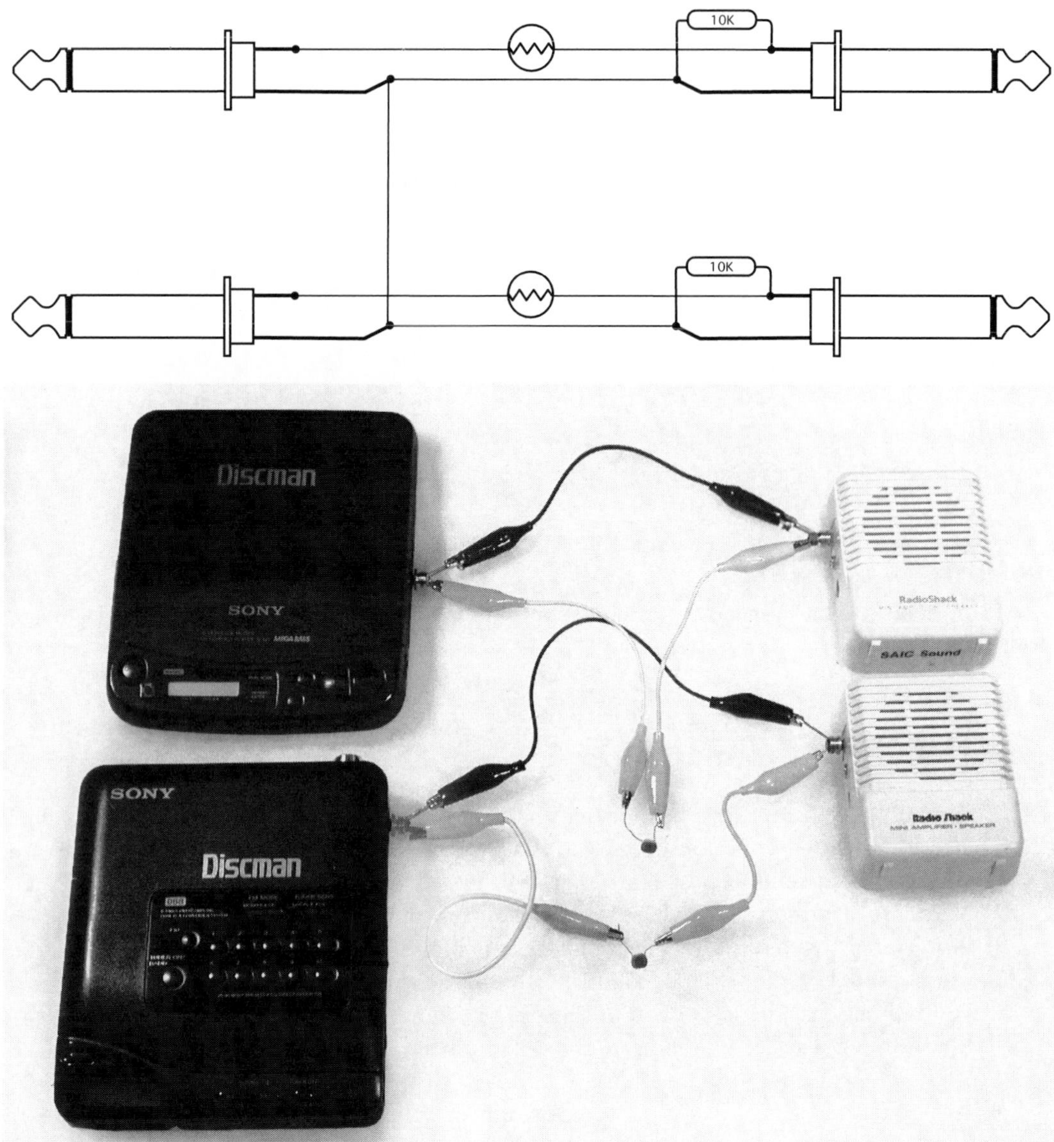

Figure 21.5 Patchable optical panner/mixer.

Flashlight-controlled circuits like these gates, panners, and mixers (as well as our photoresistor-controlled oscillators) occupy a distinguished place in the early history of live electronic music: similar circuits can be heard in the work of David Tudor, Lowell Cross, and other early hacker-composers. Figure 21.6 shows a four-channel panner designed and built by composer and pianist Frederic Rzewski in 1967, for example.

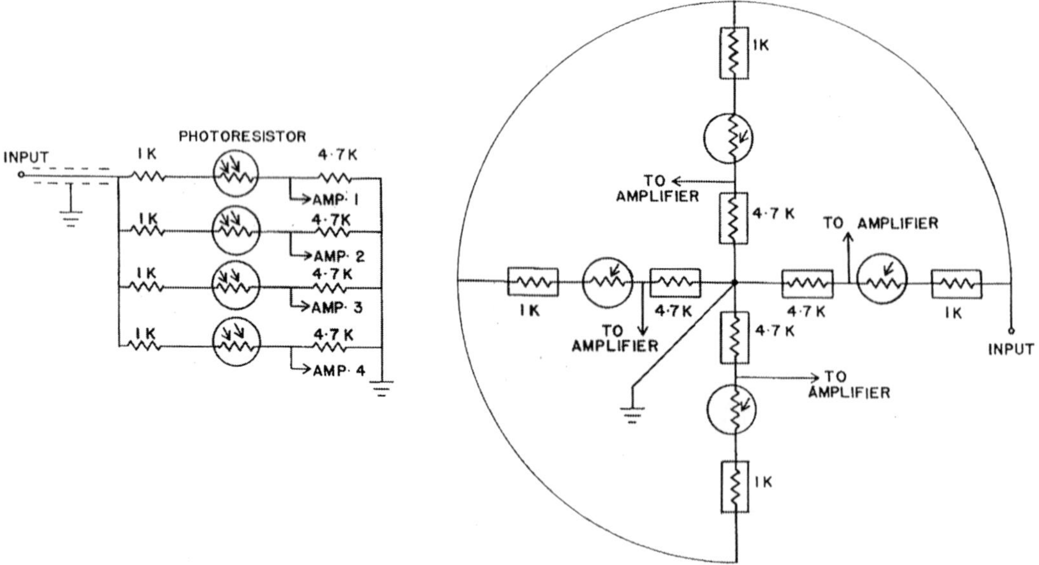

Figure 21.6 Four-channel panner by Frederic Rzewski (1967): schematic diagram (left) and physical arrangement (right).

BLINKIES

An LED (Light Emitting Diode) is a small cheap source of light that can be controlled electronically. Get one (or, better yet, two), in the color of your choice—blue ones are very cool, but tend to cost more than red, green, or yellow. LEDs are polarized, like batteries, electrolytic capacitors, and the ordinary diodes we used to mix our oscillators in Chapter 18; they only light when current flows through in one direction, not the other. Notice that one leg of the LED is shorter than the other, and if you look closely you will see that the lower rim of the LED is slightly flattened above the shorter leg—the short leg and flat side indicate the "–" connection of the LED, the other leg is the "+" connection (see Figure 21.7).

Breadboard the circuit shown in Figure 21.8. It should light up. Swap the polarity of the LED and observe that it only lights in one orientation. Substitute different values for the resistor and note the change in brightness: the smaller the resistor, the brighter the light, but only to a point, after which the LED might burn out, or the circuit driving it (our next step) will start to misbehave. Don't use a straight wire—1 kOhm is a good value to start with if you are using a 9-volt battery.

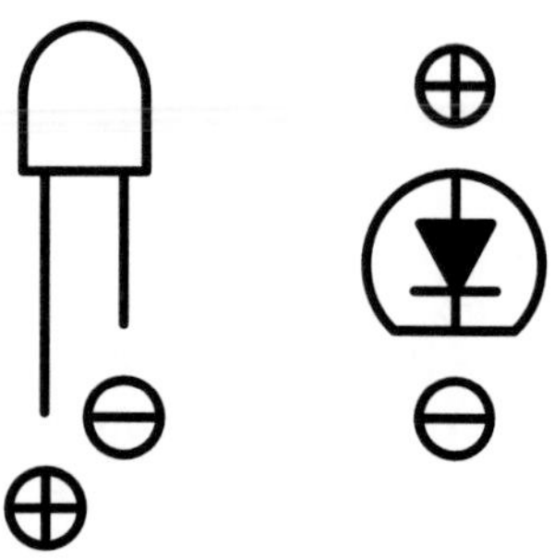

Figure 21.7
LED orientation.

Rule #22: Always use a resistor when powering an LED, otherwise the circuit and/or LED might blow out.

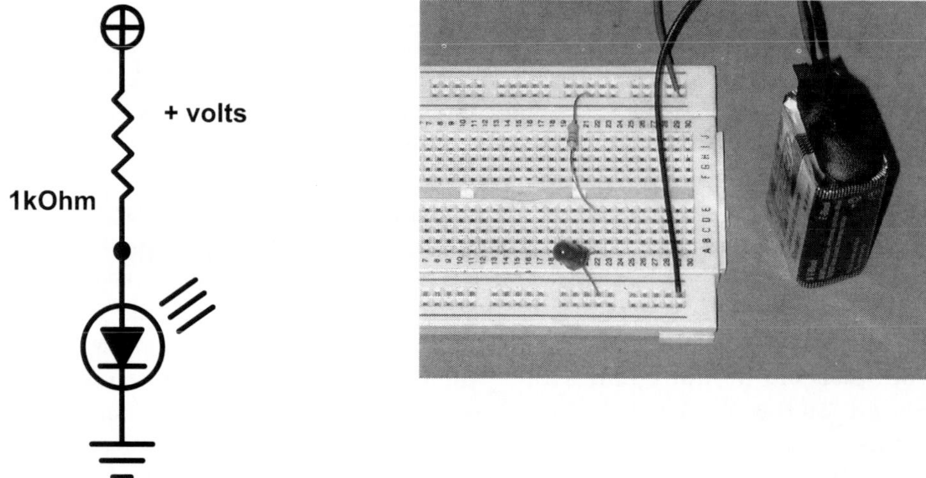

Figure 21.8 Lighting an LED.

Breadboard the circuit shown in Figure 21.9. This uses the simple oscillator from Chapter 18 (you could use the basic 4093-based design in Figure 20.2 of Chapter 20 if you prefer), but now we are connecting the oscillator's output to an LED instead of to an amplifier. The circuit should now blink the light instead of clicking the speaker (to confirm this try listening to it). Since we want it to blink at an observable rate, we use a reasonably large capacitor (2.2–10 uf) and pot (1 megOhm) to keep the oscillator in the metronome range, rather than an audio frequency. If it doesn't blink you probably either used too small a pot or capacitor, put the LED in backwards, or have omitted some connection. Vary the speed and watch the effect. Fun enough just to look at, but wait—it gets better!

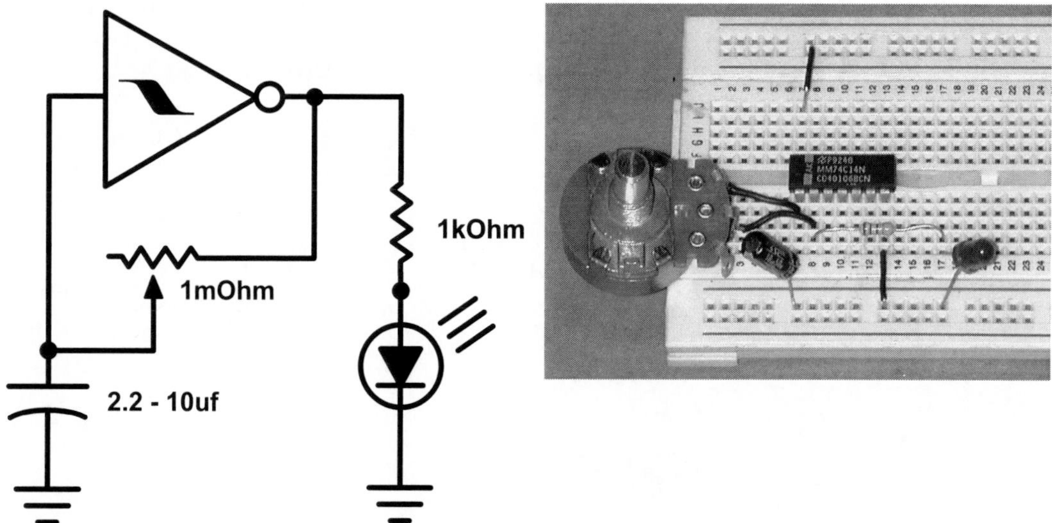

Figure 21.9 Blinking an LED with an oscillator.

Now take the LED and nuzzle it up against the photoresistor as shown at the top of Figure 21.10. Spread the leads of the photoresistor and LED apart so they do not touch each other, then wrap the photoresistor and LED in electrical tape so that they are sealed from outside light (lower photo). Don't let any of the legs short together or the circuit will not work—you can wrap some tape on each of the four wires if this helps. Replace the LED in the Figure 21.9 circuit with this bundle, making sure that you have the LED's polarity right—check the placement of that shorter leg, since if you've done the wrapping right you won't be able to see the light blink (shades of Schrödinger's Cat and Heisenberg's Uncertainty!). Connect the hot/tip of an input and output jack to the photoresistor legs as we did in Figure 21.2, link the audio jack grounds with another clip lead or solder some wire between them, as shown in Figure 21.11. **Note that you do not connect the ground of the breadboarded circuit to the grounds on the audio jacks—this is important for maximizing the audio quality of this circuit!**

Connect an audio source to input jack, and connect the clamped output jack to an amplifier. Vary the speed of the oscillator. You should hear your sounds get chopped on and off as the LED blinks. As you speed up the oscillator the distinct on/off rhythm is replaced by a kind of a wobbly modulation. Experiment with different size capacitors until you find a workable range—you may want to add a resistor in series with the pot to limit the maximum speed, as we demonstrated in Chapters 14 (Figure 14.1) and 18 (Figure 18.12).

If you want to *see* what's happening as well as hear it you can add a second LED in parallel as an indicator light as shown in Figure 21.12.

We can extend this basic oscillator/LED/photoresistor gate design to create an automated version of our flashlight-controlled panner/mixer. The circuit shown in Figure 21.13 uses one stage of the Hex Schmitt Trigger to make a low-frequency oscillator, while a second stage simply inverts the clock signal generated the first. The jacks have been omitted from the schematic for clarity—just expand the hookup arrangements from the previous two circuits. By connecting a separate LED to each inverter, each via its own resistor, one LED is off when the other is on—remember how an inverter always outputs the opposite state of the input signal? Note that when wiring a stage of the 74C14 as a simple inverter you need no capacitor or feedback resistor, as you did in your oscillator circuits.

Use a Y-cord or clip leads to connect an audio source to both circuit inputs and connect each photoresistor output to a separate amplifier (as we did with the circuit in

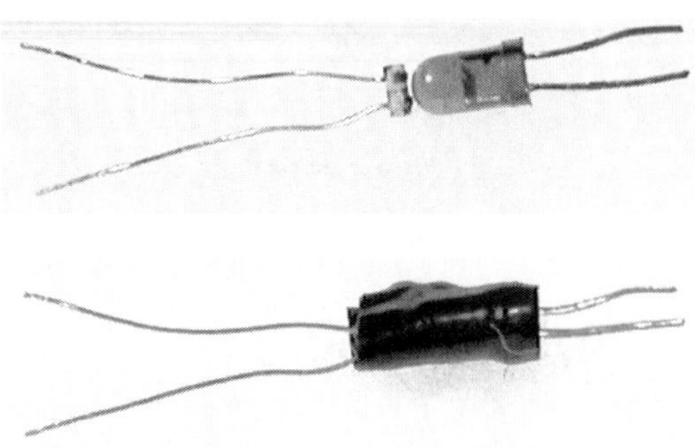

Figure 21.10
LED and photoresistor kissing (top) and bundled in electrical tape (bottom).

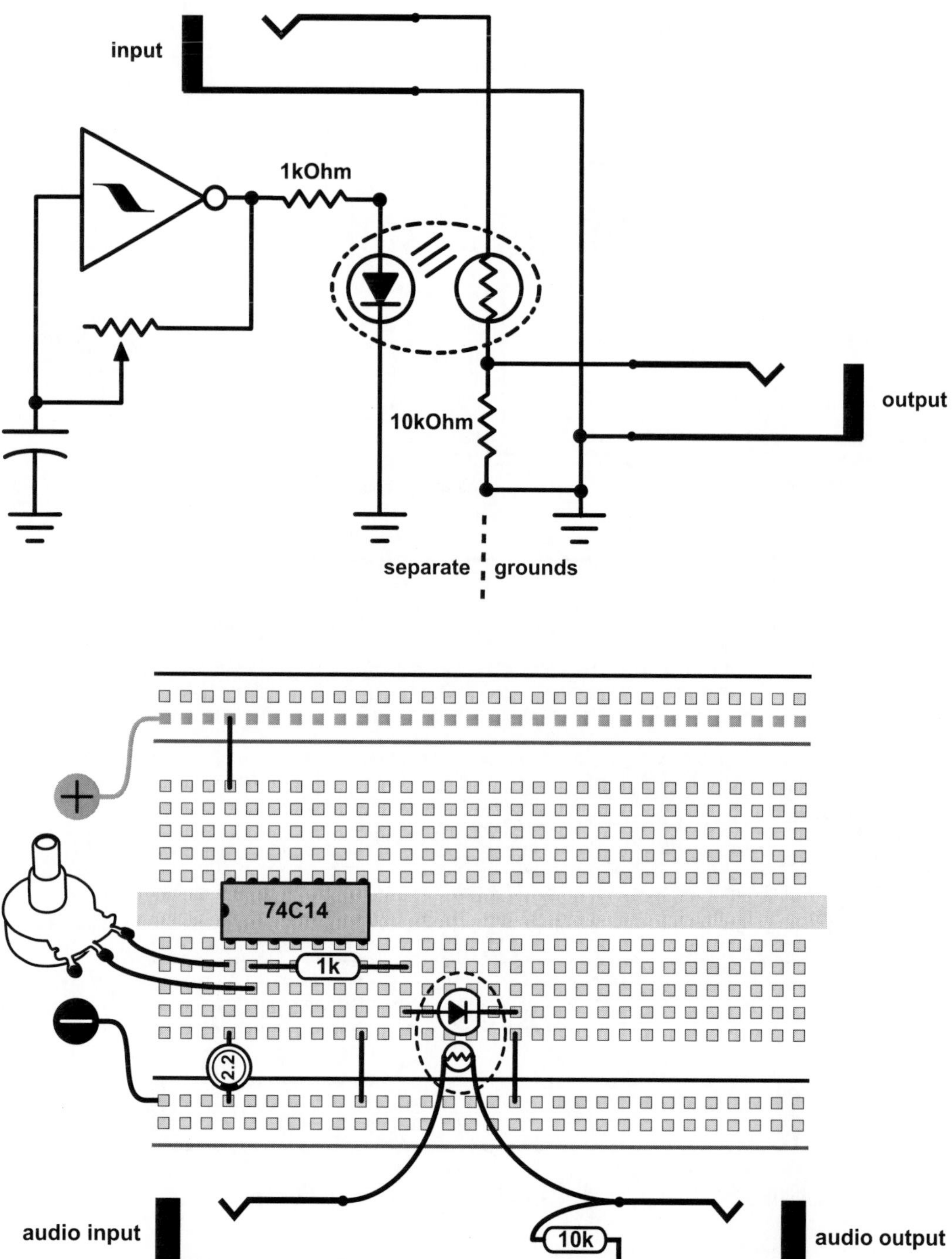

Figure 21.11 A blinking-LED-controlled gate.

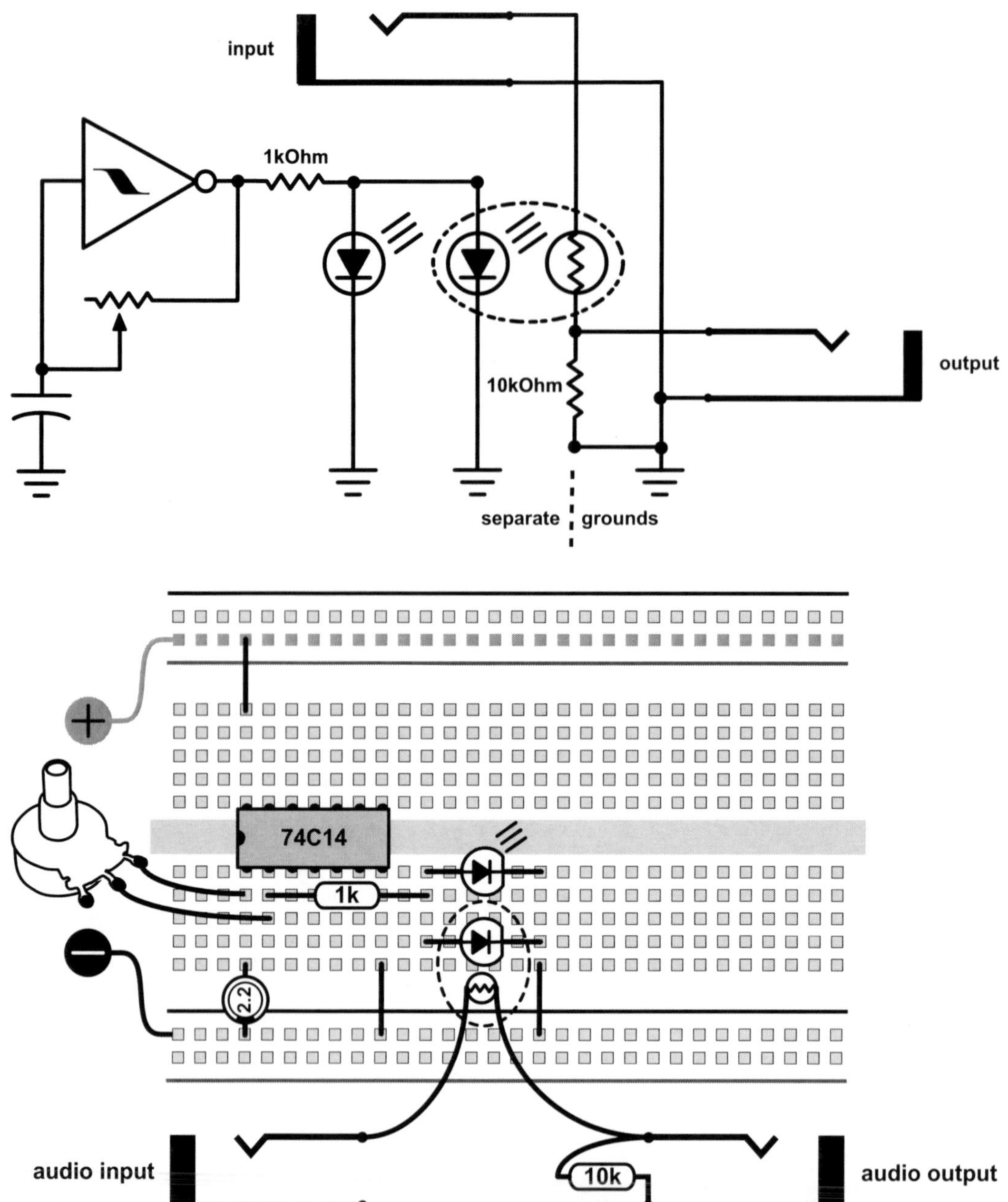

Figure 21.12 A blinking-LED-controlled gate with second indicator LED.

Figure 21.13
A blinking-LED-controlled panner/mixer. Boxed version by Nicolas Collins, from the collection of Robert Poss.

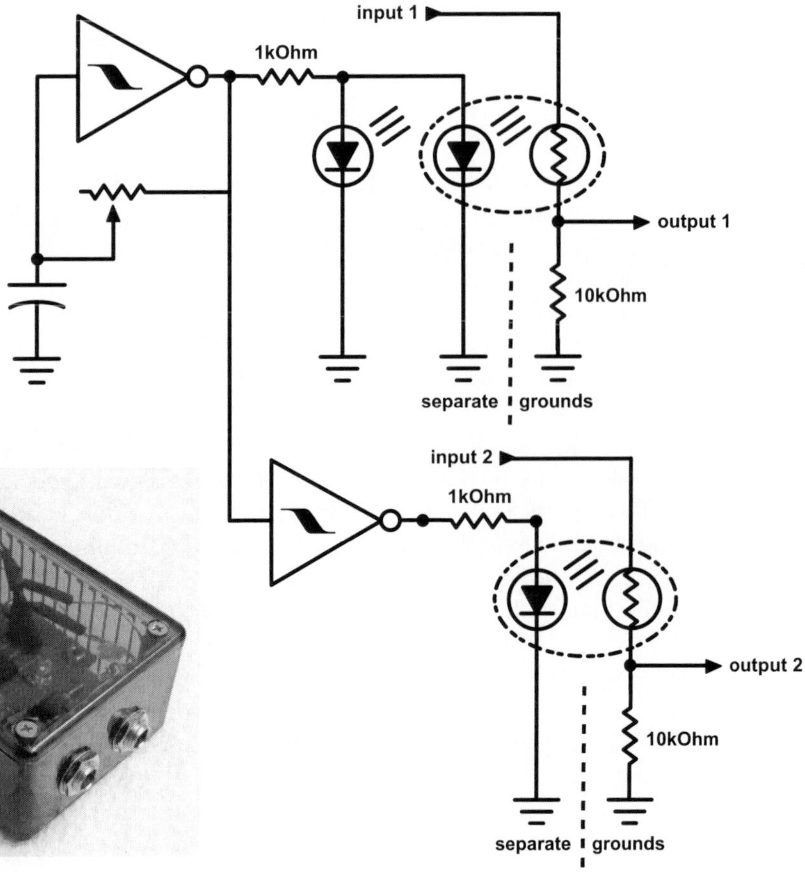

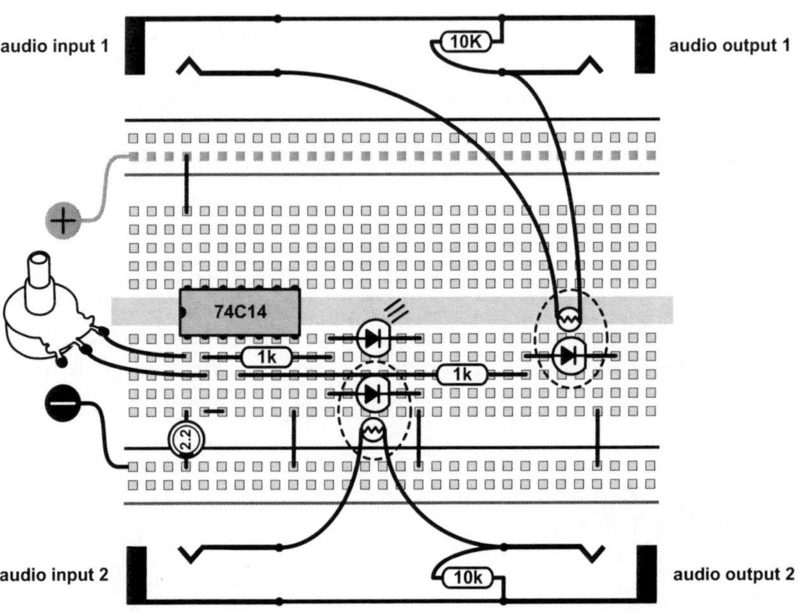

Figure 21.5); adjusting the oscillator frequency changes the panning speed—Psycho-Pan-Scan! Now hook up two different audio sources to the two circuit inputs, mix both circuit outputs to one amplifier input (Y-cords or clip leads), and adjust the oscillator to cut between the two signals—Super Crab! Remember to link all the input and output jack grounds together, but (as with the other gate circuit) you do not connect the oscillator's ground bus to the ground of the audio jacks.

The basic concept of the blinking LED chopping audio can be extended from simple oscillators to more complicated control circuits—driving the LED with the output of the cascaded gated oscillators we made in Chapter 20 yields weird rhythms. Just don't forget to include a resistor before the LED.

As with the flashlight-in-the-closet experiment, these circuits do not produce a total mute when off—some of your audio signal will continue to bleed through even when the LED is off, and even if you use the 10 kOhm clamp resistor. The amount of bleed depends on the specific photoresistor used and how effectively it is shielded from outside light, and the intensity of the LED (a brighter LED will give a wider dynamic range). The photoresistor should have as large a difference as possible between "on" and "off" resistance—they're usually best picked by ear through substituting different choices into the circuit, although a data sheet (if available) can help. Bear in mind that the slight leakage of audio during the "off" state will most likely be masked by other sounds in your mix unless this circuit is being used all by itself in a very quiet setting.

If you're really obsessive about silence take the panner circuit shown in Figure 21.13, connect input of the second channel to the output of the first, and connect the output of the second channel to ground, so that the second photoresistor replaces the clamping resistor at the output of the first channel. This way whenever the first channel is off the second channel connects the channel one output firmly to ground, greatly increasing the depth of the mute.

If the masses of electrical tape offend your sensibilities, you can put the photoresistor and LED inside an opaque soda straw, or the plastic sleeve of a mini plug or guitar plug—you may want to put some BluTak or opaque silicon sealant into the ends of the tubes to prevent light leakage. Once again, be careful to avoid shorting the leads against one another.

Heat shrink tubing is another tidy solution to light isolation (see Figure 21.14). Slide narrow pieces around the legs of the LED and photoresistor to insulate them from shorting against one another, if you wish. Slide a wider piece over the LED and cell, nuzzle the

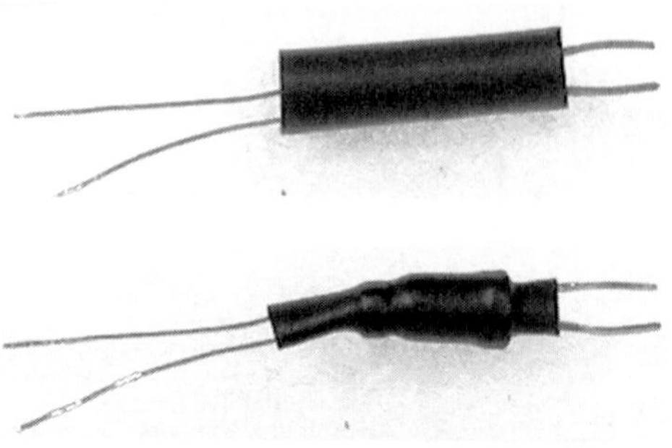

Figure 21.14
LED and photoresistor in heat shrink tubing, before shrinking (top) and after shrinking (bottom).

Figure 21.15 Pre-packaged audio opto-isolators.

two components tightly together, and apply heat from a hair-dryer to shrink the tubing tight around them. Voilá! A microelectronic "Bruit Secret."

I should mention that one can buy pre-packaged LED-photoresistor modules that are well suited for insertion into the circuits we've just built (see Figure 21.15). The most widely used is the "Vactrol" by PerkinElmer, but they're not always easy to find, and rolling your own can be oddly satisfying anyway. Caution: the Vactrol is sometime referred to as an "opto-isolator"; many, many versions of opto-isolators are available from many sources, but most of them are intended for digital applications and do not employ the photoresistor necessary for controlling an audio signal.

Optical gating and dynamic control is much prized by audiophiles for its sonic purity. The complete separation of the audio ground from the oscillator's ground contributes to the inherently high quality of the sound (which is why I stressed it earlier). Only a few small, if confusing, additions stand between these simple circuits and some very expensive studio noise gates, compressors, and limiters. We'll return to these devices in the next chapter.

OTHER USES FOR PHOTORESISTORS

As you should grasp by now, the photoresistor is a resistor like any other, but for its Nosferatu-like response to light. You can substitute a photoresistor for almost any resistor or pots, in almost any circuit, and then modulate that resistance with light—either performed (flashlights, shadows, etc.) or automated, as we showed in this chapter using blinking LEDs. If you already have a toy whose pitch is controlled by a photoresistor, wire up an oscillator-driven LED, press the blinking light against the cell, and listen to what happens (similar to the toy cross-modulation experiments we tried at the end of Chapter 15).

Modulating the pitch of a photoresistor-controlled audio oscillator with an LED blinking at a genre-appropriate beats-per-minute yields a pleasingly disco-tinged "Syn-Drum" swoop. Just take any of the audio oscillator designs from the previous two chapters, and replace the photoresistor in the feedback loop with one of our LED-photoresistor bundles. Drive the LED from a second, low frequency oscillator (try the design in Figure 21.9). Adding a pot in series or parallel with the photocell makes it possible to adjust the sweep range.

Sometimes, if its "on" resistance is low enough, a photoresistor can be substituted for any low-current switch. For example, if your toy has switches to trigger sounds or enable functions, try paralleling one of those switches with a photoresistor: connect the two photoresistor legs to the points on the toy's circuit board that are joined when the switch is closed. This is very effective when the toy has the kind of switches that consist of trace patterns on the circuit board shorted by a rubbery switch (see Figure 28.9).

Shine a flashlight at the photoresistor, or link it to an LED driven by a slow oscillator. The function associated with the switch should be triggered when the flashlight hits the photoresistor or when the LED is on. If it works, you've got a simple solution to automating some of the toy's functions; if not, try another switch or another toy. Don't use the photoresistor as a substitute for the power-on/off switch in a circuit, or between the circuit and its speaker, since it can't pass enough current. But the photoresistor switch is a convenient way to extend the duration of toy playback, especially of slowed down samples, by automatically and repeatedly "pressing" the same button; they can also be used to press various switches at different rates to produce quasi-random results.

CHAPTER 22

Amplification and Distortion: A Simple Circuit That Goes From Clean Preamp to Total Distortion, and an Envelope Follower

You will need:

- Something to amplify: an electric guitar works best; a contact mike, telephone pickup or CD/MP3 player are also useful.
- One of your oscillator circuits.
- A breadboard.
- A CD4049 (or CD4069) CMOS Hex inverter.
- Assorted resistors, capacitors, and pots.
- Some solid hookup wire.
- An amplifier.
- Assorted jacks and plugs, to match your amplifier and sound source.
- A 9-volt battery and connector.
- Hand tools.

In addition to turning sounds on and off, as we did in the previous chapter, there are many occasions when we want to make something LOUDER (see the Second Law of the Avant-Garde). Loudness comes in different flavors, and a little experimenting with the CD4049 Hex inverter demonstrates several of them. This is yet another example of a digital logic chip being "misused" for analog purposes.

The chip's internal configuration and pinout are shown in Figure 22.1. The 4049 is a rare exception to the general rule of corner pins for power hookup in CMOS chips (as in the 74C14 Hex Schmitt Trigger and the 4093 NAND Gate we used in previous chapters). It also has two more pins: 16, versus the 14 of our earlier chips. If you are at all dyslexic, now is the time to hold onto your hat with both hands: although the ground connects to pin 8 as expected, + volts connects to pin 1, rather than the anticipated pin 16. The "NC" by pins 13 and 16 indicates that they make "no connection" to any internal circuitry. Note also that the 4049 inverters face the opposite direction than those in the 74C14. IMPORTANT: do not substitute the 74C14 Hex Schmitt Trigger for

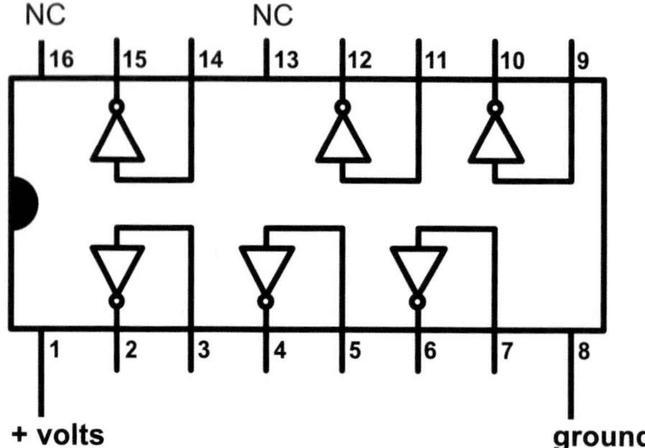

Figure 22.1 CD4049 Hex Inverter pinout.

the 4049 in the examples in this chapter—the 74C14 *is* an inverter, but has a different internal circuit design that won't work properly in our new configurations (if I may wax technical, the 4049 omits the Schmitt Trigger circuit essential to making our oscillators snappy, but incompatible with the designs in this chapter).

At the head of this chapter I've listed another chip, the 4069, as an alternative to the 4049. It has the identical pinout, voltage requirements, etc. In most cases the two parts can be substituted for each other with carefree impunity. However, because they differ slightly in their internal circuit topology, periodically some silicon sage makes the argument that one of them is preferred over the other for particular audio designs. I've had similar results with both, so although I specify the following circuits as employing the 4049, you should feel free to experiment with either chip.

PREAMPLIFIER

Hook up the circuit shown in Figure 22.2. This is a general-purpose preamplifier circuit, useful for boosting the signal of low-level sound sources (such as microphones, contact mikes, guitar pickups, and coils) to the line-level signal strength typical of CD players, cassette decks, etc. After preamplification these signals can be sent to simple powered speakers, such as those used for computers or iPods, or intermixed with line-level sources using simple, passive mixers (see Chapter 26). A preamplifier is not the same as a power amplifier—this circuit cannot be connected to a loudspeaker directly—for that you need another kind of design, discussed in Chapter 27.

This preamp circuit has five basic components:

1. The CMOS inverter stage. As with our oscillator circuits, the six sections of the 4049 chip are interchangeable.
2. The input resistor, RI, generally around 10 kOhms.
3. The feedback resistor, RF, generally larger than RI.
4. The input capacitor, CI, generally around 0.1 uf.
5. The output capacitor, CO, generally around 10 uf.

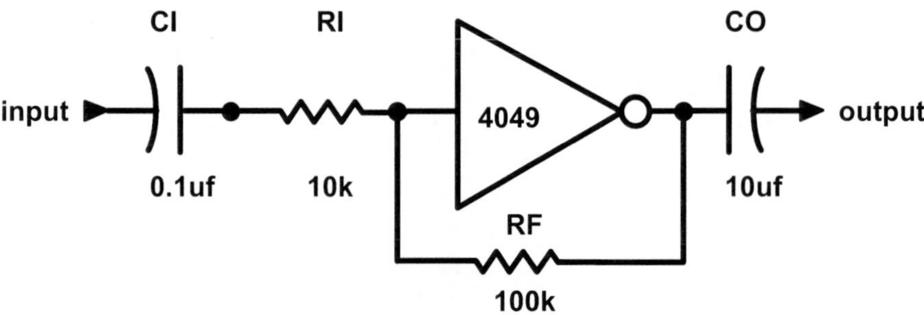

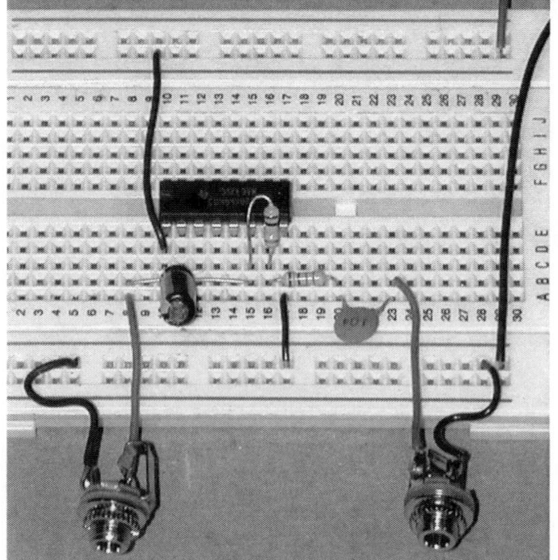

Figure 22.2
A basic preamplifier.

Your guitar (or other sound source) connects to the jack at the *right* hand side of the breadboard as shown, while the output emerges from the *left* jack—a much needed Semitic twist on the right-to-left orthodoxy we've been observing in our circuits so far.

The values of CI and CO are not critical, but when using electrolytic capacitators please observe the polarity shown in the schematics. RI can usually be set at 10 kOhm. The only real thought goes into selecting RF. The gain—how much the circuit amplifies the incoming signal—is determined by the ratio of RF/RI. So, if RI is 10 kOhms and you use a 100 kOhm resistor, the gain will be 10, which means that any signal you plug into the circuit comes out 10 times larger. If RI = 10 kOhms and RF = 10 mOhms, the gain is 1,000, which makes your input much MUCH louder.

[An aside about loudness. A gain of 10 means the voltage amplitude of the signal is 10 times larger than before, but that doesn't mean that it *sounds* 10 times louder. Our ears have a logarithmic, rather than linear, response to loudness. A gain of 10 translates into a change of 20 decibels (dB), the unit of measurement used for loudness. A change of 6 dB is perceived as a doubling of loudness, roughly. So a gain of 10 equals an increase of 20 dB, which would sound about three times louder. A gain of 100 equals a 40 dB boost, or about seven times louder. But, as usual, I suggest that you don't obsess too much about the math, and evaluate gain values by listening.]

By substituting a large pot (i.e. 1 megOhm) for the fixed feedback resistor RF we can vary the gain of the circuit, as shown in Figure 22.3. For a typical preamp (as you might use for a contact mike or coil) you may wish to wire up a 10 k resistor in series with a 1 megOhm pot: this lets you adjust the amplification smoothly from unity gain when the pot is at one extreme of rotation (signal out = signal in) to a gain of just over 100 (output = 101 × input). Reduce the size of the pot to 100 kOhm or so (or increase the size of RI) if you are getting too much gain. You can also use a photoresistor for RF to make a light-controlled amplifier that provides more gain as the photoresistor gets *darker* (the reverse of the photoresistor-controlled gates in the previous chapter).

Besides the different power supply connections, the unused pins, and the "backwards" orientation of the inverter stages, this circuit uses more parts than anything we've done so far. It might seem confusing at first, and with each additional component the chance of making a hook-up mistake increases. It may take a few tries before you get the circuit working properly. Remember that the most important detail of this design is the choice of the feedback resistor, RF. The input and output capacitors (CI and CO) block the DC voltage present in the circuit from reaching whatever you're plugging into—as with the electret mike circuits in Chapter 10 and our triangle wave output in Chapter 20. Don't think about them too much now if they bother you, just put them in with the right orientation. They are necessary for the stability of the circuit, and usually don't affect the sound much.

TONE CONTROL

You may notice some noise or high frequency oscillation at very high gain. You can minimize these unwanted artifacts by putting a *very* small feedback capacitor (CF) in parallel to the feedback resistor as shown in Figure 22.4. Try values in the range of 10–100 pf (picofarad)—the "small" 0.1 uf capacitors we have been using in our oscillators are *way* too big.

Beyond getting rid of noise or unwanted oscillation, this "feedback capacitor" sets the upper limit of the frequency response of the circuit. By substituting slightly larger

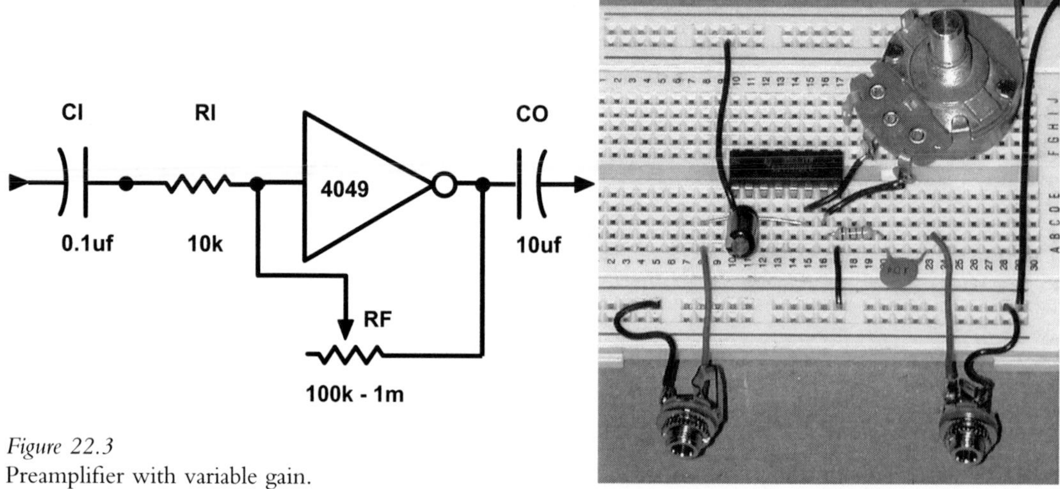

Figure 22.3
Preamplifier with variable gain.

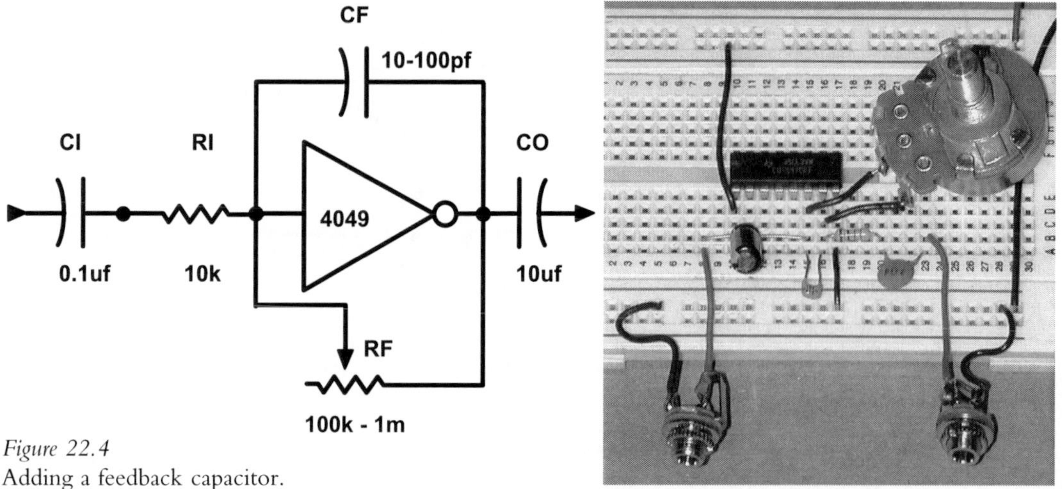

Figure 22.4
Adding a feedback capacitor.

capacitors you can transform our preamp into a simple "low pass filter" tone control—one that blocks all frequencies above a certain pitch, while letting those that are lower pass through.

Connect a sound source to the circuit and adjust it for relatively low gain, so as not to distort the listening amplifier. Try different value feedback capacitors: 10 pf, 100 pf, 0.001 uf, 0.01 uf, and finally 0.1 uf. As you increase the size of the capacitor you should notice the treble rolling off and the music getting bassier. By the time you reach 0.1 the circuit will probably be rolling off *all* audible frequencies; you'll hear almost nothing and probably think the circuit is mis-wired. You will need to patch the preamp into a better amp and bigger speaker than the mini-amp we've been using, if you want to hear the detail of these changes.

Given the extremely bright high end of our square wave oscillators, some high-frequency roll-off can be welcome now and then, as we demonstrated with our filter add-ons in Chapter 20 (Figures 20.12 and 20.13). Patch any of your oscillators through a preamp set for unity gain (RI = 10 k, RF = 10 k) and try the capacitor substitutions as above. The waveform should mellow out into something more like a triangle wave or sine wave, bringing relief to canine and human cohabitants alike.

Unlike the low pass filter on a synthesizer, the High Frequency Equalizer on a mixer, or even the simple treble control on a home stereo, the amount of roll-off cannot be continuously adjusted with a pot in this design. This circuit is best suited for fixed settings, such as softening the oscillators into pseudo-triangle waves, or mellowing a distortion circuit (see below). You could add a switch to select between two or three different capacitors if you want some variation (rotary switches can select amongst several different values). Before you sneer at switched EQ, remember that vintage Neve, API, and Pultec equalizers command stratospheric prices despite (or because of) such switching—and a few other factors, admittedly, such as brilliant design, beefy transformers, luscious knobs, rarity, etc.

But if you insist on continuously variable EQ, try the circuit in Figure 22.5, which is a variation on the swept low pass filter we introduced in Chapter 20 (Figure 20.13). Here, instead of being tacked directly onto the output of an oscillator, a similar

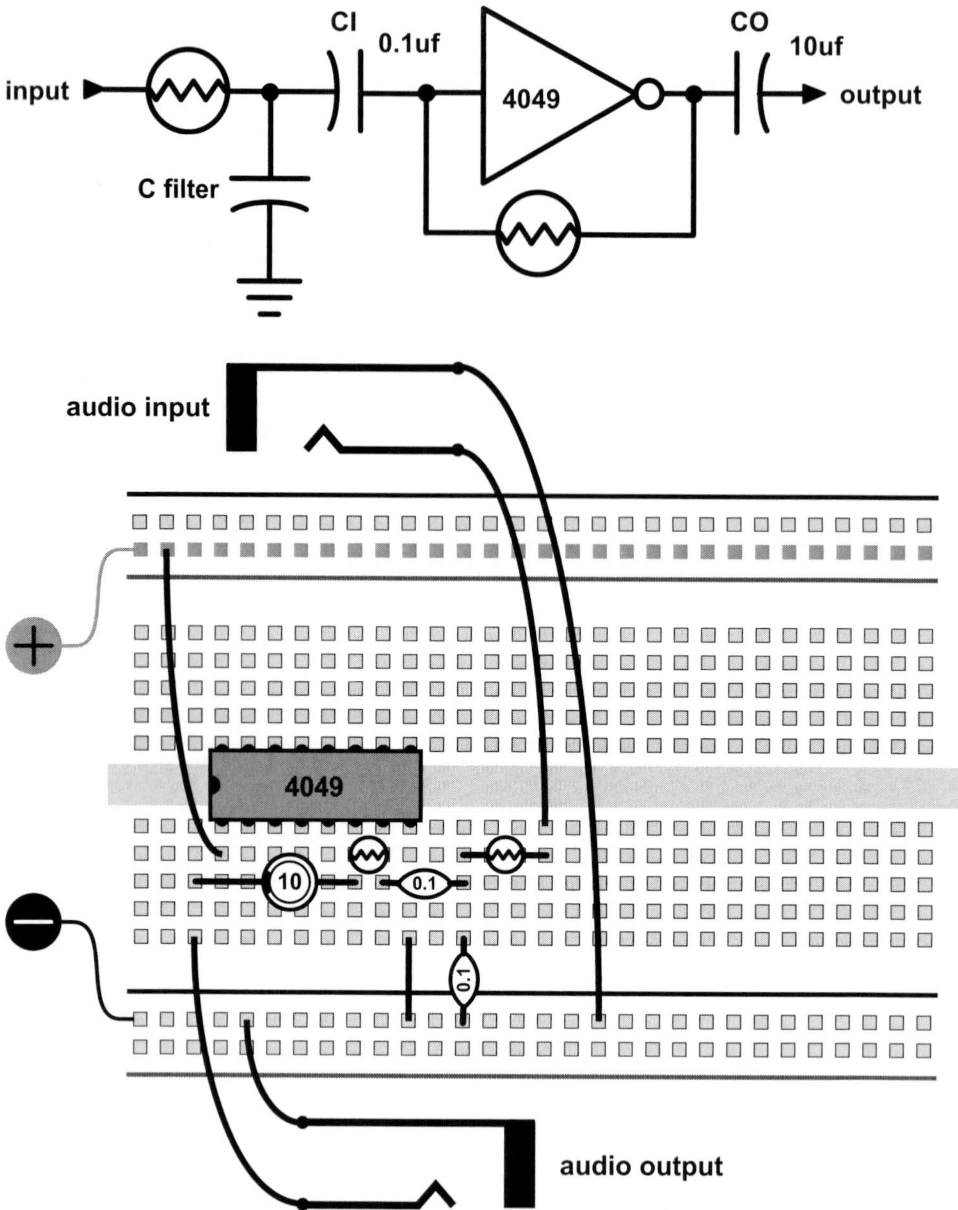

Figure 22.5 Swept low pass filter.

arrangement of a capacitor and two photoresistors is added to a stage of a 4049. The two photoresistors should be close together so they receive the same amount of light and "track" each other with the same resistance value, otherwise you will hear a significant change of loudness in addition to the sweep of the filter. Experiment with different values for C-filter—0.1 uf is a good place to start. CI and CO are needed, as in our basic preamp circuit, to block DC.

If you want more accurate control of this filter you can substitute a dual potentiometer for the two photoresistors. A dual pot consists of two potentiometers controlled by a single shaft (see Figure 26.6). Thanks to this mechanical coupling, they track each other very closely, and eliminate the problem of trying to keep the light level the same on the two photoresistors.

Remember the Input and Output Capacitors I told you not to worry about? If you enjoy worrying, try capacitor substitutions on CI. As you *decrease* its value from the nominal 0.1 uf down to 0.01 uf, 0.001 uf, 100 pf, and 10 pf, you should notice that the bass frequencies begin to vanish and the sound gets tinnier—we've made a simple "High Pass Filter." With careful selection of the optimum capacitor size, you can make a useful filter for rolling off low frequency wind noise or handling noise and rumble from microphones and contact mikes.

The six inverter sections in the 4049 can be used interchangeably. You can make a six-channel contact mike preamp with one chip, for example (just watch out for the illogical pinout of the various inverter sections). Each preamp can be wired to its own output jack for connection to an external mixer, or you can sum them together with 10 k resistors as we did with the multiple oscillators in Chapter 18. You can build them into the mixer designs we will introduce later, in Chapter 26.

DISTORTION

Amplifier sections can also be cascaded to produce greater gain: by putting two stages with 10x gain in series you get a net gain of 100. But this simple approach to cumulative amplification is not "perfect," and by adding a lot of gain in series we introduce distortion, the guitarist's friend. The circuit in Figure 22.6, based on a venerable design by Craig Anderton (the godfather of musical hacking), is simple, versatile, and sounds great.

Plug in a guitar (always the best instrument for evaluating distortion) and turn the pot. As the gain increases the sound should shift from clean amplification through tube-like "overdrive" into distortion and, eventually, uncontrollable noise and oscillation. Ahh, bliss! As Robert Poss says:

Rule #23: Distortion is Truth (Poss's law).

But truth and distortion both come in many flavors. If you are really interested in distortion, you should spend some time substituting different components throughout this circuit until you find perfection. In particular, try:

- Various resistors, from 100 kOhm to 10 mOhm, for RF2.
- Substituting a pot for RF2, for separate control of gain at each stage.
- Adding a "clamp resistor," around 10 k to 100 kOhm, from the input to the second stage to ground—this sometimes reduces noise and oscillation at high distortion/gain settings.
- Inserting 10 k input resistor between CI and the first inverter (as we had in our preamp circuits).
- Various feedback capacitors, from 10 pf to 100 pf, for CF1 and CF2, in parallel to RF1 and RF2. These values can be set with fixed values or made switch selectable, as we suggested above for our simple EQ circuit (see Figure 22.7).

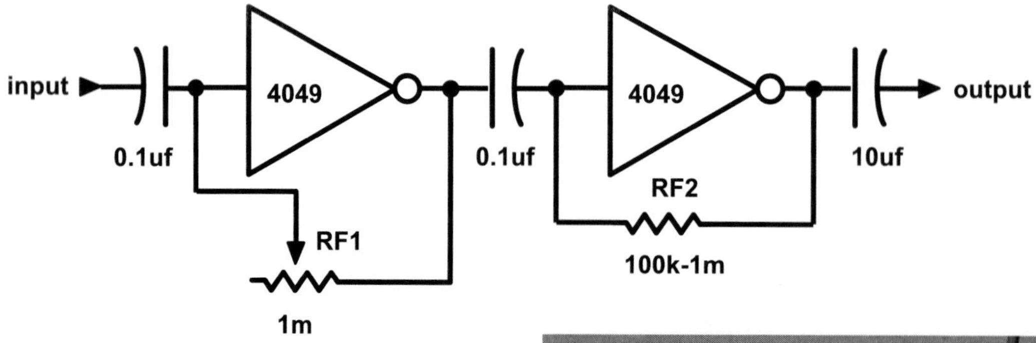

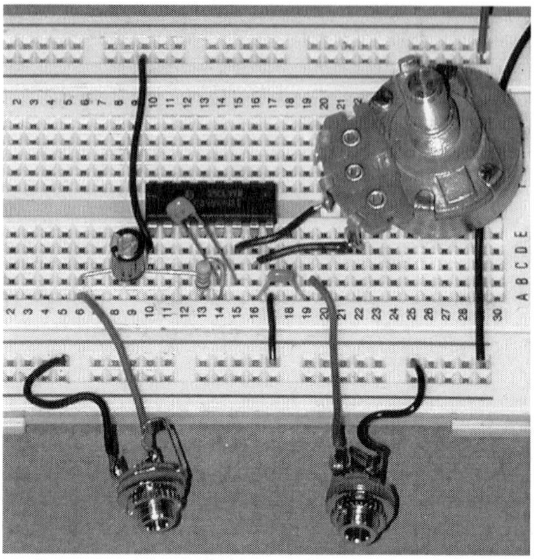

input
4049
4049
output
0.1uf
0.1uf
10uf
RF2
RF1
100k-1m
1m

Figure 22.6
Craig Anderton's basic distortion circuit.

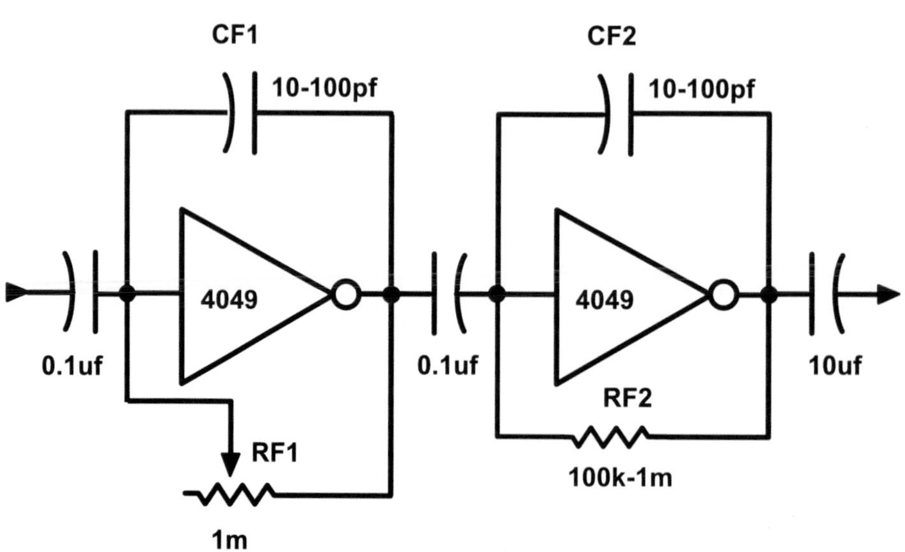

Figure 22.7 Distortion with feedback capacitors added.

- Adding a stage of swept low pass filtering as shown in Figure 22.5 above.
- Adding an additional gain stage, which makes the distortion more extreme (see Figure 22.8).
- Following the circuit with a Schmitt Trigger inverter from a 74C14, which clips the signal into a buzzing pulse wave in the style of a classic 1960s fuzztone (see Figure 22.9). You will probably need a resistor of about 100 kOhms between the input of the Schmitt Trigger and ground to keep the circuit from oscillating in the absence of an actual input signal. Sometimes removing the capacitor marked with the asterisk and replacing it with a straight wire also improves stability.

When running at high distortion the output signal of these circuits reaches 9 volts peak-to-peak, like our oscillators. You may want to add an output volume pot (similar to Figure 20.8 in Chapter 20) to lower the output level independently of the amount of distortion before you plug it into a guitar amp—most commercial distortion circuits have separate controls for "distortion" and "level." Another common feature in stomp boxes is a bypass switch that allows you to select between the processed output of the

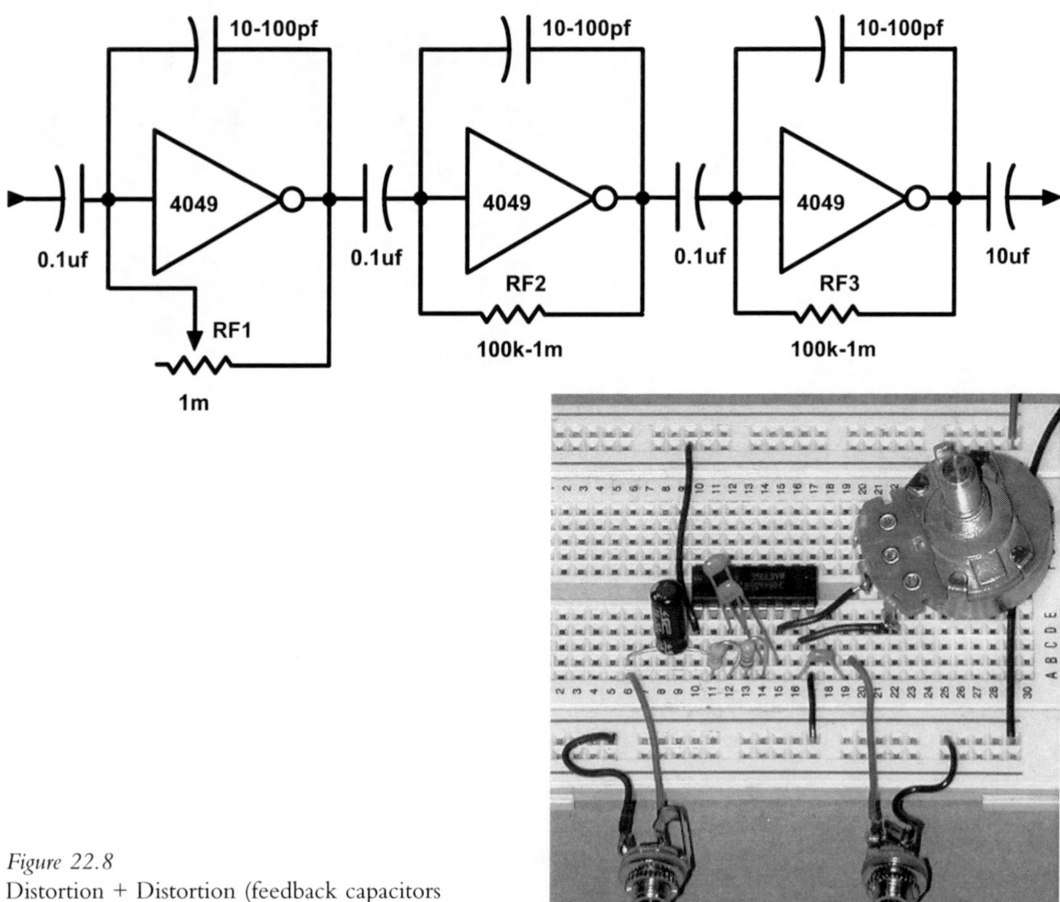

Figure 22.8
Distortion + Distortion (feedback capacitors omitted from breadboard for clarity).

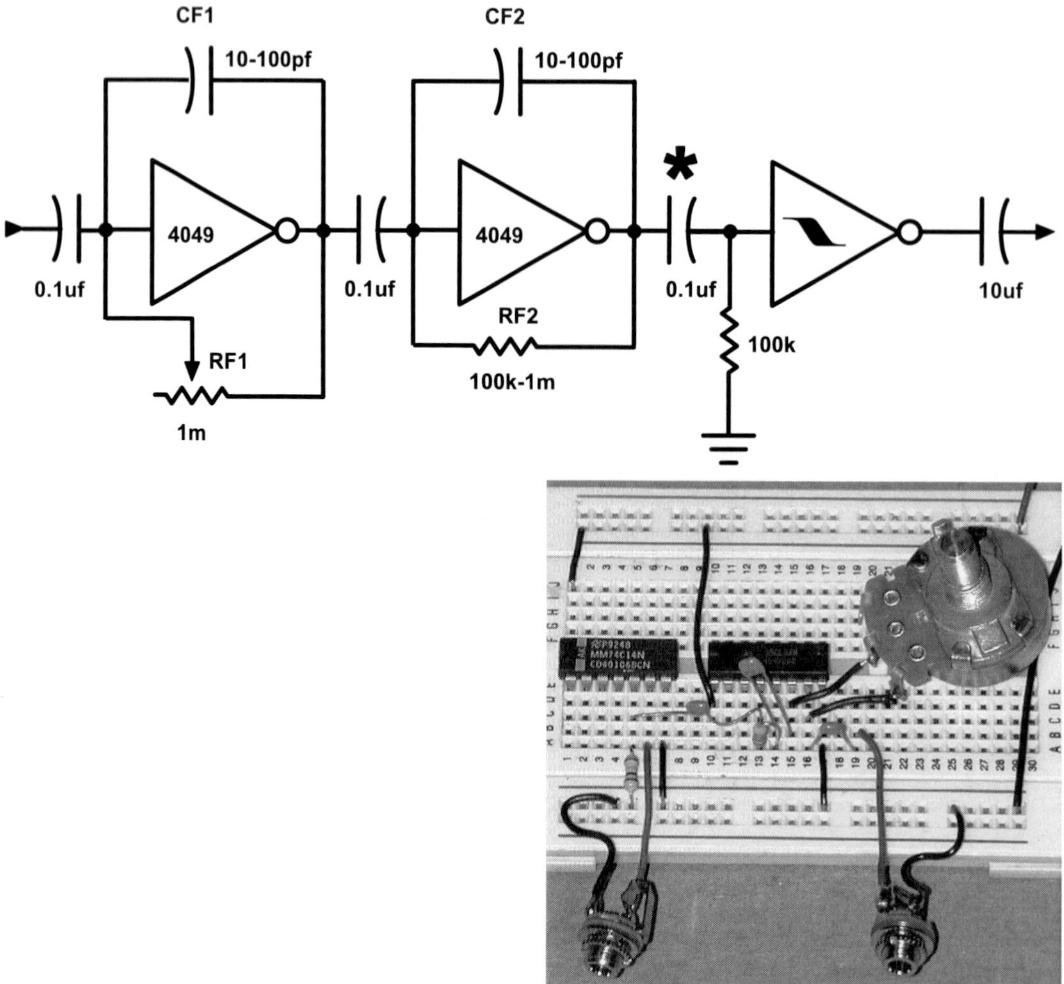

Figure 22.9 Distortion + Fuzz (feedback capacitors omitted from breadboard for clarity).

circuit and the unprocessed input. Adding this function simply requires a SPDT switch (usually a heavy duty push-on/push-off switch that alternates positions on each pressing) with the NC terminal connected to the input to your circuit, the NO to the circuit output, and the C to the output jack's tip/hot. A toggle switch might be more useful than a pushbutton if you're operating the circuit by hand on a tabletop, rather than with your foot on the floor. Figure 22.10 shows both these features added to the basic distortion circuit, but you can include them in any of the variations we discussed above.

If you haven't started breadboarding these designs yet let me forewarn you, and if you're already deep in it let me reassure you: as I mentioned earlier in this chapter, these circuits—with the odd pinout and increased parts count—are inherently more confusing than anything else we've done in this book. Moreover, as you raise the overall gain, especially in the distortion designs, your circuit starts to behave irrationally: it's supposed to be amplifying your guitar, but it seems to be oscillating all by itself. Why? Basically,

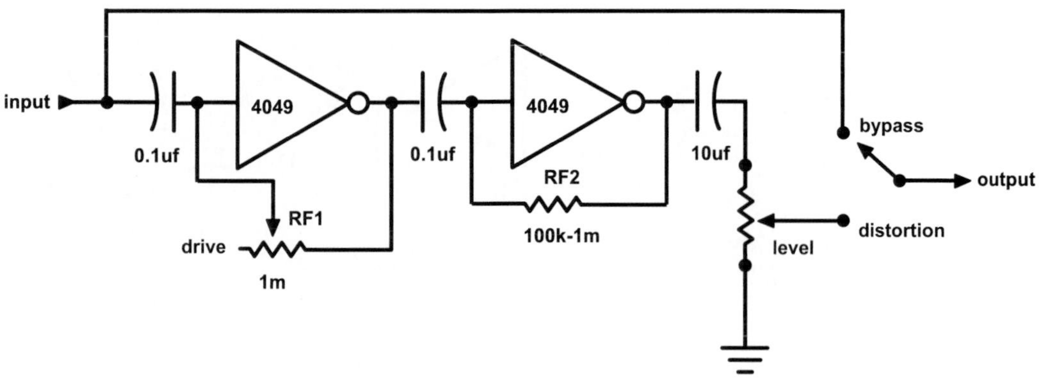

Figure 22.10 Output level and bypass switch added to basic distortion circuit.

at high gains the circuit isn't just amplifying the signal you think you're sending it (the guitar), but it acts as though you've plugged in a loose coil pickup as well: it boosts all manner of spurious noise and hum to the point that there is no more "off" or silence. In the battle of the "signal to noise ratio," noise is favored by a large margin. To improve the signal's odds:

- Keep all jumper wires as short as possible—on the breadboard, and between the pots and jacks and the breadboard.
- Use shielded cable to connect to and from the circuit.
- Add tiny capacitors in the various feedback stages, and a 0.1 uf cap between + power and ground.
- Try clamping resistors to ground at the various inputs.
- Lower the gain of each stage.
- Build the circuit up one stage at a time, testing after each addition.

It's worth persevering. As frustrating as it can feel sometimes, designing and debugging by ear is essential to building good-sounding audio circuitry (oh no, you mutter, a *learning experience*!). Preamplifier and distortion circuits are very useful in their own right. And if we can successfully boost ordinary analog audio to the near-square wave signals put out by our more extreme distortion circuits, we open the door to a host of very unusual signal processing, as you will see in the next chapter.

ENVELOPE FOLLOWER

Sometimes it's useful to be able to trace the loudness of an audio signal, like the meters on a mixer do. The circuit in Figure 22.11 translates loudness of any signal into the brightness of an LED. And as you know by now, when we nestle the LED up to a photoresistor, we can control all manner of things.

I know this one looks complicated, but fear not: it is built up from modules that should be quite familiar to you by now. If you study the design for a moment you'll notice that it starts out sort of like a distortion circuit, which then dumps the highly

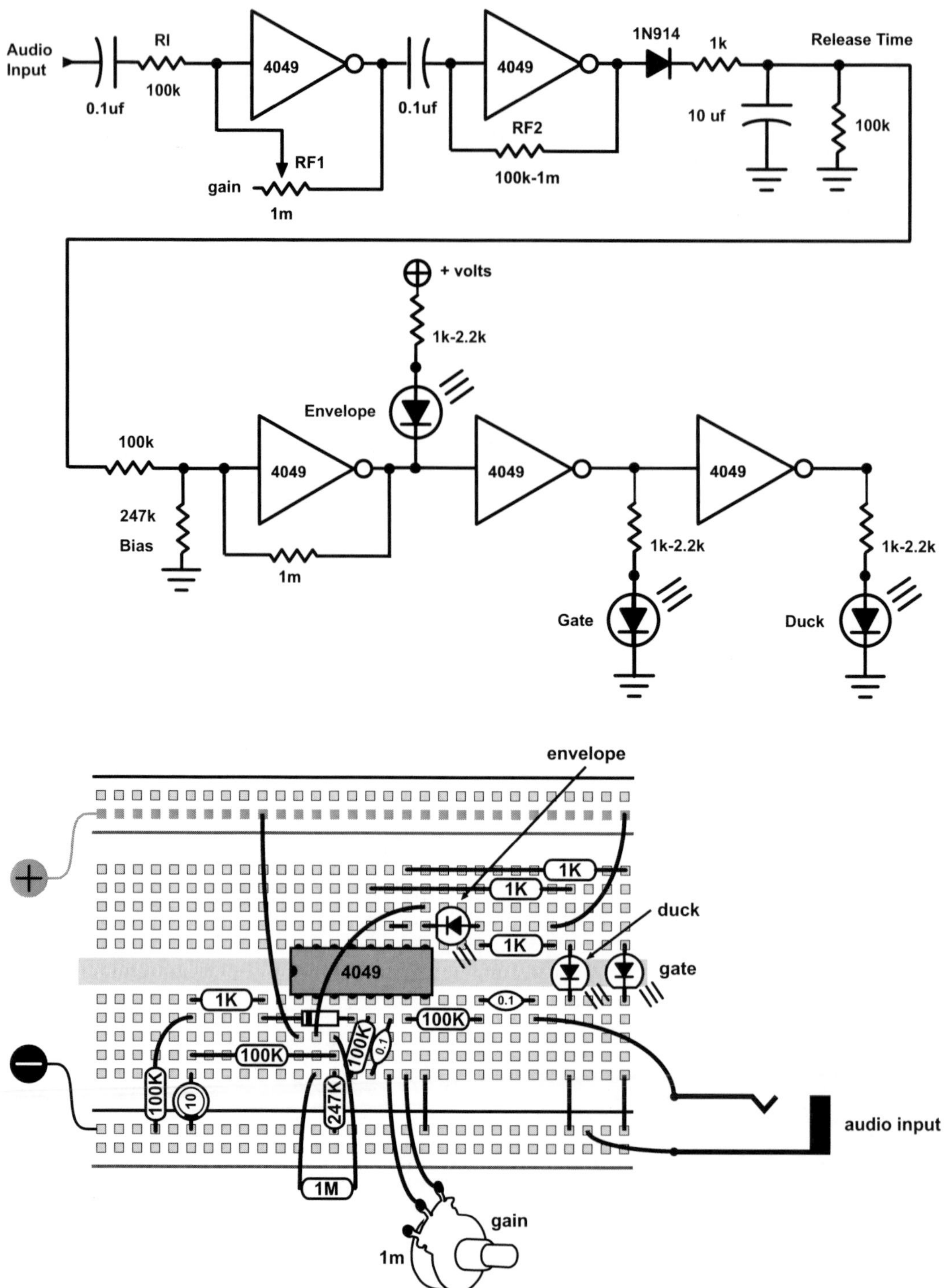

Figure 22.11 Envelope Follower with Gate and Duck outputs.

amplified audio through a diode into a 10 uf capacitor. The brightness of the Envelope-LED that follows (after another inverter, used here as one more stage of amplification) essentially indicates the "water level" in that capacitor—i.e. the loudness of the signal. The capacitor slowly drains off between audio peaks through the 100 k "release time" resistor (think back to the bucket-capacitor and rusty pipe analogy we used to explain the 74C14 oscillator in Chapter 18). The 4049 stage after the Envelope-LED switches state between on and off when the envelope crosses one-half the supply voltage, thereby translating the fluctuating analog voltage back into a binary signal. The Gate-LED that follows it lights whenever the audio signal is louder than a certain threshold. The next (and last) inverter does what the second inverter did in our panner circuit in Figure 21.13 of the previous chapter: it outputs the opposite of its input, so the Duck-LED lights when the signal is *below* the threshold, whenever the Gate is off.

The main control for this circuit is the 1 megOhm pot in the first gain stage—all the LEDs respond to the level coming through from this point. This may be enough variation for you, but you can replace the 100 k Release Time resistor and/or the 247 k Bias resistor with 1 megOhm pots if you want to vary the speed of the envelope follower and adjust the Gate/Duck threshold (see Figure 22.12). If the circuit doesn't seem responsive enough you can increase the overall gain by lowering RI from 100 k to 10 k.

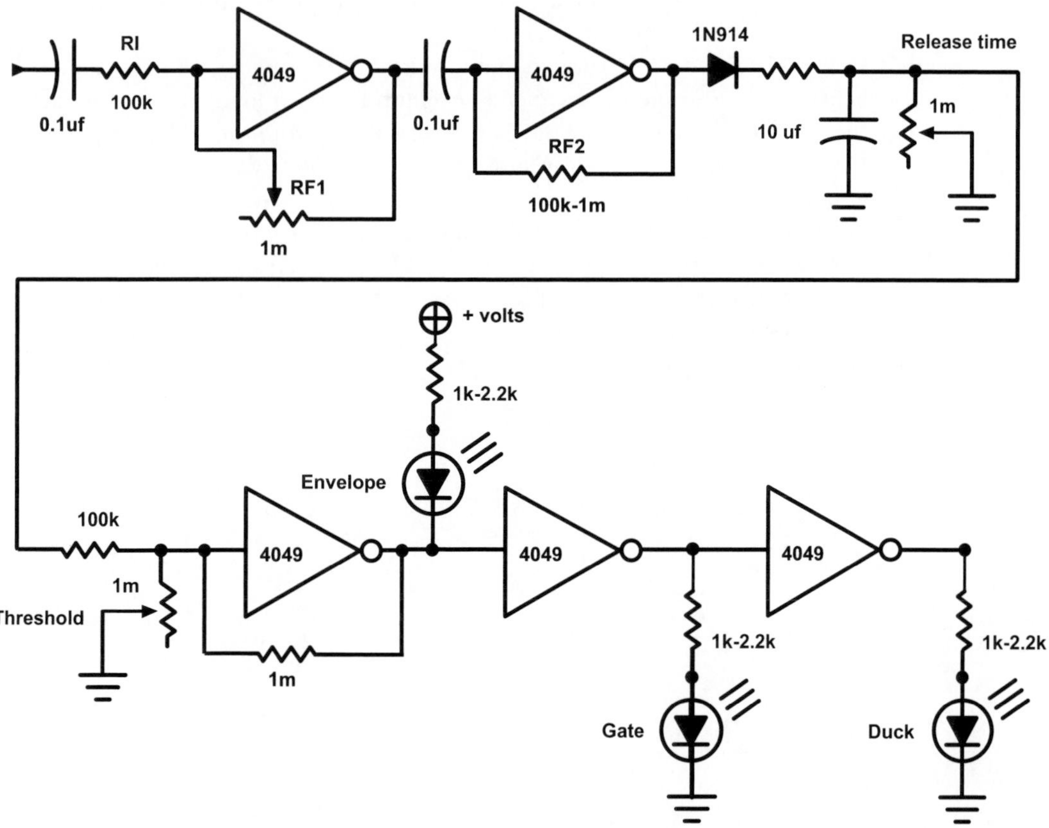

Figure 22.12 Envelope follower with variable release time and gate threshold.

What can you do with this circuit? You can substitute its various LEDs for those in any of the light-controlled circuits we created in the previous chapter. To start with, send an audio signal (such as a music recording) through the Envelope Follower and adjust the level so the LEDs blink in response to peaks in the sound. Couple a photoresistor to the Envelope-LED, using some electrical tape or a bit of heat shrink tubing (you can always add second LEDs in parallel for indicator lights, as we did in Figure 21.12 in the previous chapter).

- If you wire this photoresistor as the pitch control resistor in a 74C14 oscillator, its frequency will sweep in response to the loudness of the Envelope signal: plug in a contact mike, tap it, instant electronic drums; plug in a music file you'll hear an odd sort of bouncing ball effect following the dynamics of the track.
- If you send a *second* audio signal through the photoresistor it will be articulated by the loudness of the *first* signal you connected to the input to the Envelope Follower—like the flashlight experiment at the beginning of the last chapter: plug a kick drum into the Envelope Follower, run a bass guitar into the photoresistor, and your bass line will be accented by the kick—an old Motown trick, by the way. This latter effect is essentially a simple "externally keyed expander."
- The Gate and Duck LEDs can replace the complementary blinking LEDs in our panning and mixing circuit: just substitute them for the two LEDs in Figure 21.13, and you have an externally keyed noise gate and ducker—another great way to synchronize disparate rhythmic elements.
- When the Gate and Duck-driven photoresistors are used as timing resistors in oscillators they will cause the frequency to jump between high and low pitches in response to peaks in audio—like our Syn-Drum circuit from the end of the previous chapter. By soldering the photoresistor to a toy's switch you can trigger it with audio signals.
- If you split one audio signal and send it both into the Envelope Follower input *and* into photoresistors controlled by the various LEDs, you can arrange the components to create a simple self-keyed compressor/limiter, expander, gate, or ducker. These circuits will require tweaking before they replace that classic Fairchild in your rack, but the essentials are all there. Stay up late.

CHAPTER 23

Analog to Digital Conversion, Sort Of: Modulating Other Audio Sources with Your Circuits, Pitch Tracking, and Sequencers

You will need:

- Something to amplify: an electric guitar is best.
- A breadboard.
- Distortion circuit from the previous chapter.
- CD4093 Quad NAND Gate Schmitt Trigger.
- CD4040 Binary Counter/Divider.
- CD4046 Phase Locked Loop.
- 74C14 Hex Schmitt Trigger.
- CD4017 Divide-by-Ten Counter.
- Assorted resistors, capacitors, and pots.
- Some solid hookup wire.
- Assorted jacks and plugs.
- A 9-volt battery and connector.
- An amplifier.
- Hand tools.

The preamp and distortion circuits in the previous chapter are useful on their own to boost a low-level signal, or make an electric guitar sound legitimate but also in conjunction with other circuits. As I've already mentioned several times, an ordinary line-level audio signal (such as the output of a CD player) measures a bit less than 1 volt peak-to-peak, and fluctuates in a curvy, Baroque way. The circuits we've been making with CMOS digital chips put out 9 volts peak-to-peak (if powered by a 9-volt battery), and their outputs snap between 0 and 9 volts with Modernist decisiveness. They won't react to any puny 1-volt wigglings at their inputs. Until a signal pokes its head up over one-half the supply voltage it always feels like 0 to the chip, no matter how much it jumps around. But with enough gain and distortion, any analog audio starts to look like one of our digital square waves. And the more it resembles a digital signal, the easier it is to fool our CMOS chips into accepting it as kith and kin. This deception lets us interface analog sounds from the real world to our digital circuits for some unusual signal processing.

THE FUZZY DICER

The Distortion+Fuzz circuit in the previous chapter (Figure 22.9) was the first step in this conversion of snakey analog waveforms into digital signals: we boosted our guitar high enough that it could trigger a 74C14 to produce pulse waves. The next step is the circuit shown in Figure 23.1, which is a variation on the gated oscillator in Chapter 20. Instead of using one oscillator to modulate another, however, here the almost-square wave output of the distortion circuit is connected to the control input of the NAND gate oscillator. When the oscillator runs at low speeds (1.0–10 uf timing capacitor) it effectively "chops" the output of the distortion circuit on and off—like the photocell gate in Chapter 21, but with a much sharper edge and a complete muting in the off state. At higher speeds (0.01–0.1 uf capacitor) the oscillator interacts with the distorted signal to create sounds reminiscent of a ring-modulator or (somewhat inexplicably) a wah-wah pedal.

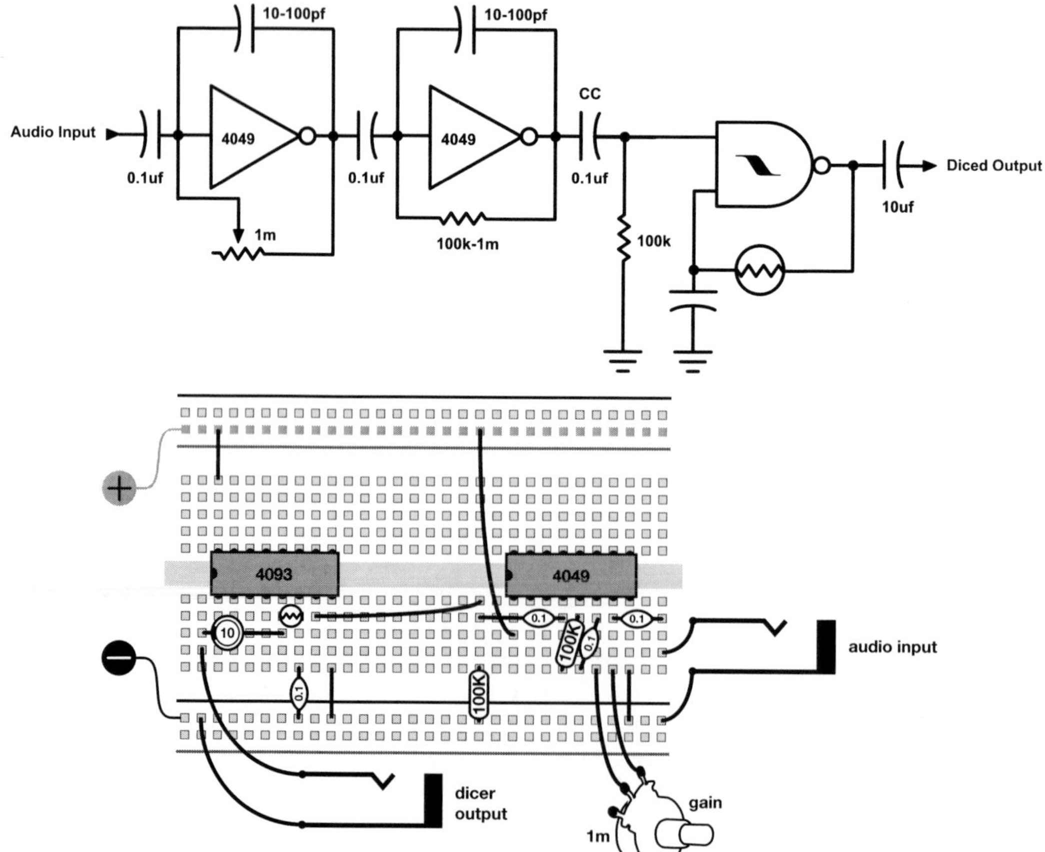

Figure 23.1 The Fuzzy Dicer.

Try this with your favorite version of the distortion circuits from Chapter 22. Don't include any output volume control or bypass switch between the distortion section and the NAND gate; if you want you can put it after the oscillator instead. The circuit may be more stable without the coupling capacitor CC—try both ways. A "pull-down" clamping resistor between the 4093 input and ground is usually needed to mute the circuit when you are not sending it an input signal (i.e. not playing the guitar). If it seems insensitive, only triggering on very loud signals, you may need to increase the gain of your distortion circuit—try the triple-inverter version shown in Figure 22.8, or the fuzz circuit in Figure 22.9.

If it seems *too* sensitive, and prone to fits of high frequency squealing, try putting larger capacitors in the feedback loops of any or all of the 4049 stages of the distortion circuit. This will roll off the higher frequencies, remove excess noise, and send the subsequent circuits a signal that emphasizes the fundamental pitch—the output of your 4049s might sound unacceptably muffled for use as distortion, but will do a better job of feeding the NAND gates a stable signal to track. This circuit may take more tweaking than the ones we've made before, but the effort is worth it—and you can't buy one anywhere (yet).

You can modulate a second NAND gate with the output of the first, as we did in Chapter 20 (Figures 20.4 and 20.5), when we introduced cascaded oscillators. If you use the same value capacitor for both stages the cool flangey effect interacts with whatever you've plugged into the circuit. You'll need to add the coupling capacitor and clamping resistor shown in Figure 23.2. You can hear this circuit on audio track 17.

You can also use the NAND gate as a kind of digital mixer. Breadboard two distortion circuits (there are enough inverters in a 4049 for three) and send one into each input of a 4093 gate, as shown in Figure 23.3. Now instead of an oscillator modulating the external signal, two external signals cross-modulate each other in a vaguely ring modulator way. Try a guitar in one channel and a toy or recording in the other. Plug the same signal into both inputs. Experiment.

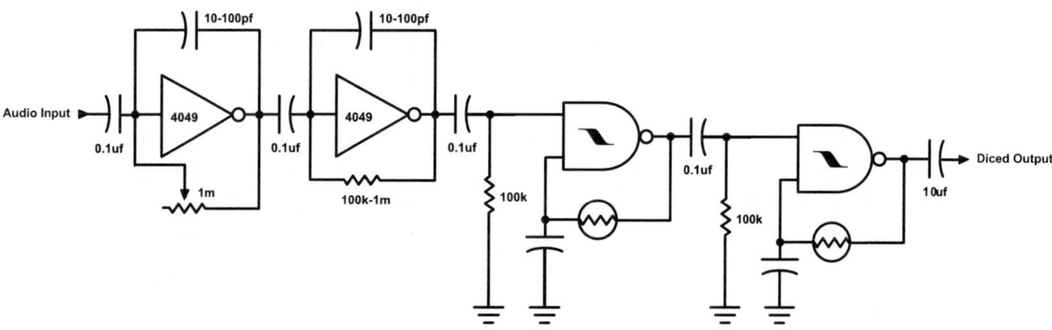

Figure 23.2 The Super Fuzzy Dicer.

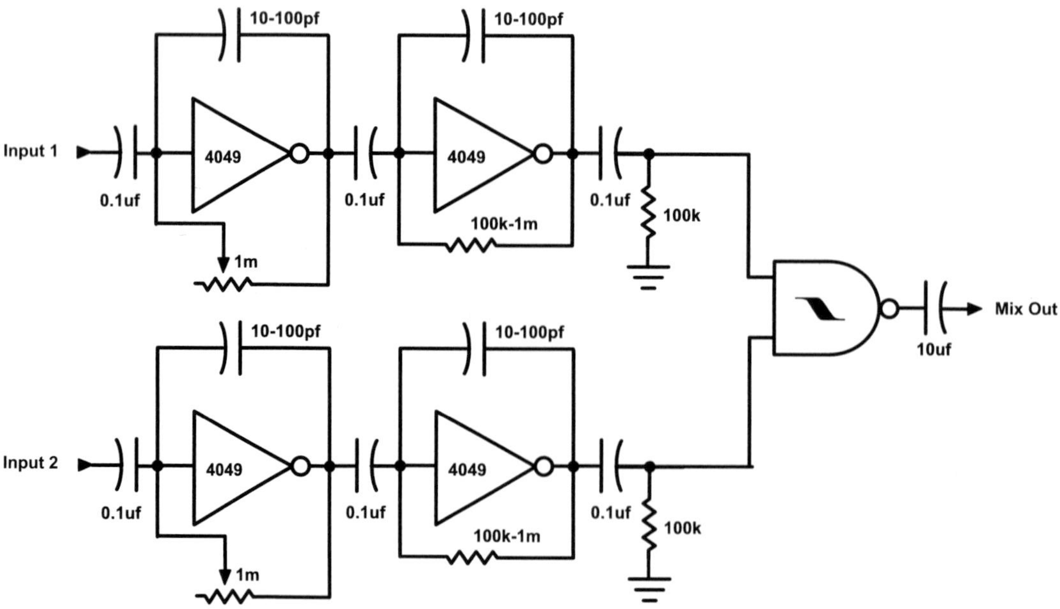

Figure 23.3 Using a NAND gate as a "digital mixer."

DIVIDERS—AN INTRODUCTION

We've been squeezing the maximum number of applications out of the absolute minimum number of chips to save you money, avoid clutters and as a classic minimalist exercise in permutational strategies for a finite set of materials. As loath as I am to add more chips to your modest collection, some of the ones in this chapter are just too cool to pass up. The musical applications of digital logic circuits go well beyond simple oscillators and extreme distortion. Once you accept that an audio square wave can be regarded as a simple alternation between two binary numbers (0 and 1), the logical and arithmetic functions on which all digital calculations are based can be seen as potential sound transformations. For example, a chip that performs division, such as the CD4040 12-stage binary divider, can be used to generate several harmonically related pitches from a single master oscillator. Figure 23.4 shows its pinout and internal configuration.

Breadboard the circuit shown in Figure 23.5. Note that the 4040 has 16 pins like the 4049 in the previous chapter, but the power connections have a similar diagonal arrangement to the 14-pin chips we've used so far: ground connects to pin 8, +9 volts to pin 16. You also need to connect ground to pin 11, designated "rst" (for reset), or the chip will not run. A master oscillator, built from one section of a 74C14 or 4093, is connected to pin 10, the clock input of the 4040. Don't forget to hook up +9 volts and ground to the 74C14/4093 as well. The 4040 has 12 cascaded stages, each of which divides the frequency by 2; outputs are provided for the clock signal divided by 2, 4, 8, 16, etc., all the way to 4096.

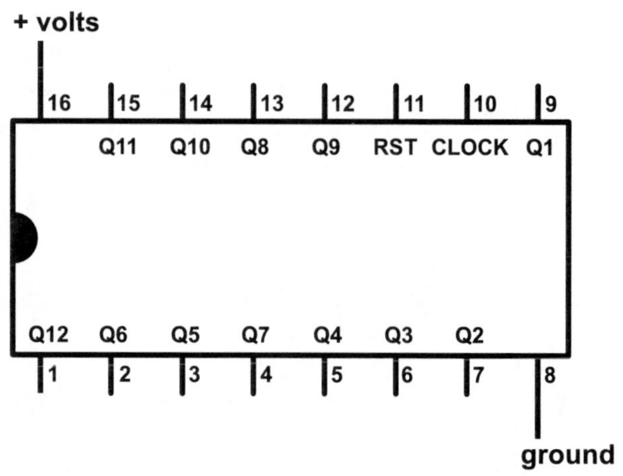

Figure 23.4
CD4040 12-stage binary divider pinout.

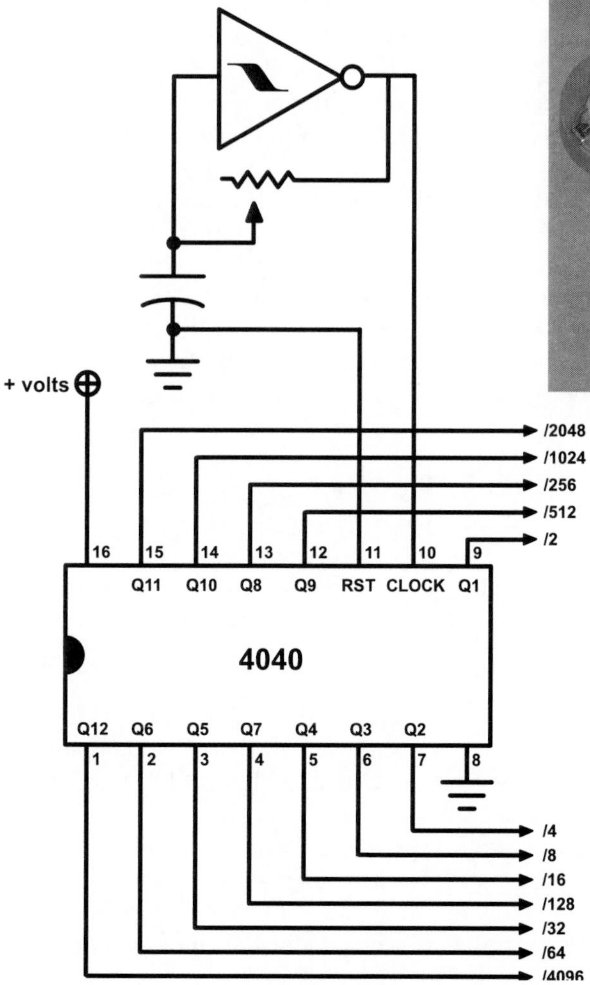

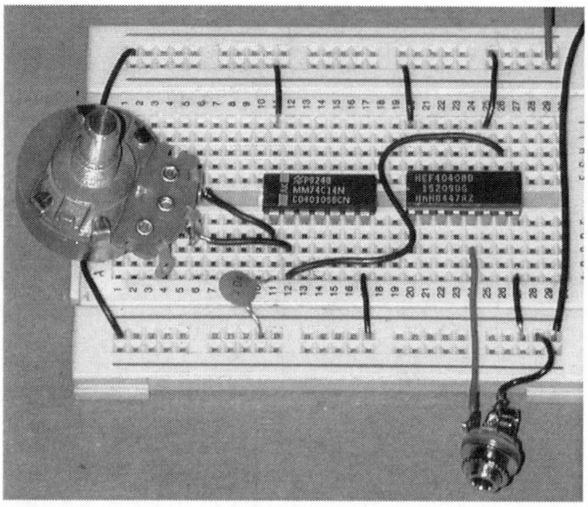

Figure 23.5
Divided oscillator.

Connect the shield of your audio jack to the ground bus, and attach a few inches of solid hookup wire to the hot connection. Connect this hot lead first to the output of the clock oscillator (master output) by inserting it into the vertical bus at pin 10 on the 4040, and tune the clock to a high frequency audio pitch. Now listen to Output Q1 (clock frequency/2) by pressing a wire into the bus at pin 9, and note that it sounds an octave lower. Output Q2 sounds an octave below that (clock/4), Output Q3 an octave below that (clock/8), etc. (Note also the irrational relationship between pin number and divisor value.)

You can use switches to select amongst these different subharmonics. The switches can be anything: momentary pushbuttons, toggle switches, our homemade tilt-switch from Chapter 16, or just some wire jumpers on the breadboard. Because of the thick texture and low frequencies of this circuit, you may want to listen to it over a larger loudspeaker than that of the tiny test-amplifier. If you want to *mix* multiple outputs to build up a rich waveform, make sure you send each output through a resistor before tying together (as we did with the multiple oscillators in Chapter 18, Figure 18.18). Because of the simple harmonic relationship of the octaves you will note that the differences in mixes are subtle: a slight shift in overtone balance, rather than an impression of distinct voices being added.

If you slow the master oscillator down to the rate of a tempo, rather than a pitch, the various outputs become subdivisions of the beat—good for setting up nested rhythmic patterns. You can connect the various outputs of this divider circuit to multiple LED/photoresistor gates (like those in Chapter 21) to chop multiple sound sources in rhythmic patterns—Hacking Dub!

Disconnect the oscillator circuit from the clock input (pin 10) and connect a short piece of solid wire sticking up into the air from the breadboard. Sometimes your body carries enough of an electrical charge that if you touch the end of the wire the noise of your flesh will trigger the divider. Listen to the different divisor outputs as you experiment with brushing and squeezing the wire. Sometimes it helps to connect large resistors (100 kOhm–1 mOhm) between the clock input pin (pin 10) and ground and/or +9 volts to stabilize the circuit when you are not touching it. An excellent ghost detector, by the way.

THE LOW RIDER

Substitute the output of a distortion circuit for the clock/oscillator driving the clock input of the 4040 divider circuit (pin 10), and the divisor outputs become subharmonics of whatever you play into it—a "Rocktave" box, to use the vulgar industry parlance (see Figure 23.6).

You can use switches to select different subharmonics (how about a tilt-switch on the headstock of your Fender Mustang?) or set a fixed mix. Divided down far enough, an E-chord becomes a rhythmic pattern, which can be used to blink an LED to control a photoresistor to gate on and off . . . whatever.

As with the Fuzzy Dicer, you may need to experiment with the coupling capacitor and the pull-down resistor; adding the Schmitt Trigger buffer as we suggested for the Fuzzy Dicer will likely make the pitch tracking more stable. Likewise, larger feedback capacitors in the distortion stages will strengthen the fundamental pitch of the incoming signal and minimize the circuit's tendency to fly off.

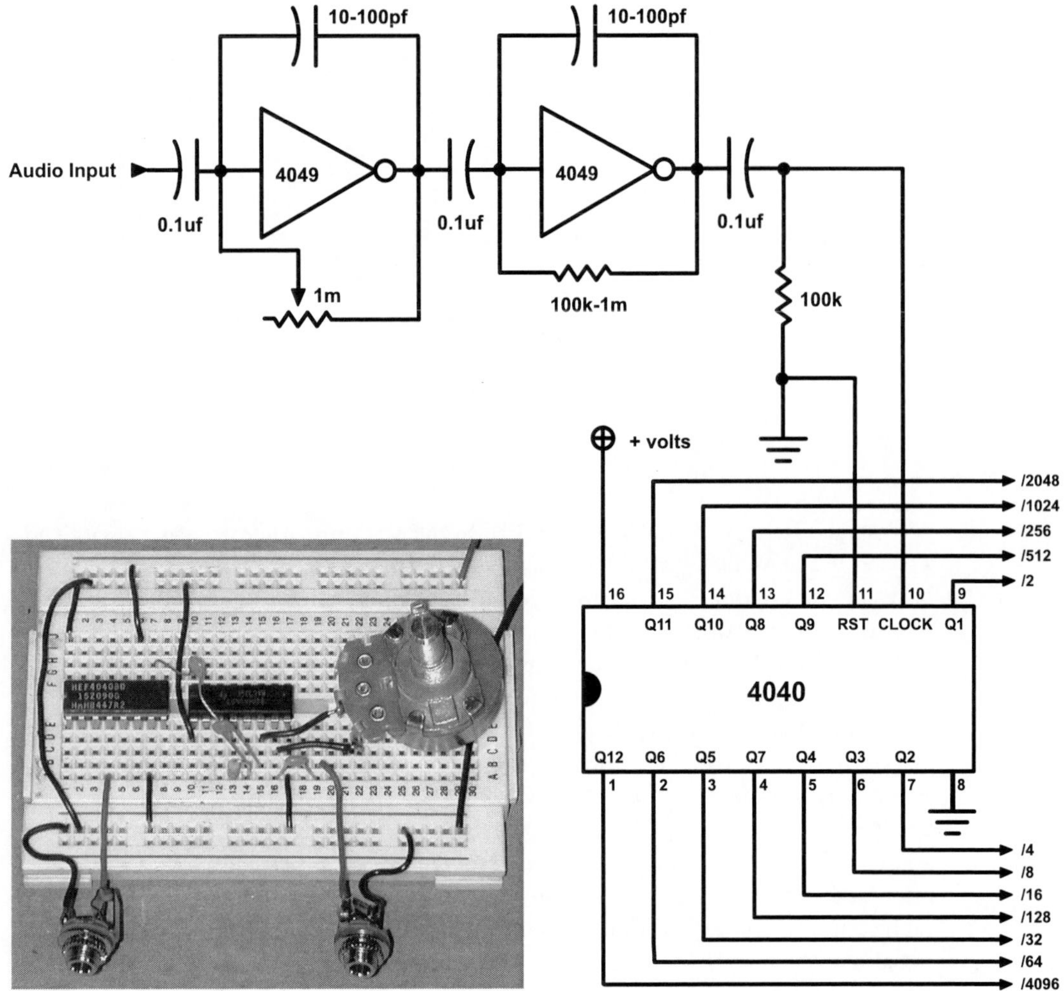

Figure 23.6 The Low Rider.

POOLS OF PHASE LOCKED LOOPS

Phase Locked Loops are versatile pitch tracking devices. They can be used to detect specific frequencies, to convert a complex signal into a simple square wave, to multiply or divide pitches by a factor, and more. The first widespread consumer application of PLLs (as they are commonly abbreviated) was in Touch Tone Telephone networks, converting the merry melodies generated by punching buttons back into numbers to direct your call.

Figure 23.7 shows the pinout and internal configuration of the 4046, with all its inscrutable acronyms and abbreviations. Relax. The two key components of a PLL are:

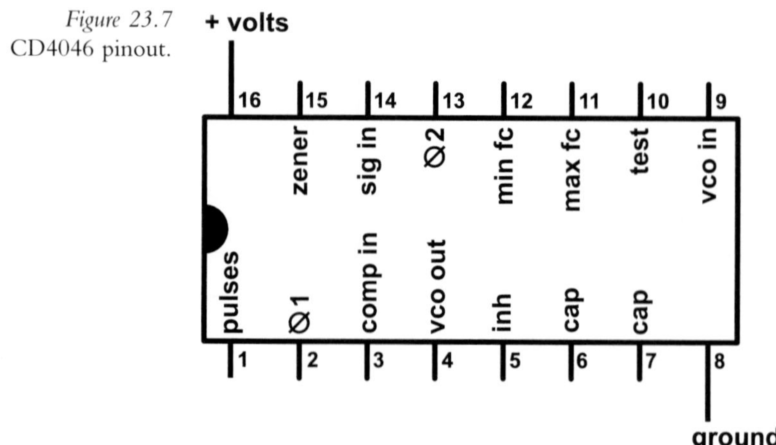

Figure 23.7 CD4046 pinout.

- A Voltage Controlled Oscillator (VCO). Similar to our earlier designs, the VCO is a square wave oscillator whose overall range is similarly set by a capacitor connected between pins 6 and 7 (rather than going from one pin to ground). But instead of controlling the frequency with a resistor, a *voltage* is applied to pin 9 ("VCO in"). Voltage Control greatly increases the versatility of an oscillator, as we shall see—it is the functional underpinning of traditional analog synthesizers. The output signal of the VCO appears at pin 4.
- A Phase Comparator. This module compares the frequency of an external signal (the pitch we want to track) with that of the internal VCO and generates an "error voltage" proportional to the difference between the two. If this error voltage is connected to the VCO the resulting control feedback loop forces the VCO to match pitch of the external signal. The 4046 contains two different styles of Phase Detector, each suited to a different kind of input signal (we'll elaborate later): both track the signal applied to pin 14 ("signal in"); the error voltage from Phase Comparator 1 appears at pin 2 and the voltage from Phase Comparator 2 is at pin 13.

Power hookups are "normal": ground to pin 8, + voltage to pin 16. "Inhibit" (pin 5) must be connected to ground for the circuit to run (connecting to + voltage mutes the output).

PITCH TRACKING

Tracking the pitch of a real world sound—such as a guitar, music file, or ambient recording—with a square wave oscillator may sound like a stupid idea. But the very fallibility of the PLL produces wonderful artifacts: by adjusting a few parameters we can go from extreme distortion, to watery burbling, to swooping glissandi chasing our original sound. The first step is to condition our input, as we did to feed the circuits earlier in this chapter. We preamplify and distort the incoming signal until it resembles a square wave, as before: connect the output of a basic distortion circuit to pin 14 of the PLL, as shown in Figure 23.8; a 100 kOhm clamping resistor to ground keeps the circuit

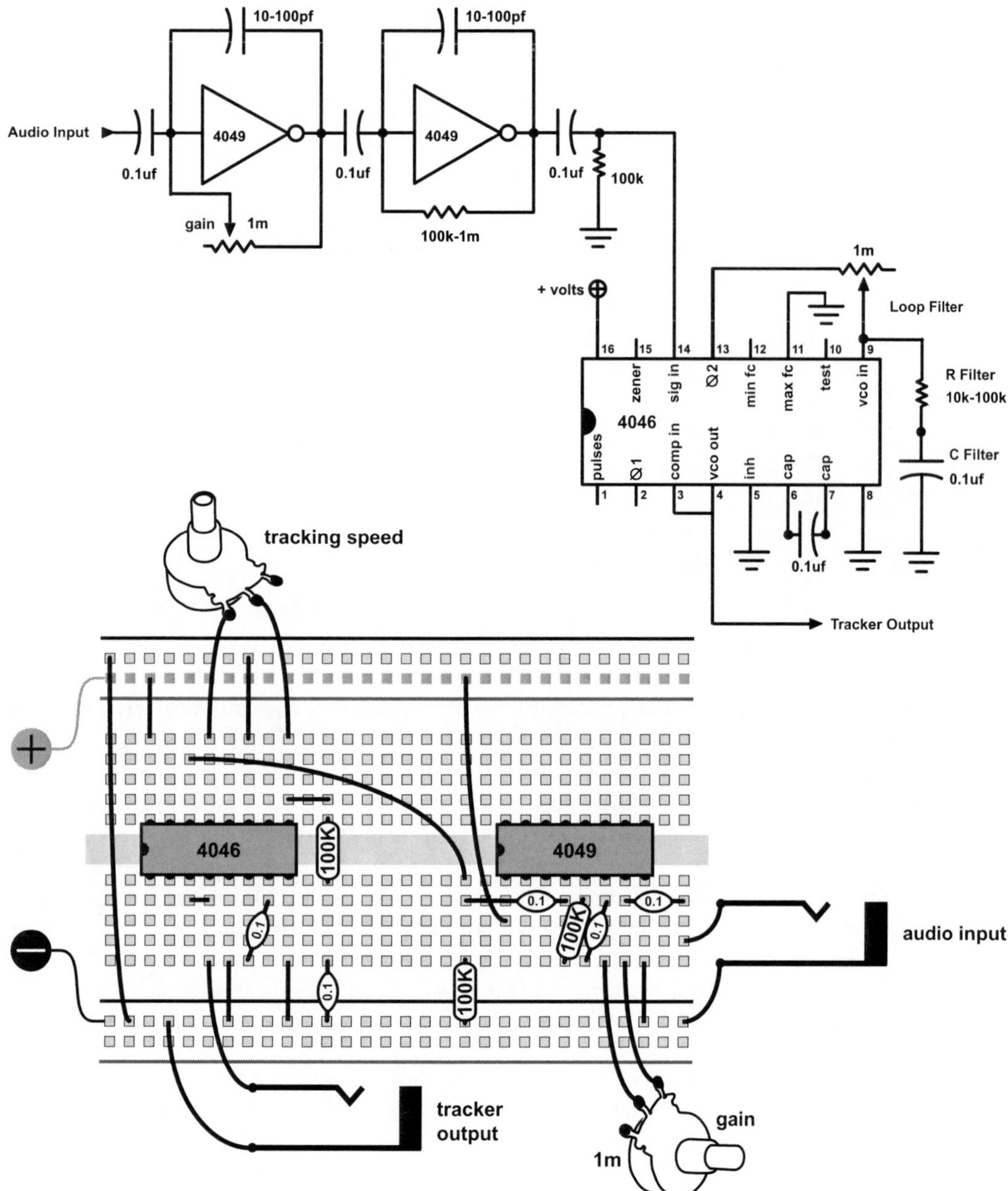

Figure 23.8 4046 Pitch follower.

from sputtering when no signal is present. We connect "VCO out" (pin 4) to "Comp In" (pin 3). "VCO out" is also the pin we listen to, by connecting pin 4 to our amplifier.

The critical variables are:

- "VCO Cap" (pins 6 and 7): the capacitor we insert between pins 6 and 7 determines the pitch range of the VCO. Start with 0.1 uf.

- "Max Freq" (pin 11): a resistor between pin 11 and ground sets the highest frequency the VCO will reach. Start by linking it to ground, which sets the highest pitch to the maximum possible with the VCO Capacitor you've chosen; later you can substitute progressively larger resistors to lower the maximum frequency.
- "Min Freq" (pin 12): similarly, a resistor between pin 12 and ground sets the *lowest* frequency. Start by leaving it unconnected (a very large resistor indeed), which sets the lowest pitch to the minimum possible with VCO Capacitor. Later you can substitute progressively *smaller* resistors, starting with a value around 1 mOhm, to raise the minimum frequency.
- Choose which "Phase Comparator" you will use: for general audio tracking "Phase Comparator 1" (pin 13) is usually more effective, but go ahead and experiment with "Phase Comparator 2" (pin 2) if you wish.
- "Loop Filter": the Phase Comparator output you've chosen connects to the VCO through a Loop Filter, which determines how quickly the VCO tracks the input pitch. As shown in the figure, connect one ear of a 1 megOhm pot to your Phase Comparator output (here pin 13, for Comp. 2), and connect the nose to VCO In (pin 9). Now connect pin 9 through a resistor (R Filter) and capacitor (C Filter) to ground—good starting values are 47 kOhm and 0.1 uf.

Connect a signal to the distortion input (a guitar, as usual, works best for initial tests). Raise the gain of the distortion circuit until you begin to get a response at the VCO output. You should notice that adjusting the loop filter pot affects the speed of the tracking, from so fast at one extreme as to sound like noise, and progressively smoother as you rotate it. Substituting other values for R Filter and C Filter will also change the circuit's behavior. Keeping the capacitor the same, increasing R Filter to 100 kOhm will make the tracking more accurate, but slower to move to a new frequency, with swooping glissandi; reducing R Filter to 10 kOhm makes everything faster and noisier (a great way to turn any music file into delirious noise). You can replace R Filter with a second potentiometer of an appropriate size (probably with a limiting resistor in series) if you want maximum flexibility. Increasing C Filter similarly slows things down and smoothes them out, while decreasing it dirties stuff up.

Once you've got the basic circuit running you can:

- Try the other Phase Comparator.
- Try a larger or smaller VCO Capacitor to shift the tracking range.
- Substitute progressively larger resistors between pin 11 (Max Freq) and ground, to limit the highest frequency to within the "musical" range.
- Substitute progressively smaller resistors between pin 12 (Min Freq) and ground, to limit the lowest frequency similarly.
- Try listening to pin 1 ("Phase Pulses") and the unused Phase Comparator output—these signals reflect the weird interaction between the input signal and the VCO as the VCO tries to lock on the input. You might like this sound.
- Breadboard a simple 74C14 oscillator running at a low frequency (2.2 uf capacitor and 1 megOhm pot). Connect its output to pin 5 of the 4046 (Inhibit) in lieu of the jumper to ground. Adjust the rate of the 74C14 to gate the 4046 on and off; at higher frequencies you get nice modulation effects.
- Add a simple low pass filter (such as shown in Figure 22.5 of Chapter 22) to the output if the square wave becomes too harsh.

- Drive the PLL with one of the divisor outputs of the Low Rider instead of the output from a simple distortion circuit. Now the tracking will be transposed down by the interval you select.
- Transpose the tracking VCO *up*, instead of down, by inserting a 4040 between the VCO out and Comparator In. Connect pin 4 on the 4046 (VCO out) to pin 10 of the 4040 (clock input). Then connect any divisor output of the 4040 to pin 3 on the 4046 (Comparator In). Linking 4040 pin 9 (divide-by-2) to the 4046, for example, will cause the PLL to track an octave above the actual input signal. Don't forget to hook up power to the 4040 (pins 8 and 16) and tie Reset (pin 11) to ground.

To any of these new signal processors (Fuzzy Dicer, Lower Rider, PLL Tracker) you might want to add the bypass switch and output volume control we discussed at the end of the previous chapter. If you use a single distortion circuit as the "front end" for all three of them you may want to build in a simple mixer that lets you blend any combination of the clean audio input, the output of the distortion stage, the Fuzzy Dicer output, a few divisors of the Low Rider, and the Phase Locked Loop—skip ahead to Chapter 26 for mixer designs.

OTHER APPLICATIONS OF EXTREME AMPLIFICATION

These audio processors get us into a very woolly area of circuit conglomeration, never anticipated by the original designers of these chips—trial and error is the best working method. This distortion-based pseudo-analog-to-digital conversion can be coupled with many other chips in the CMOS family (time to start downloading those PDFs or order a data book—see Appendix A), some of which will yield exquisite signal transformations.

You can also use signal boosting and distortion to inject external signals into your hacked toys, keyboards, or radios—just reverse the technique we introduced in Chapter 17 for extracting additional signals from a toy: link the grounds between the distortion circuit and the toy or radio, then use another clip lead to connect the distortion output to any location on the other circuit board. You can find points in toys and music keyboards that will wobble the clock, simulate key closures, or disrupt the LCD display (see Chapter 25). You can modulate or mix with radio signals. But please observe our No-AC-Power safety practice and use these techniques on battery-powered circuits only.

LOVE PARADE

The 4046 PLL's VCO has applications beyond its bumbling attempts at whistling along with your work. For the next project we'll have to introduce one more chip, however. The CD4017 is a counter/divider, like the 4040. But where the 4040 performs *binary* division (each output is 1/2 the frequency of the previous one) the 4017 is a *decade* counter: it counts to 10, over and over, like a counselor ticking off campers on his fingers (see Figure 23.9).

To familiarize yourself with the 4017 breadboard the circuit shown in Figure 23.10. The power hooks up to pin 8 (ground) and 16 (+9 volts). A 74C14 is configured as a simple clock, connecting to pin 14 ("clock") of the 4017. Pin 15 ("reset") and 13 ("enable")

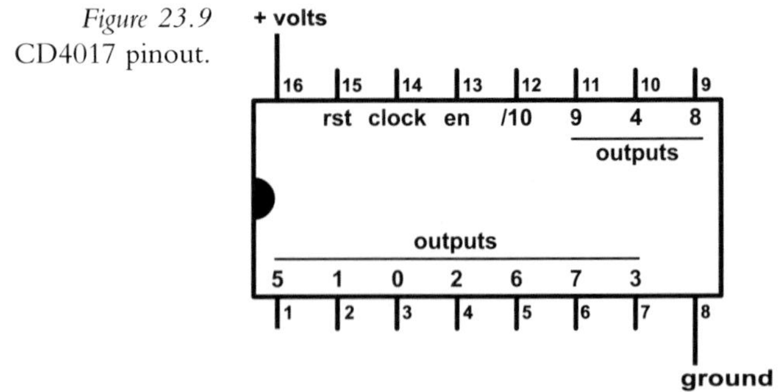

Figure 23.9
CD4017 pinout.

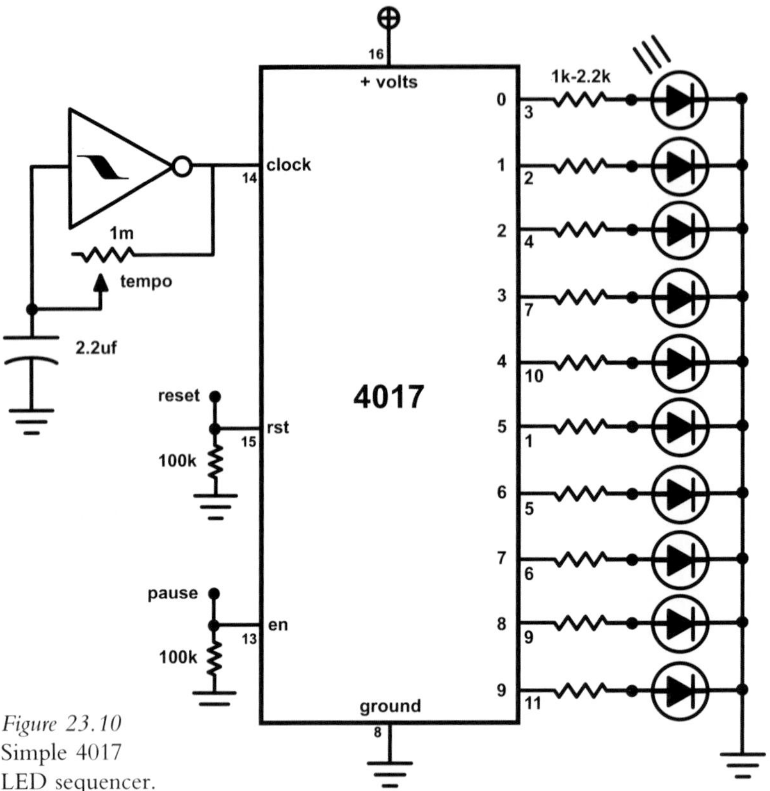

Figure 23.10
Simple 4017
LED sequencer.

are both tied to ground through 100 kOhm resistors. Each of the outputs, 0–9, passes through a 2.2 kOhm resistor to the + side of an LED; the – side of each LED goes to ground. When you connect the battery the LEDs should go on, one at a time, cycling from 0 to 9 before starting over (digital engineers love to start counting at 0, instead of 1—get used to it). This chip is the heart of a lot of pointless, if captivating, "LED Chaser" kits—remember this next time you're stuck for a gift.

I've not included a breadboard layout for this circuit because the snarl of all those resistors and LEDS is almost impossible to reduce to two dimensions. But by now you should be pretty used to translating from schematics to the breadboard, so this is as good a time as any to take off the training wheels.

Momentarily connecting pin 15 (reset) to +9 volts through a jumper (leaving the 100 kOhm resistor in place) immediately resets the count to zero; when the jumper is removed the 100 kOhm resistor pulls the pin back down to ground and the count starts up again. Jumping pin 13 to +9 volts pauses the count, which resumes when the jumper is removed and the pin is pulled back to ground by the resistor.

By coupling a photoresistor to each LED we can turn on and off multiple audio signals sequentially, or pan a signal amongst ten speaker channels—just follow the guidelines for gates and panners we gave in Chapter 21.

By adding a potentiometer and diode to each output of the 4017 we can construct a simple analog sequencer that generates a control voltage for each of its ten steps. This connects neatly to the VCO in the 4046 (see Figure 23.11). The pitch of each step of the sequence is adjusted with a 10 kOhm pot: rotating the shaft sets the voltage anywhere from 0 to 9 volts, which should cover the full frequency range. We've kept the LEDs from the previous circuit so you have some visual feedback on where you are in the sequence. The resistors on pins 11 and 12 of the 4046 set the maximum and minimum frequency of the VCO as before—I've shown good starting values; you can experiment with other values to limit the VCO's range to useful audio frequencies, and maximize the resolution of the pitch pots. You can use other size pots, but its best if they all have the same value. The Reset and Inhibit pins can be momentarily connected to + voltage to reset and pause the sequence as we did above with the LEDs.

If you want a sequence shorter than 10 steps simply take the output 1 greater than your last step and connect it to Reset: i.e. for a 4-step pattern connect pin 10 to pin 15. You can wire up a ten-position rotary switch that selects any output to reset the count and make the sequence length easily adjustable. Add some pushbutton switches for Reset and Pause and you'll be in techno heaven.

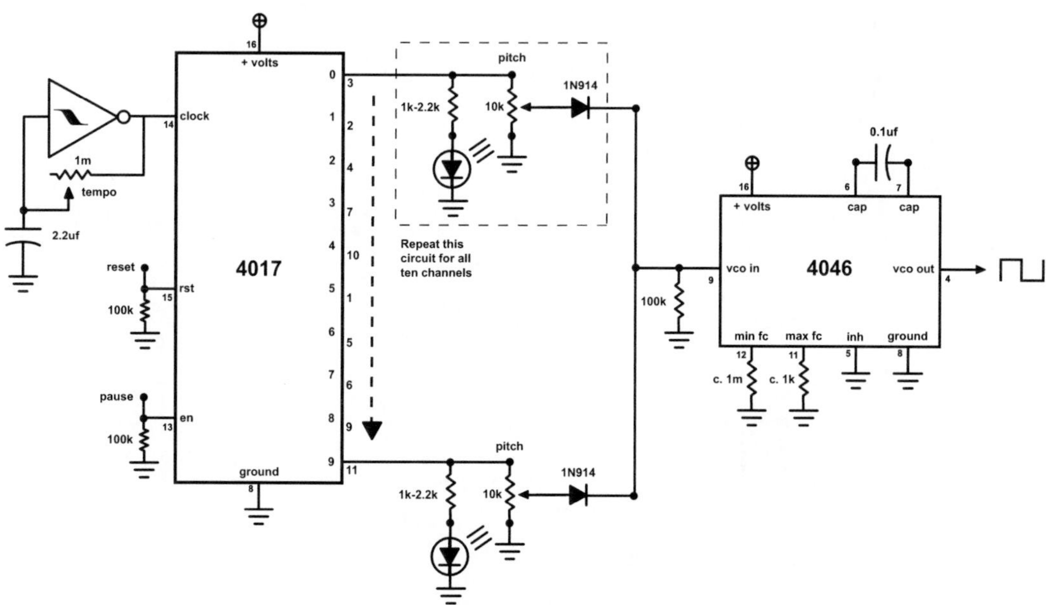

Figure 23.11 Ten-step analog sequencer driving a voltage-controlled oscillator.

As before, the profusion of pots makes a breadboard drawing impractical. Note also that, in the interest of clarity, the sub-circuit that generates the control voltage and lights the LED for each stage has been shown only for the first and last steps of the sequence

By the way, it is possible to cascade multiple 4017s to make longer sequences, but the circuit is a bit complicated to go into here. If you're driven, download a PDF data sheet on the 4017 and there should be a schematic that will do the trick. You can also replace the simple 74C14 clock with any other source of square waves: chained 4093s replace the steady Kraftwerk beat with off-kilter rhythms (see Douglas Ferguson's video on the DVD); track something with another 4046, divide down its pitch with a 4040, and the sequencer speed follows the pitch of your source (more or less).

For an experiment in "wavetable synthesis," remove the 4046, the LEDs, and their resistors from the circuit. Replace the diodes with 10 kOhm resistors. Tie the free ends of these resistors together and send them to an amp. Replace the 2.2 uf capacitor in the clock with a smaller one (0.1 or 0.01 uf) so that it runs in the audio frequency range (see Figure 23.12). Now the pots adjust the levels of individual segments of a ten-stage waveform generator, rather than control voltages going to the VCO in the 4046. Varying the step levels changes the timbre of the waveform. Follow this circuit with a simple low pass filter (like those in Figures 22.4 or 22.5 in the previous chapter), or a basic graphic EQ pedal, and you've got a very flexible, multi-timbre oscillator. The pot in the clock circuit can be replaced by a photoresistor, electrodes, etc., to make it more performable; and cascaded 4093s can be substituted for the simple 74C14 clock (as described above) for more variation.

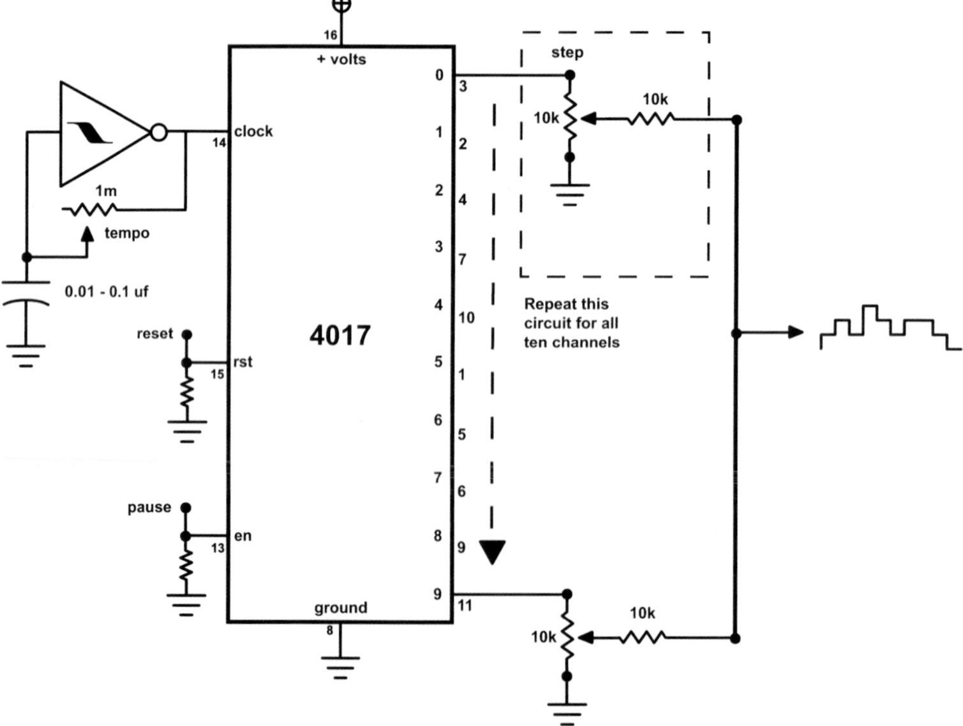

Figure 23.12 Wavetable synthesizer using 10-step sequences.

PART V

Looking

CHAPTER 24

Video Music/Music Video: Translating Video Signals into Sound, Hacking Cheap Camera Circuits, and Extracting Sounds From Remote Controls

You will need:

- A video camera or camcorder (it will not be destroyed).
- A video monitor.
- A cheap, hackable CCD video camera circuit board.
- Some phototransistors.
- Some photoresistors.
- A 74C14 Hex Schmitt Trigger.
- A CD3093 Quad NAND Gate.
- An infrared remote control from a TV or other appliance.
- An audio amplifier.
- Some raw speakers of various sizes.
- Some small mirrors, a laser pointer or flashlight.
- A piezo disk.

Various ingenious software tools exist for translating pictorial data into sound and vice versa: Soundhack's "Open Any . . ." turns any computer file into a sound file (i.e. a Photoshop-to-hit-record converter), STEIM's "Big Eye," and Max's "Jitter" track moving objects in a video image and extract MIDI or audio information. But here are a few simple hardware approaches to the same task that bypass the computer.

LIGHT AND SHADOW

Several artists have translated images directly to sound by placing photoresistors on video monitors or projection screens (see Art & Music 10 "Visual Music"). Wire up a few photoresistor-controlled oscillators (see Chapter 18), using long sections of stranded wire to connect the photoresistors to the circuit board. Place the sensitive side of a photoresistor

ART & MUSIC 10

VISUAL MUSIC

Electronics have pervaded and altered our visual world as profoundly as our sonic one and, furthermore, allowed us to link the two in peculiar, causal ways. In his 1965 work "Magnet TV," Nam June Paik sat a large magnet on top of a television set to distort its image; although technically rather crude, this piece presaged the considerably more "sophisticated" electronic image processing that would come to typify much subsequent video art. "Magnet TV" established a hacker precedent that would remain a consistent presence in Paik's work, as well as in that of many multimedia artists who followed him.

Before lightning-fast personal computers with massive amounts of memory made digital video processing as commonplace as word processing, Paik-like hacks were the only affordable way to manipulate visual images in real time, or to create linkages between video and audio. Video feedback was as popular a tool for early video artists as audio feedback was for electronic music composers: Bill Viola (USA) made extensive use of it in the 1970s; more recently Billy Roisz (Austria) has VJ-ed with video feedback, modifying it through video mixers and keyers, and splitting the video signal to feed the PA as well, so that the bursts and jitter of the images are heard in parallel as glitches and hums (see Figure 24.1).

"Cloud Music" was a video/music installation developed by David Behrman, Bob Diamond, and Robert Watts between 1974 and 1979. In the earliest version, a camera

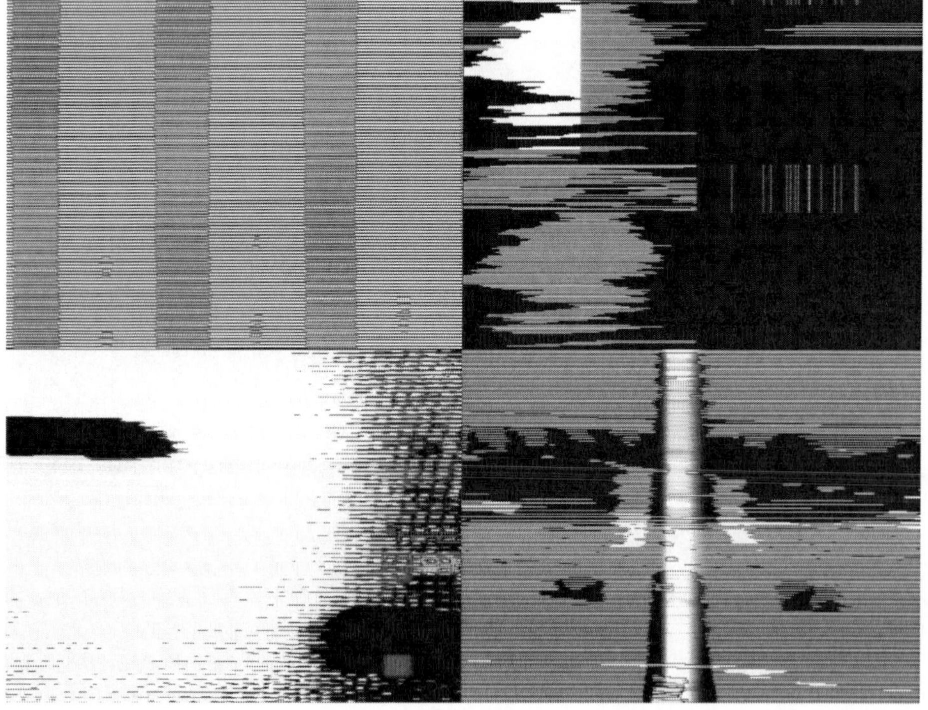

Figure 24.1 Four stills from video feedback performance by Billy Roisz.

was pointed at the sky and connected to a video monitor. A number of photoresistors were affixed to the screen and connected to circuits. The light values of the passing clouds changed the resistance of the photoresistors, and, in turn, affected the sound score. Yasunao Tone (JP/USA) used a similar approach in his "Molecular Music" (1982–85): photoresistors were taped to the surface of a screen onto which a film was projected; each photoresistor controlled the pitch of an oscillator (similar to those described in Chapter 18), and the resulting sound mass responded directly to changes in the projected images (see Figure 24.2). Today Tone is best known as the "grandfather of glitch": he began "wounding CDs" in 1985 by applying Scotch Tape, punctured by pinholes, to the underside of the disks; the resulting frenetic digital error-fest was the first documented music made with intentionally damaged CDs (see audio track 20). The intertwining of light and sound are central to Tone's work: the deflection of lasers through pinholes is a miniaturized, but nonetheless logical, extension of film interrupting the projector's light before it strikes the photoresistors. Similar experiments in controlling circuits through photoresistors reacting to projected light have been done more recently by Jeffrey Byron and Jay Trautman, Joe Grimm, and Kyle Evans (see their videos on the DVD).

In 1969, long before planetarium laser shows, Lowell Cross (USA), a frequent collaborator of John Cage and David Tudor, created the first sound-modulated laser projections for his work "VIDEO/ LASER II": the laser (enormous at the time—see Figure 24.3) was reflected off mirrors mounted on transducers called galvanometers, which vibrated in response to sound input to create curving Lissajous patterns on the wall. (Lowell Cross also built a photoresistor-based matrix mixer embedded in a chessboard for the famous 1968 John Cage/Marcel Duchamp chess-playing performance, "Reunion".)

In 1999, when Stephen Vitiello had an artist's studio on the ninety-first floor of the World Trade Center in New York City, he and Bob Bielecki (see Art & Music 11 "The Luthiers," Chapter 28) hooked up a photoresistor to a battery (as shown in Figure 24.12), placed it on the eyepiece of a telescope, aimed it down at New Jersey, and sat together listening to the flashing lights on a police car across the Hudson. Vitiello has made a series of recordings using this "audio-telescope" (see audio track 18). Norbert Möslang and Andy Guhl of Voice Crack (see Art & Music 8 "Composing Inside Electronics," Chapter 14), have used similar circuits to extract surprisingly rich rhythmic

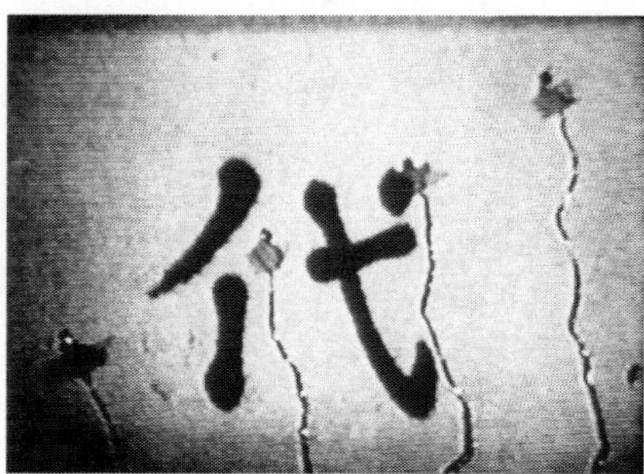

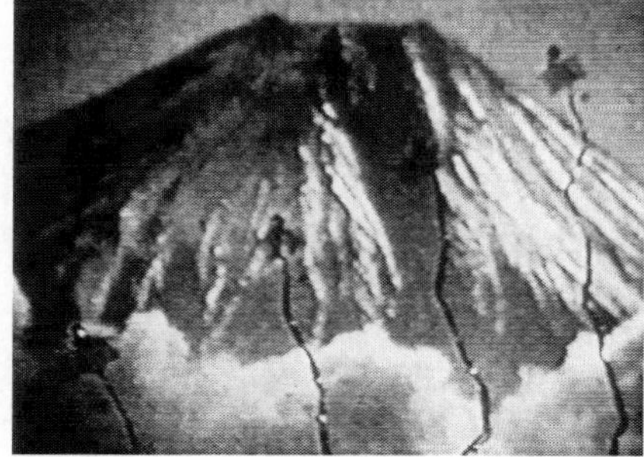

Figure 24.2 Two stills from "Molecular Music," Yasunao Tone.

Figure 24.3a Lowell Cross (left), Eugene Turitz (center), and David Tudor (right) setting up for the first laser light show to use x-y scanning, Mills College, Oakland, California, May 9, 1969.

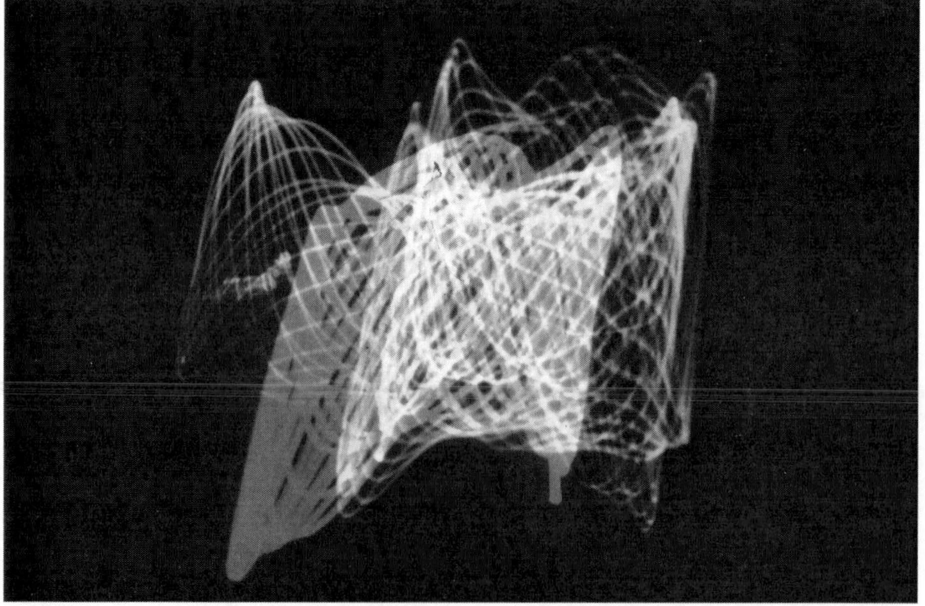

Figure 24.3b Laser projected image from "VIDEO/LASER II" (December 1969), Lowell Cross.

and harmonic textures from the light patterns of bicycle flashers and LEDs on toys (see audio track 19 and Andy Guhl's video on the DVD).

Computers finally caught up with video, but visual hacking hasn't stopped. The disparity between the $100 portable LCD TV and the $5,000 video projector offended the sensibility of the Dutch electronic performance trio BMBCon (Justin Bennett, Wikke't Hooft, and Roelf Toxopeus), so in the mid-1990s they took the screens from damaged stadium TVs (which have the same dimensions as 35 mm slides) and dropped them into old slide projectors from the flea market—voilá: the home-made, low-budget video projector (see Figure 24.4 and their video on the DVD). In my installation "Daguerreotypes" (2006) high intensity LEDs shine through LCDs from toys and games, projecting a sort of miniature *wayang* shadow play onto the walls of a gallery (Figure 24.5).

Figure 24.4 Homemade LCD projector, BMBCon.

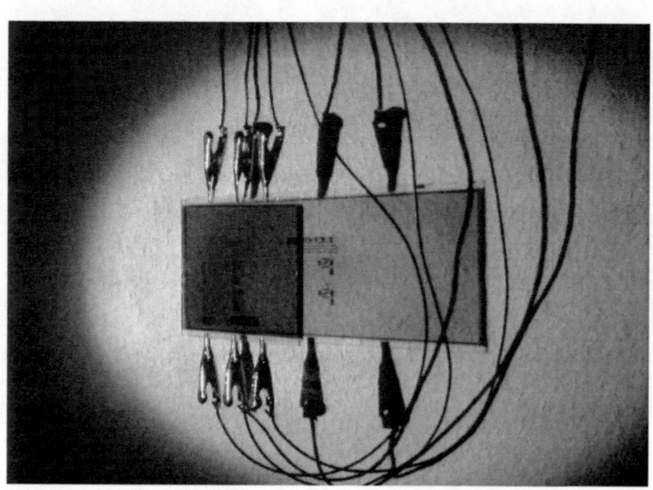

Figure 24.5 Detail from "Daguerreotypes" installation with LCD screens and LEDs, Nicolas Collins.

Jon Satrom (USA) has built his VJ career on transforming a child's "video paint box" into an instrument he calls the "Vitch" (see Figure 24.6 and his video on the DVD). By inserting circuit bending-style jumpers between various points on the circuit board, Satrom is able to disrupt the toy's functions to produce a remarkable range of fragmented, frozen, superimposed, and digitally warped images (essentially a video equivalent of the keyboard malfunctions described by Phil Archer in Art & Music 9, "Circuit Bending," Chapter 15). Similar video circuits have been bent by Jordan Bartee (USA), J. D. Kramer (USA), Phil Stearns (USA), and the trio of Abbot, Archer, and Tombs (UK)—see their videos on the DVD. Tali Hinkis and Kyle Lapidus of the video hacking duo LoVid (USA) have created wonderful homemade video synthesizers, occasionally built into soft sculpture and wearables. Their "Kiss Blink Sync Vessel" is a collection of modules, embedded in tabletops, that can be patched together to synthesize both video and sound (see Figure 24.7 and their video on the DVD).

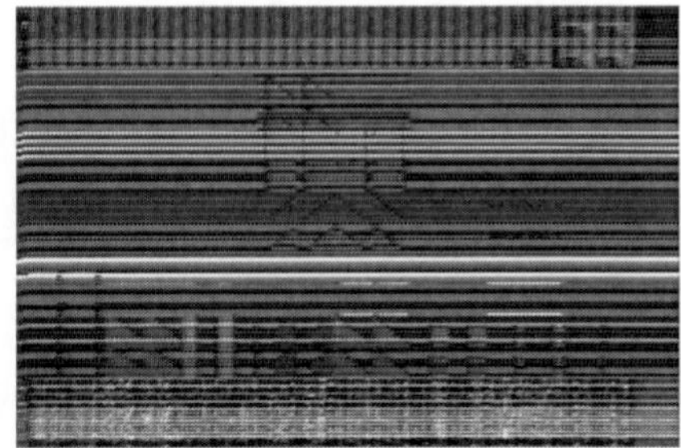

Figure 24.6 The "Vitch," Jon Satrom (left). Video image from performance with the "Vitch," Jon Satrom (right).

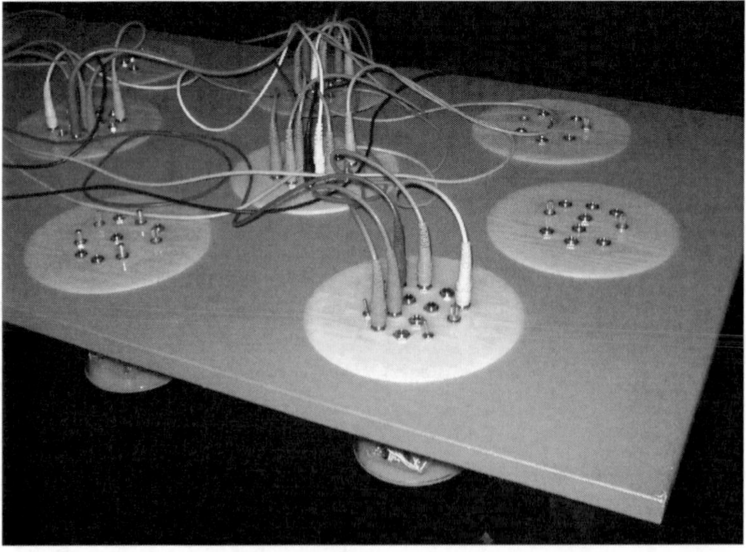

Figure 24.7 "Kiss Blink Sync Vessel" by LoVid.

And in a pseudo-Victorian twist that would make John Bowers proud, Dutch artists Arthur Elsenaar and Remko Scha attach electrodes to Elsenaar's face and electrically stimulate the muscles of expression to provide an "emotional display" for their computer (see Figure 24.8 and their video on the DVD).

Figure 24.8 Portrait of Arthur Elsenaar's face displaying an artificial facial expression.

against the screen of a video monitor and use a thin strip of opaque electrical tape across the back to hold it in place; repeat for each photoresistor, distributing them across the screen. Connect a camera or other video source to the monitor and listen as you sweep the camera across the room or play back a tape. Action! Instant soundtrack! You can do this on a projection screen as well—the effect is stronger with film projection than video because of the increased contrast between light and dark. LCD screens on laptops work but similarly have less contrast than ordinary TVs.

You can also use the photoresistors to adjust the *loudness* of any audio signal (CD, computer, microphone, etc.) in response to fluctuations in the image, by adapting the gating and panning circuits from Chapter 21 to work with photoresistors affixed to a monitor or projection screen. You can adapt the Theremin circuit in Chapter 20 (Figure 20.9) for video response as well—using the image to control both pitch and volume of multiple oscillators gives the resultant texture considerably more variety.

FRAME RATE MUSIC

We can also listen directly to a video signal itself. Use a Y-cord to split the analog video output of a camera. Connect one leg of the Y to a video monitor and the other to an amplifier and speaker—that's right: the *video* output to the *audio* input. As luck would have it, a camera puts out a video signal that is approximately the same amplitude as a line-level signal from a CD, etc. Pan the camera around the room as you listen. You should hear a steady drone whose overtones fluctuate in response to the image content and brightness. The fundamental pitch is a function of the video frame rate (between B♭ and B♮ with NTSC video, between G and G♯ with PAL), and therefore unwavering if the camera is functioning normally, while the overtone balance directly represents the image data, line by line. Very nice, if you like drones.

Aim the camera through a rotating fan; vary the fan speed and you should hear interference patterns between the frame rate and the fan speed. Focus on a white card off-center on a black turntable mat, and switch between 33 and 45rpm. Aim the camera at the monitor and look and listen as you experiment with video feedback. A video mixer, keyer, or special effects box introduces audible artifacts as well as visible ones. With a Y-chord splitter you can see and hear the effects—this is a technique used by several experimental VJs and video artists, including Jon Satrom, Billy Roisz, and LoVid (see Art & Music 10 "Visual Music"). Aim an infrared remote at the camera (most video cameras detect infrared light and display it as hot white) and listen to the burst pattern of the encoded data (see Channel Surfing Music below).

You can similarly listen to the video output of your DVD player—fanning it will have no effect, but it's an easy way to generate an automatic soundtrack.

The frame rate is fixed, and normally doesn't budge unless you move between NTSC and PAL. But if you invest in a cheap black and white CCD camera circuit board (scrounged from a surveillance camera, or available from most electronic surplus outlets for a modest price), you can experiment with tickling the clock frequency by a laying of hands (as we did in Chapters 11 and 12) or replace the clock crystal with a variable oscillator (as discussed in Chapter 20). The crystal is usually pretty conspicuous on the circuit board—often a metallic-silver small cylinder or block (see Figure 24.9).

Figure 24.9 Camera board with switch for disconnecting crystal (circled, right) and electrodes for tickling clock frequency (left, visible below switches).

Split the camera output between a video monitor and amplifier using a Y-cord as before, so you can see as well as hear the effect of your hack. The video image produced by a tickled camera is reminiscent of 1960s "scratch animation" films, and the sound is somewhat meatier than the typical hands-upon-radio swoops. Sometimes lifting one leg of the camera's crystal time base makes it just unstable enough to produce a coherent image when left alone, but jitter like crazy when touched (I have no idea how this can possibly work, but it does). If you replace the crystal with your own adjustable clock circuit you can transform the video camera into an oscillator whose pitch is controlled by a pot, photoresistor, sequencer-driven 4046 VCO, etc., but whose timbre is a function of what it sees. To the best of my knowledge no one has built a synthesizer with such a hacked camera as its basic oscillator module, so jump on this one.

The hacked camera will not generate a stable sync signal when tickled. Most video monitors will continue to display scratchy video in the absence of a stable sync, but most video projectors are too "smart": they will interpret intermittent or erratic sync as an indication that there is no video signal at all, and will display that irritating blue screen with the legend "no video input." A circuit that provides a proper sync under scratch video is beyond the scope of this book, sorry. Use an old TV instead: focus a video camera at the screen and send *that* signal to the projector. Or invest in a cheap video mixer, keyer, or other device that generates its own sync or lets you patch in a second, "normal" camera for a sync signal.

As long as we are on the subject of old TVs, I would be remiss if I did not remind you, the reader, of the beautifully liquid image distortion that results from putting a hefty magnet in close proximity to a television picture tube (ineffective on modern LCD screens). Take an old TV. Tune it to any station or even inter-station static. Move a big magnet over the top and sides, and watch the image wiggle—a gift from Nam June Paik (see Art & Music 10 "Visual Music").

I must warn you once again of the electrocution hazards posed by all of the above projects. The fingers-on-circuit activities have the usual risk:

EXERCISE EXTREME CAUTION WHEN CONNECTING THE CARESSED CAMERA BOARD TO ANY AC-POWERED VIDEO MONITORS OR PROJECTORS!

But the greatest danger with the magnet-on-TV experiment is that the circuitry inside old-fashioned TVs and video monitors (the kind that use "picture tubes" instead of LCD screens) actually steps *up* the wall voltage from a deadly-enough 120/240 volts to several thousand volts (kilovolts, as they are known—although we might dub them "killervolts"). Please, no matter how much you want to get the magnet closer to the tube:

DO NOT OPEN UP THE TV!

(Or you're in for a nasty shock.)

VIDEO-FREE VIDEO

Visual display of sound patterns can be accomplished without video cameras and monitors, of course. As we mentioned at the end of Chapter 5, you can take a large raw loudspeaker, fill it with sand or talcum powder, connect to an amplifier, play some sound, and watch the dancing dust. Coat the inside of the cone with paint or rubber cement, fill it with water or oil, and repeat the experiment; you can reflect a focused light or laser pointer off the water's surface onto the wall or ceiling (see Figure 24.10). This works best with low frequency sound. A mirror glued to the center of the cone also reflects a laser nicely. Planetarium laser-shows use electromagnetic transducers called "galvanometers" to deflect mirrors on several axes—sometimes these gizmos turn up on surplus sites, but the old speaker-and-mirror technique works pretty well.

CHANNEL SURFING MUSIC

In Chapter 3 we used coils to pick up the electromagnetic signals given off by various appliances and electronic devices. We can also eavesdrop on light signals of various kinds by using a specialized type of light sensor: the *phototransistor.* A phototransistor is the heart of any infrared remote control receiver circuit, such as that in your TV. It looks

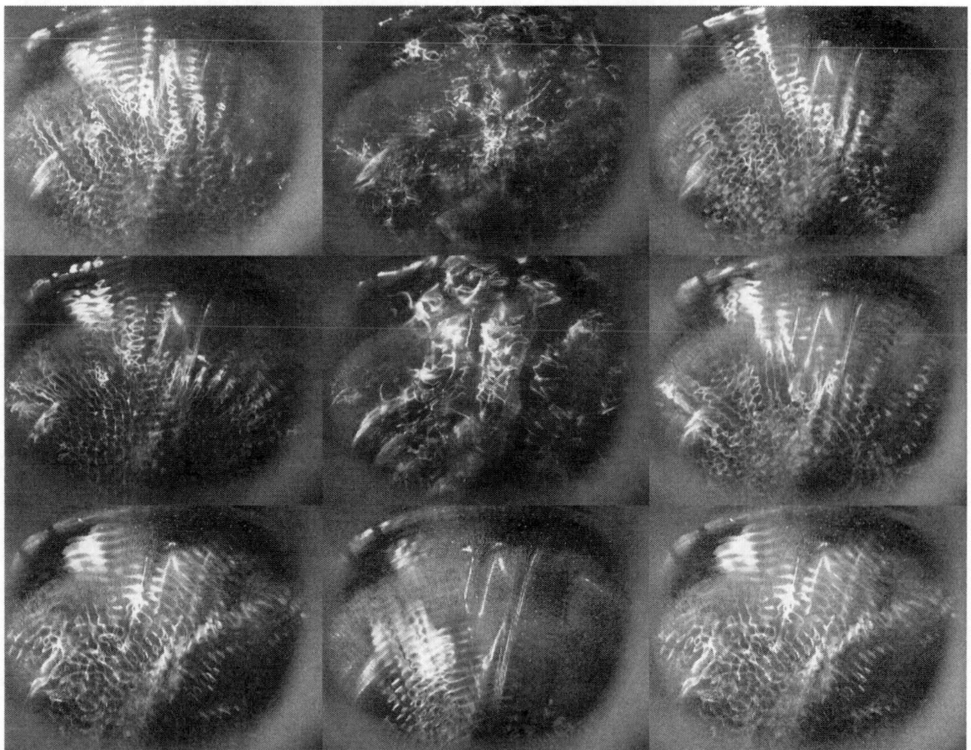

Figure 24.10 Water-filled speaker, showing ripples produced by low-frequency sound.

just like an LED, so make sure you keep them clearly separated in your parts collection. You can find them quite easily, even at Radio Shack. Phototransistors detect the pulses of infrared light sent by your remote control and convert them into a stream of binary pulse waves that are, in turn, translated back into digital data by the microprocessor in the TV. Earlier in this chapter we detected these data burst using a video camera, but phototransistors are cheaper (and smaller).

Aim a remote control at the simple circuit in Figure 24.11. Keep it close, and you should hear pulse trains as you press the buttons. If not, reverse which leg of the phototransistor connects to +9 volts and which connects to the load resistor. Normally the phototransistor is "off" and the 2.2 kOhm resistor holds the output to ground (0 volts). When infrared light strikes the phototransistor it turns on and effectively connects the output to +9 volts. So bursts of light from the remote cause the output to switch between 0 and 9 volts, just like our old friend the CMOS oscillator, only here the waveform is not a simple square but instead a cycling sequence of pulse waves of various duty cycles, which has rather a different timbre.

The variation between one button and another may sound pretty subtle: although the encoded numbers are different, the base frequency remains the same—the effect is similar to listening to video, where the constancy of the fundamental sometimes overpowers the variations in image-dependent overtone content. Try several different remotes. You'll notice that the loudness of the signal falls off pretty sharply as you pull

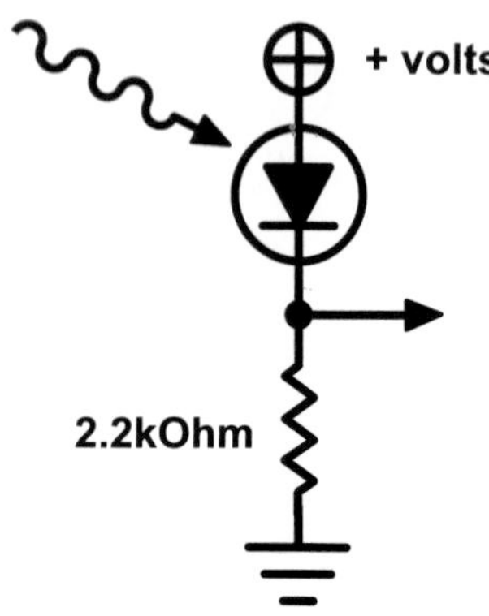

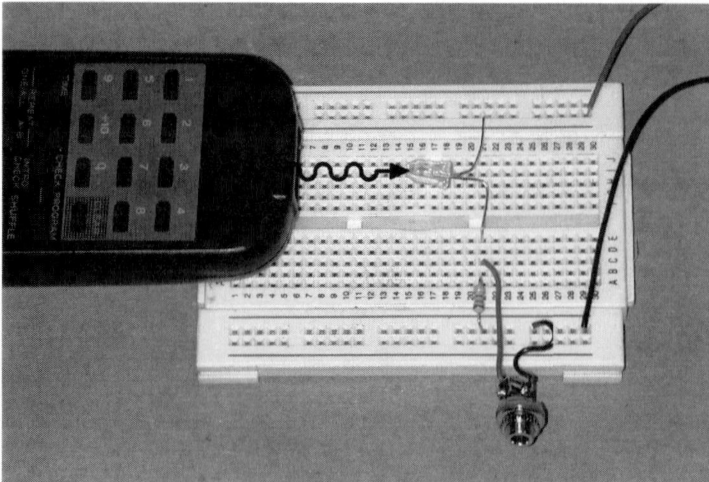

Figure 24.11
Simple infrared detector circuit.

the remote farther from the circuit, so you do have some dynamic control over this instrument.

You can substitute an ordinary photoresistor for the phototransistor. You may need to increase the size of the load resistor from 2.2 kOhm to 10 kOhm or larger, as shown in Figure 24.12. Because photoresistors are sensitive to light across the spectrum (not just infrared), you will get much more interference from the power grid's AC frequency present in incandescent and fluorescent lighting (60 hz in the United States, 50 hz in Europe), resulting in an underlying drone. But you may find this interesting rather than irritating, so try it.

Although the indicator lights on many electronic devices look steady, most are in fact "scanned" by the central processor unit. You can use our light-detector circuits to extract unexpected sound patterns from almost any gizmo with LEDs. Try it on bicycle flashers (see audio track 19), toys with blinking lights, the front panels of studio gear, TV screens, computer monitors. Certain bicycle lights and blinking toys sound astonishingly much like heavy metal chord progressions.

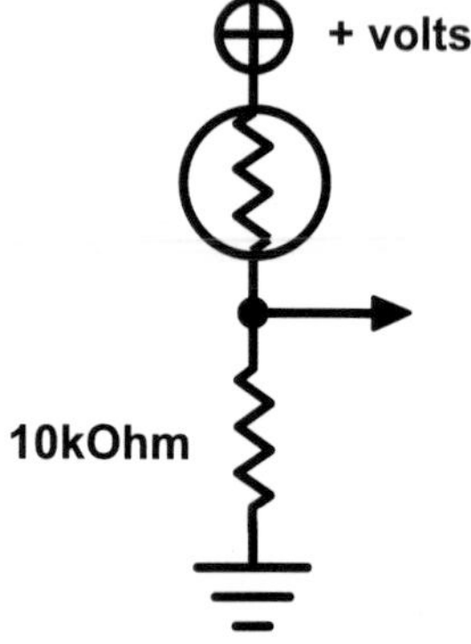

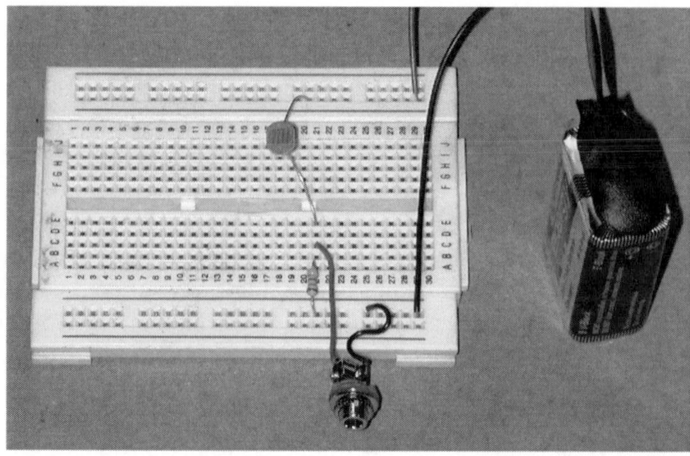

Figure 24.12
Simple photoresistor light-to-sound converter.

For greater variation try using the phototransistor or photoresistor circuit above as the control input to the basic 4093 gateable oscillator circuit from Chapter 20. If you also use a photoresistor for the oscillator's frequency control, you get a pretty expressive "multi-phase" light-to-sound converter that responds to both ambient and modulated light sources (such as remote controls) (see Figures 24.13 and 24.14). Add a Theremin-style control of output volume (see Figure 20.9 in Chapter 20) for additional expression.

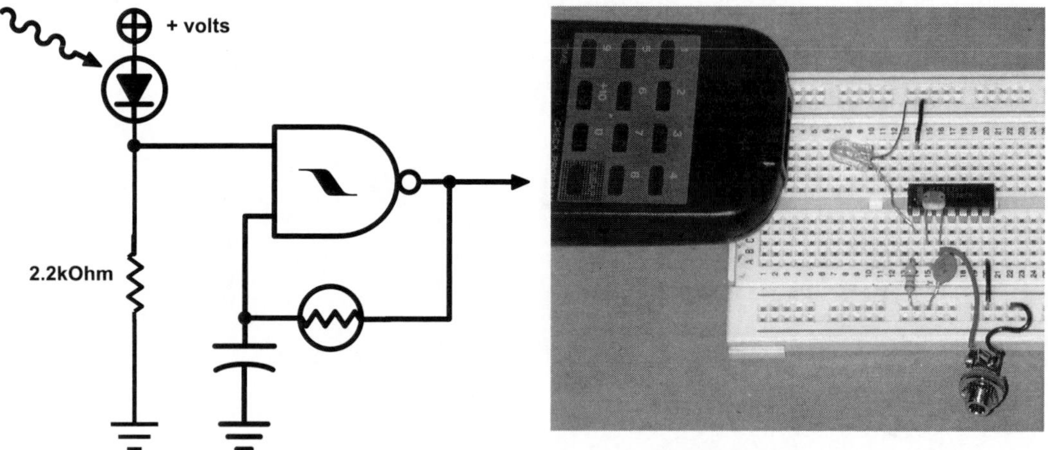

Figure 24.13 Infrared-gated oscillator with photoresistor-controlled frequency.

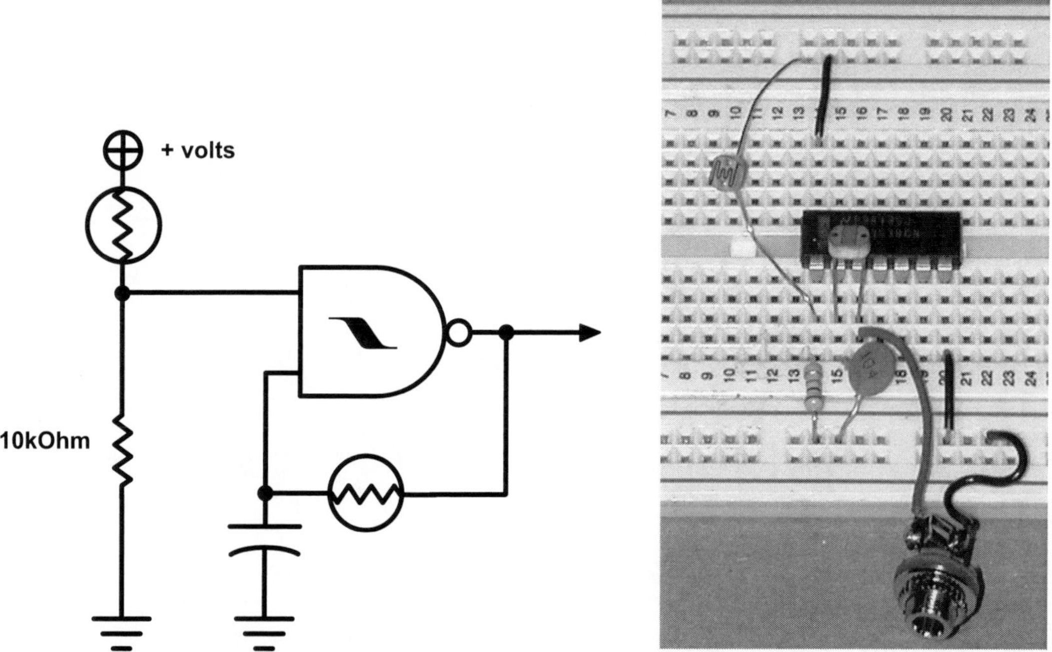

Figure 24.14 Photoresistor-gated oscillator with photoresistor-controlled frequency.

For a battery-free alternative you can connect a dozen or so phototransistor in series and hook either end of the chain up to the leads on a piezo disk (see Figure 24.15). Aim an infrared remote at your circuit and you should hear a quiet, tinny refrain of the familiar pulse train. As we suggested at the end of Chapter 20, clamping the disk to a resonator of some sort (matchbox, pie tin, etc.) will increase its loudness. Engineers at the Information Technology Research Institute at the National Institute of Advanced Industrial Science and Technology (Japan) used a similar passive design for an installation by Laurie Anderson in a Japanese Garden at Expo 2005 in Aichi, Japan. Visitors participating in "Walk" could listen to poems in four languages, transmitted with infrared light, and picked up by handheld "Aimulets" (see Figure 24.16).

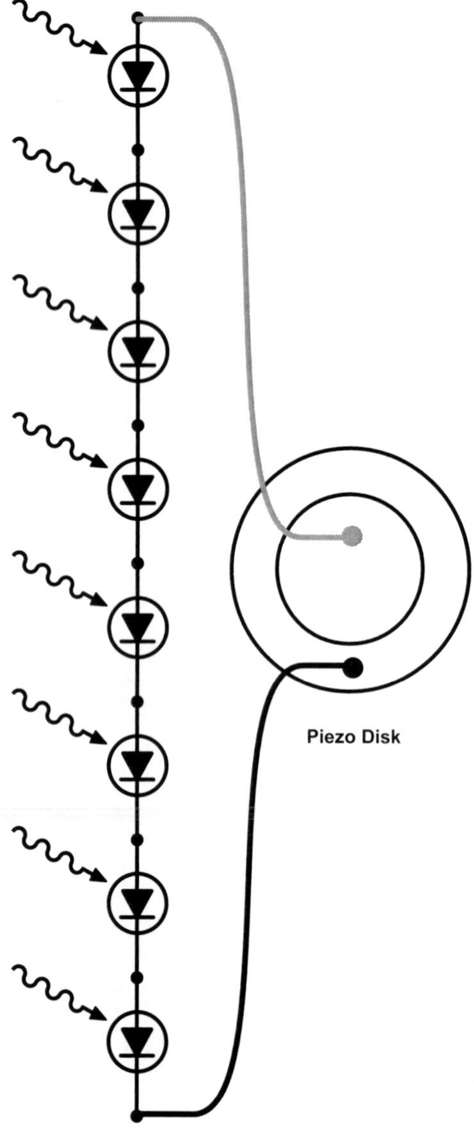

Figure 24.15
Phototransistor ladder driving a piezo disk directly.

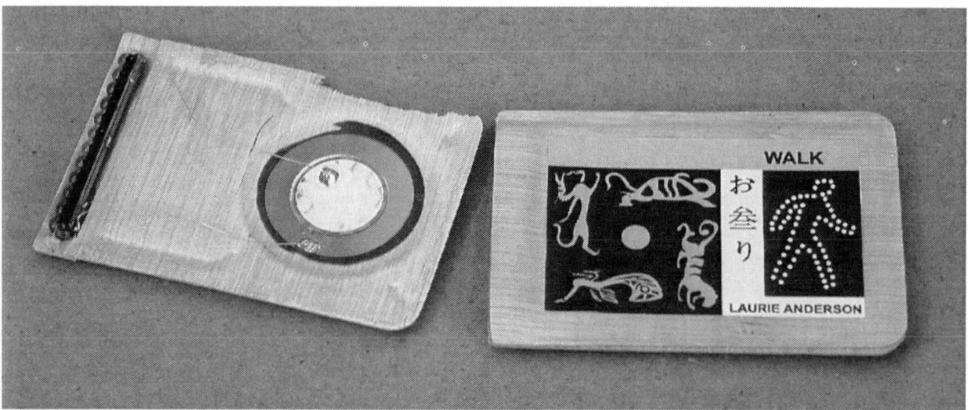

Figure 24.16 "Walk" by Laurie Anderson: photodiodes driving piezo directly, with molded bamboo resonator; designed by the Information Technology Research Institute at the National Institute of Advanced Industrial Science and Technology, Japan.

BAR CODE

Infrared transmitters and detectors are combined in the bar code readers used in our UPC dominated world—from handheld wands to the deliriously diffracted laser beams at supermarket checkout, all work by bouncing light off packaging and detecting the difference between the light and dark stripes. Handheld "wands", and their core elements, can be found quite easily and cheaply from electronic surplus sites online.

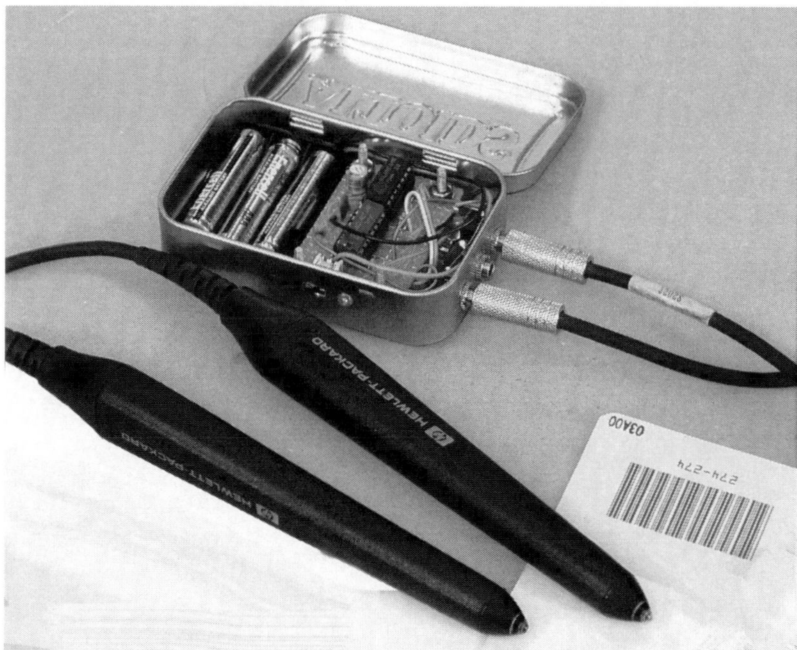

Figure 24.17 Bar code readers with power supply and planning circuit (Nicolas Collins).

Most require a 5-volt power supply (try 3 AA batteries in series). Sometime they come with wiring data, otherwise you'll have to decode it (buy a few, in case of a real flameout). Hook up power, connect the data output to an amp, pass the wand across some bar code, and you'll be rewarded with a noisy waveform similar to the tape head scratching transit cards (no surprise here, since both the magnetic tape and the UPC are encoding binary data).

Since the bar code reading mechanism detects any light/dark difference, they can be used as generic image-to-sound translators: pass them over newsprint, photographs, TV screens, Dalmations, facial stubble . . . . Figure 24.17 shows two wands with a homemade power supply. The batteries are all that's really needed, the extra circuitry in the box makes each wand's data stream to alternate between the left and right output channels with every pulse (for a little more variety).

CHAPTER 25

LCD Art: Making Animated Modern Daguerreotypes and Alternative Video Projectors

You will need:

- A toy (or other expendable device) with an LCD screen.
- Some test leads.
- A 9-volt battery and battery hook-up clip.
- Hookup wire.
- Some straight pins or short needles.
- A basic oscillator circuit from Chapter 18 or 20.
- A flashlight and some lenses.
- If possible, an old-fashioned slide projector.
- Hand tools, soldering iron, and electrical tape.
- If possible, a "video paint-box" toy with video output.
- An amplifier.

LOWER TECH

A lot of handheld toys and games incorporate small LCD screens. In general-purpose LCD displays—such as laptop screens, flat-screen TVs, Gameboys, and cellphones—a matrix of pixels is *bit-mapped* in rows and columns by a microprocessor that turns individual pixels on and off to "draw" any character or image, etch-a-sketch style. In cheap toys, on the other hand, the screen usually contains a handful of "ready-made" graphic components: lips, a nose, and pair of ears are turned on and off against a printed cardboard backdrop to add distinguishing features to Mr. Potato's otherwise generic head, for example. Although this approach severely limits the graphic options of any individual toy, it is much simpler from a programming standpoint, and much cheaper to manufacture.

These rebus-like images take on new meaning when the background is removed—leaving the body parts floating like a medium's apparitions—or superimposed on an alternative drawing or photograph that you provide (Mr. Turniphead? Baby Sister rev. 2.0?). To accomplish this gentle re-purposing, disassemble the toy carefully (don't lose those tiny screws or tear any fine wires), remove the cardboard backdrop, insert a new one of your choice, re-assemble, and prepare to amuse your friends.

The "narrative" of the game can sometimes be disrupted by shorting various points on the board, as we did in the "Almost a Short Circuit" section of Chapter 15 in pursuit of sonic effects. In some cases injecting an amplified and distorted audio signal can also confuse or modulate the imagery (see page 207 for guidance).

Although fewer in number than the tiny pixels in a bit-mapped pixel grid, these graphic elements are still arranged in a matrix: the toy's computer turns on individual images by sending logic signals through a particular row and column pair. When the screen is removed from the circuit its graphic elements can be activated with simple connections of voltages, either directly from a battery or from an oscillator. Start by tinning the tips of the red and black wires from a 9-volt battery clip to give them stiff, sharp points—you can also solder sewing pins to the wires for stronger, finer contacts.

Locate the connections to the LCD. The glass element is often connected to the circuit board through a thin rubber strip (usually pink, for some incomprehensible reason) containing a thinner strip of black conductive rubber filaments. If you look closely you'll notice that what at first may have appeared as a solid stripe actually resembles a dotted line—each dot is one end of a thread of the same material as those funny rubber hats used as keys on many toys. One end of each of these filaments presses down against a trace on the circuit board, while the other touches a narrow, ghostly grey finger on the edge of the LCD, linking the display to the circuitry. Usually this kind of LCD has connections on two or four edges of the screen. Some older LCDs have larger metal tabs that are soldered directly to the circuit board (as shown in Figure 25.1).

Poke the "+" lead from your battery against a dark point in one rubber strip, and press the ground wire to another point, on the same strip or another. If you remove the rubber strips, you can usually make direct contact with the LCD connections by pressing the wire tip directly to the glass where the rubber sat, or by clamping the jaws of a narrow clip lead to the edge of the glass. Look carefully: you should be able to see very fine lines etched on the glass—these are the electrical contacts. Move the probe or jaws along the edge and catalog the hot spots for the individual screen elements. If the screen has more substantial metal connections poke them directly (see Figure 25.1). Keep trying different pairs of contacts while watching the screen—at some point an LCD

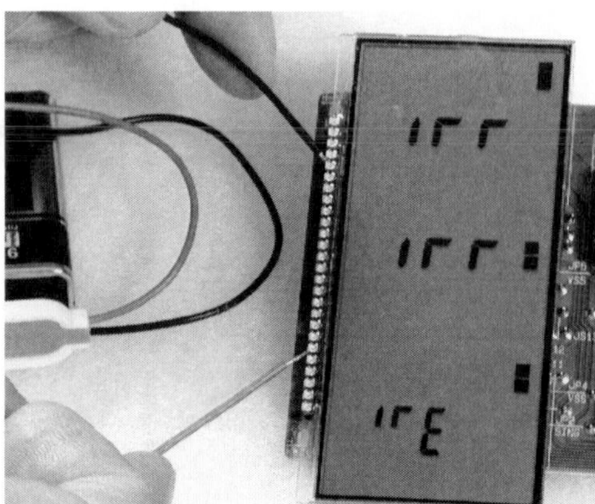

Figure 25.1
LCD elements being activated by direct battery connection.

element should become visible. Make a note of the location of the contact point pairs that enable specific graphic objects, and keep exploring. Once you find a set of images that you like, you can make the connections more stable by wedging and taping wires or pins into or against the contact strips.

You can *animate* the images by using the outputs of oscillators to blink the LCD elements instead of setting them "permanently" on with battery leads. Breadboard a simple oscillator with a 74C14; use a larger capacitor (2.2–10 uf) and a big pot so it runs at a low, metronomic speed. Take one lead from the output of the oscillator and one from the circuit ground (battery "–"). Connect the leads to points along the LCD edges that you know enable images (see Figure 25.2). Adjust the oscillator speed and watch the LCD element turn on and off. If nothing happens, swap the oscillator output and ground connections to the LCD, or try different contact points. Continue to connect more oscillators to more contacts until you achieve the visual texture you want. When more than one oscillator is connected you can usually remove the ground lead from your circuit and just use oscillator outputs—the interaction amongst low and high outputs of the various oscillators activates the matrix. Alternatively you can connect jumpers between any electronic circuit (such as a bent toy) and contact points on a device containing an LCD screen and search for interesting modulation effects (see Michał Dudek's video on the DVD).

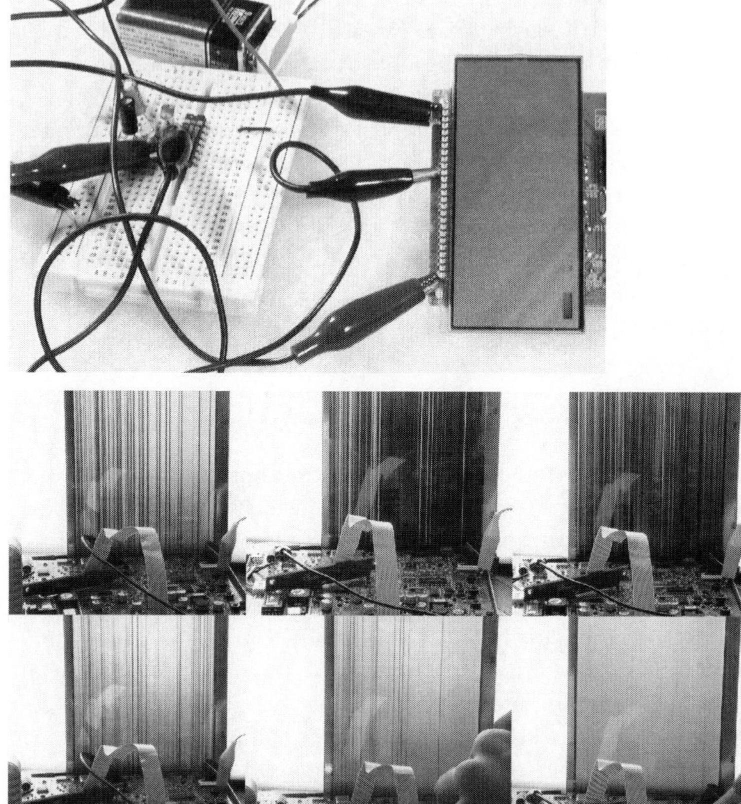

Figure 25.2
LCD elements animated by oscillator outputs.

LCDs are kind of spooky—the image often lingers on the screen for several seconds after power is disconnected before fading out. The small screens bear a resemblance to old daguerreotypes (although Berlin artist Martin Riches reminds me that daguerreotypes had resolution so fine as to be discernable on a molecular level, in stark contrast to the crude pixilation of an LCD). They have a certain charm as modern miniatures, whether superimposed on new backdrops or left in their rather ghostly, mostly transparent state as tiny digital stained glass windows. They are lovely hung in a sun-dappled window.

The current consumption of the LCD device and a low-frequency oscillator is so small that you can leave the object running on a 9-volt battery for weeks before it runs down. You could even power the whole thing with a small solar panel (especially if you *do* choose to hang it in a sunny window). If you go shopping for the latter, look for one that puts out anywhere from 3 to 12 volts, with a current capacity of 5 milliamps or greater. Some CMOS and LCD circuits can even be powered by a battery made from an apple or potato. So dust off those childhood science fair notes or skip ahead to Chapter 29 for some alternative energy sources.

With a decent light source and the appropriate lens you can project your LCD onto a wall—experiment with flashlights or bright LEDs and some lenses. Invest in one of those "third hand" devices: use the two articulated arms with alligator clips to hold the LCD and a lens to focus the image projected by a narrow-beam table lamp. You'll have to remove any printed backdrop, of course, and some LCDs without a background picture use a simple self-adhesive reflective tape that must be peeled off to make them transparent.

You can also drop the LCD screens from tiny portable "stadium TVs" into older slide projectors (see section on BMBCon in Art & Music 10 "Visual Music," Chapter 24). The screens of many older mini-TVs fortuitously have the same dimensions as a 35 mm slide. You can do the same with the displays from many cheap hand-held games—the stupidest thumb-driven hockey match looks pretty cool projected huge, upside-down and slightly fuzzy on a wall. LCDs can be damaged by excessive heat—it helps to add additional fans to cool the slide-well in the projector.

Don't let the visual charm of LCDs distract you from their sonic potential. Clip the ground half of an audio cable to the battery ground of a working LCD toy (before you do the hacks described above) and touch the hot lead to the various scan lines of the LCD display: you often hear deep, rich chords.

HIGHER TECH

Several companies specializing in high-end electronics toys, most notably V-Tech, make "video paint boxes" for children. These devices connect to a television and include a simple graphics tablet and keyboard with which the kid creates drawings and animations. Inside are some very sophisticated graphic chips that can be bamboozled into doing strange things. Following the technique described in Chapter 15, use wire or clip leads to interconnect any points on the circuit board. Occasionally you will get lucky and find connections that cause the graphics engine to freeze mid-way through drawing an image, re-color blocks of pixels, superimpose graphics from its memory, etc. Since these circuits usually use crystals for their clocks, any clock speed modification will probably require replacing the crystal with CMOS oscillator, as we discussed in Chapter 20. And while you're at it don't forget to listen to different points on the board as well. (See Art & Music 10 "Visual Music," Chapter 24.)

PART VI

Finishing

CHAPTER 26

Mixers, Matrices, and Processing: Very Simple, Very Cheap, Very Clean Mixers, and Ways of Configuring Lots of Circuits

You will need:

- An assortment of sound-making and sound-processing circuits, found or made.
- A few pots of the same value (between 10–100 kOhm), preferably "audio taper."
- Assorted resistors and photoresistors.
- An unwanted computer keyboard.
- Some solid and stranded hookup wire.
- Assorted jacks and plugs.
- Some clip leads.
- An amplifier or two.
- Soldering iron, solder, and hand tools.

MIXERS

Now that your collection of noisemakers is growing, a simple mixer might prove useful. Here are some completely *passive* circuit designs—they use no batteries, no chips, or other active components; they need no on/off switches or circuit boards. Even if you already have a "real" mixer, this is a very convenient way to expand its inputs.

The simplest mixer of all is just a Y-cord tying two signals together (see Figure 26.1). Remember earlier (Chapter 18) when I said you shouldn't do this with the output of our basic oscillator circuit because it could force the oscillator to hang up and stop oscillating? Well, professional audio designers assume users will do the stupidest thing: a Y-cord mixer is pretty high on the stupid scale (or low, depending on how you orient the chart), so they include resistors in the outputs of their circuits, so if shorted they will mix just like our oscillators did *after* we added resistors. You can keep adding channels to this mixer by buying more Y-cords and patching them together until they resemble your family tree.

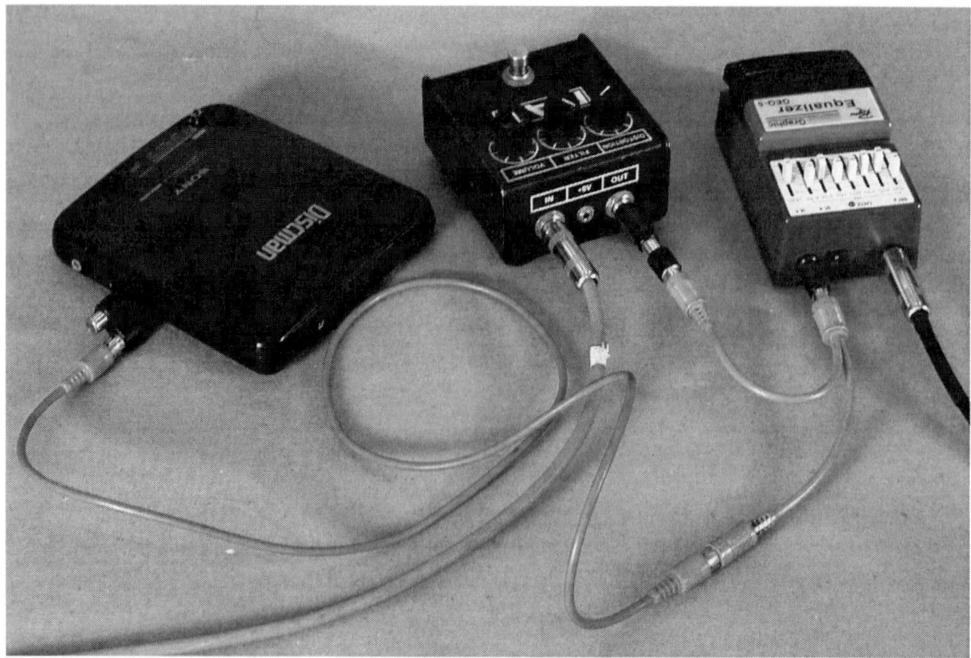

Figure 26.1 Y-cord mixing: a CD player and fuzzbox summed with a Y-cord into an EQ pedal.

If, on the other hand, you want to mix and match commercial electronic devices (with built-in resistors) and weird stuff (with unknown output characteristics) built by you and your friends, we need to increase the circuit complexity ever so slightly. We can make a nice, safe-for-all-circuits passive mixer with nothing more than a jack and a resistor for each input needed (see Figure 26.2). As you can see, each input jack connects to the output jack through its own resistor; all the grounds are tied together as well. You can expand this design for any number of inputs: just add a jack and resistor for each new channel.

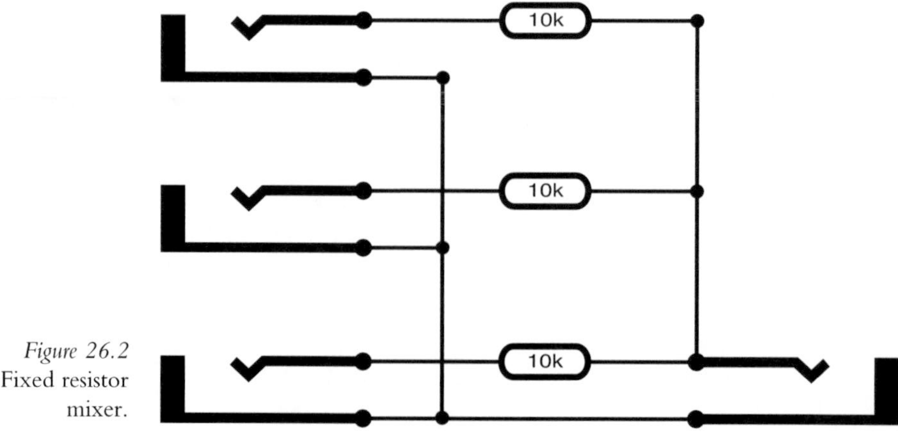

Figure 26.2 Fixed resistor mixer.

"Wait," you protest, "these aren't mixers! Where are the knobs? How can I adjust the levels?" Well, you may have noticed that the majority of audio devices, from fuzztones to CD players to iPods to laptops, have controls for adjusting their output loudness. If you plug them into a bunch of Y-cords or a fixed-resistor mixer, each device's volume control does the job of the corresponding channel fader you'd expect to see on a mixer. Tacky, yes, but poverty is the mother of invention and, as you'll soon find out, potentiometers are the most expensive part of a mixer. So if you can do without, leave them out.

On the other hand, if you are post-economy and looking to park your portfolio in potentiometers, let's advance to some more flexible, reassuringly-familiar mixer designs.

Figure 26.3 show a basic three-input mixer. The signal (i.e. tip) of each input jack is connected directly to one ear of a pot (the one designated "C" in Figure 13.6 in Chapter 13). The center tap on the pot (nose, or "B") is soldered to a "summing" resistor. The non-pot ends of all the summing resistors are tied together and connected to the signal/tip of the output jack. The grounds/shields of all the jacks are connected together with wire, and soldered to a similar wire linking the other ear ("A") of all the pots.

The summing resistors and pots should be about the same value, i.e. for 10 kOhm pots try to use 10 kOhm summing resistors, for 50 kOhm pots use 50 kOhm summing resistors, etc. The pots can be of any value between 10–100 kOhm, but it's best if all the pots are the same value. Figure 26.4 shows which ear of the pot to use for input and which for ground if you want a traditional mixer behavior (i.e. turning the knob clockwise raises the level).

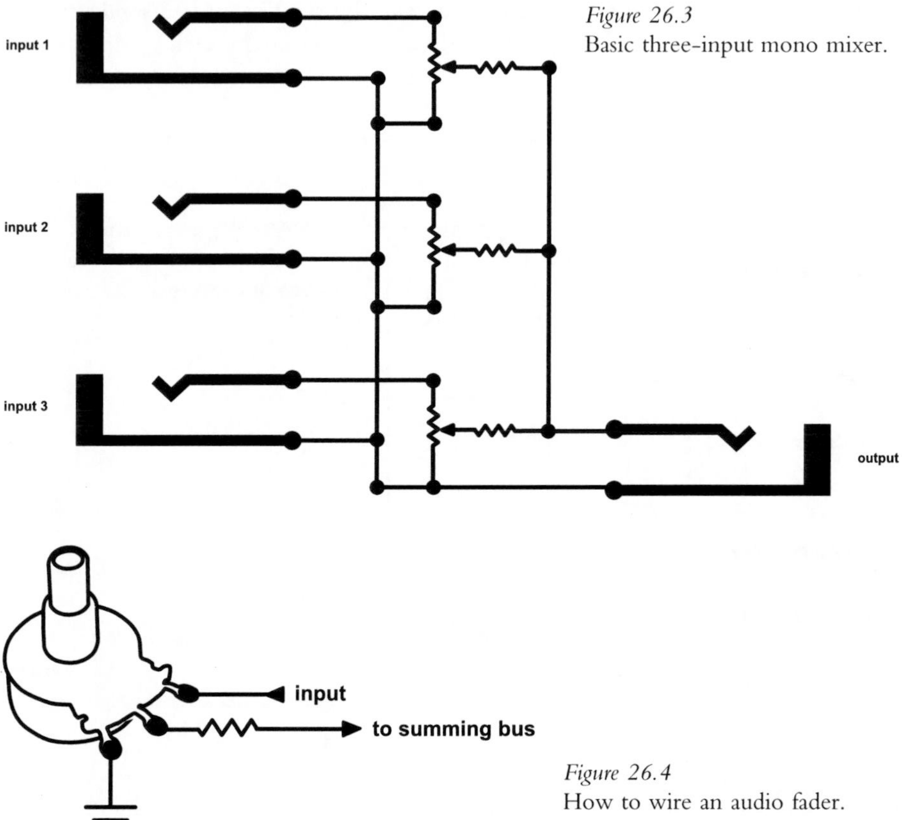

Figure 26.3
Basic three-input mono mixer.

Figure 26.4
How to wire an audio fader.

You can expand the circuit in Figure 26.3 to accommodate any number of inputs, simply by adding another jack, pot, and resistor for each new input signal. By including some switches and another output jack you can expand the output of your mixer to stereo, as shown in Figure 26.5. The ground interconnections have been omitted from the drawing for clarity's sake; instead the common ground bus is indicated by the runic ground symbol. Each summing resistor is connected to the switch's Common terminal (C); the Normally Open terminal (NO) is connected to one output jack, and the Normally Closed (NC) to the other. Throwing the switch swaps the signal between the left and right outputs. It's a kind of binary panner, alternating between hard-left and hard-right, with no gradations in between—crude, but much simpler (and cheaper) than adding a proper pan-pot.

Pots are specified as "linear taper" (what we've been using so far in this book) and "audio taper," which are optimized for adjusting audio level. You should find that raising and lowering the volume feels and sounds smoother with an audio taper pot. We respond to changes of volume on a logarithmic scale (refer back to the explanation of gain and perceived loudness in Chapter 22), and audio faders use a log curve. You can get them from the same online retailers as linear taper pots, but you may not have as wide a choice of values at budget prices.

If you are mixing a lot of stereo signals (like the outputs of MP3 players or CD players) you may want to use "stereo" pots, which are two pots coupled to a single shaft; this mechanical design lets you adjust two signals in parallel from a single knob (see Figure 26.6). One pot in each stereo pair will take the left half of a stereo input and send it to your left bus, while the other will take the right input and send it to the right bus—like our stereo mixer, but without the switches. You wire each section exactly like an ordinary single pot.

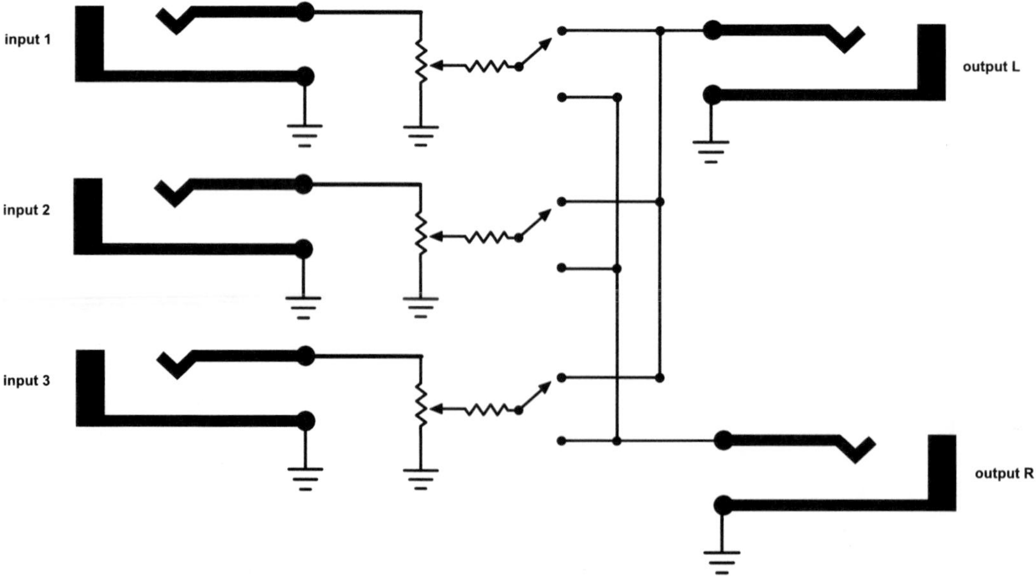

Figure 26.5 Basic three-input stereo-output mixer.

Figure 26.6
A stereo potentiometer.

You can use rotary pots for these mixer circuits, or slide pots (easy to find through online sources), which make your circuit look more like a "real" mixer, but be warned:

Rule #24: It is easier to drill round holes than slots.

Because they contain no amplifiers or other gain circuits, these circuits are best suited for mixing line-level signals of more or less similar strength (i.e. CD players, laptops, radios). They cannot *boost* a mike signal up to the line level of a CD, but can only *attenuate* line-level signal down to that of a mike—this is awkward, unless you plan to patch the output of this mixer into a mike preamp. But what these designs lack in gain and EQ they make up for in cost and audio quality—as with the optical gating and panning circuits in Chapter 21, passive mixers are quite coveted by certain audio purists.

If you want to mix low-level signals (such as contact mikes, coils, etc.) as well, you can insert a preamplifier circuit (such as those in Chapter 22) between each input and its pot. This means adding a chip and a battery, but you get six mike inputs with one cheap chip, which isn't too bad.

MATRICES

David Tudor, one of the pioneers in the field of live electronic music (see Art & Music 4 "David Tudor and 'Rainforest'," Chapter 8), used mixer matrices to combine relatively simple circuits into networks that produced sound of surprising richness and variation. Instead of simply mixing a handful of sound sources down to a stereo signal, Tudor interconnected his circuits with multiple feedback paths and output channels. The recent rise of the "no input mixing" school of internal machine feedback has exposed a new generation of musicians and listeners to the possibilities of matrix feedback. Consider the 3 × 3 matrix mixer shown in Figure 26.7. (Input and output jacks have been omitted for clarity.)

You will notice that the design is similar to that of our basic mixer in Figure 26.3, but here each input signal is connected to three pots instead of one, and we have three output buses as well. You can expand this circuit with as many pots and jacks as you need and can afford. Figure 26.8 shows some packaged matrix mixers using this design.

Connect a few circuits, including both sound *generating* circuits, such as your oscillators or toys, and some *processing* circuits, such as the photoresistor gate, the distortion circuit, or some guitar pedal (such as a delay, wah-wah, or graphic equalizer). Send one output of the matrix to an amplifier for listening, and the others can be sent to the inputs of your sound processing circuits. By adjusting the levels of the various pots you can create a pretty straightforward signal path (toy through distortion to speaker) or a more devious one (toy through distortion to speaker, distortion also to delay, which goes both to speaker and back into its own input).

The piezo-driver pseudo-reverbs we discussed in Chapter 8 work beautifully in these configurations. Some of the most unassuming rock pedals reveal astonishing musicality

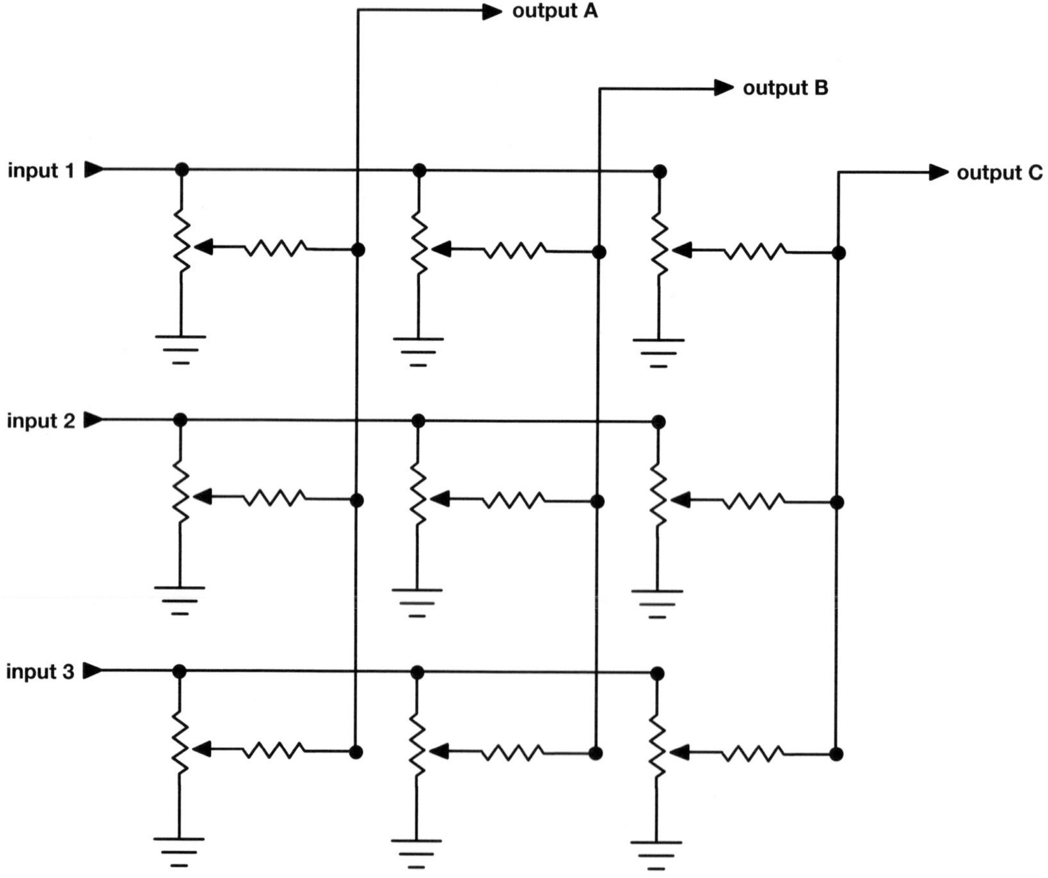

Figure 26.7 3 × 3 matrix mixer.

Figure 26.8 a, b & c
Matrix mixers by Alex Inglizian (upper left), James Murray (upper right), Douglas Ferguson (lower left) and Steve Marsh (lower right).

when placed in feedback loops. Incorporated into matrices, time-based effects (such as delays and flangers) contribute an instability that transforms a table of commonplace effects into a richly challenging performance instrument (see Figure 26.9).

If you intend to use a matrix to generate feedback you will need some gain, which you can provide with the simple 4049 preamp circuit of Chapter 22, or using effect pedals that can boost a signal (i.e. almost any). Feedback matrices benefit greatly from the inclusion of some kind of equalization, to aid in steering pitch response and nulling

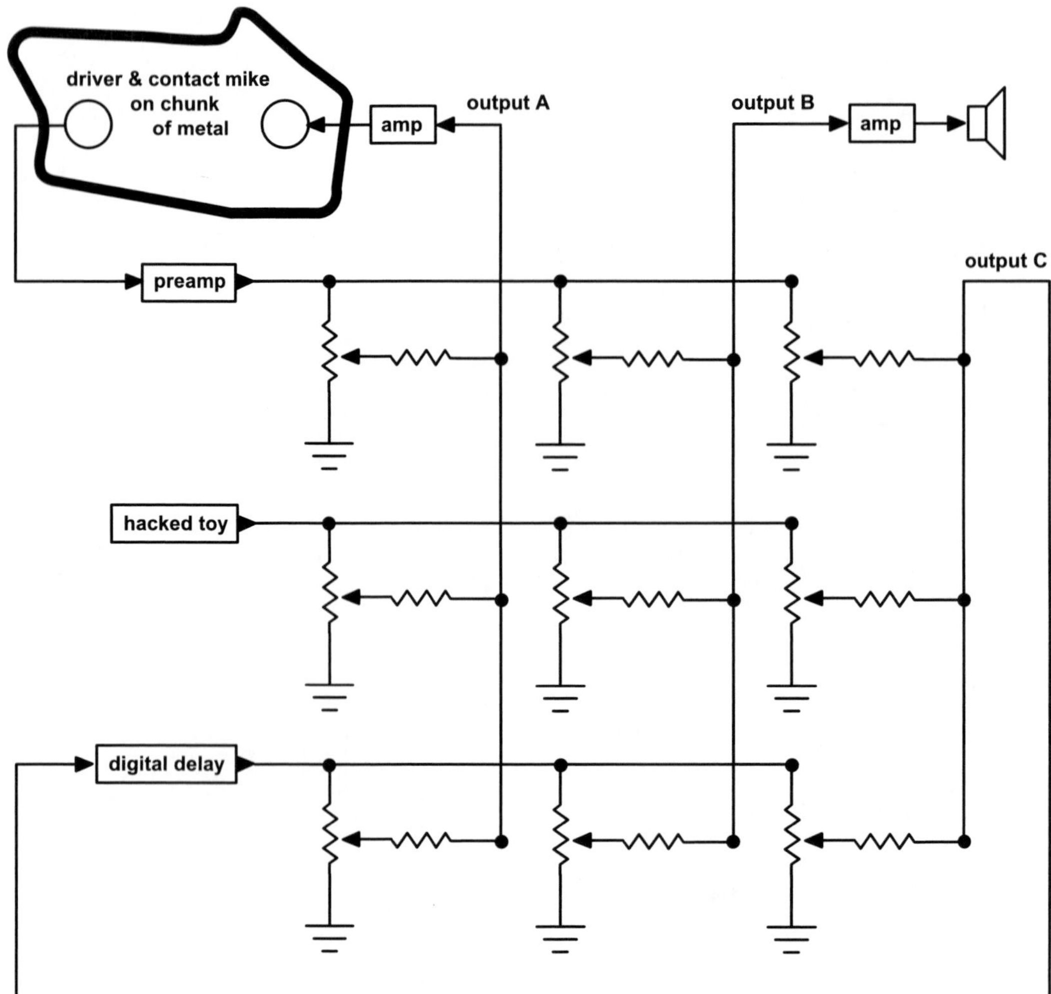

Figure 26.9 Circuits in a matrix.

out unwanted shrieks—a simple graphic EQ effect pedal provides both the requisite level boost and useful frequency shaping.

A last word of advice on matrix mixers: when using a passive such as ours it behooves you to keep any unused inputs turned fully *off*. This reduces the crosstalk between channels.

COMPUTER KEYBOARDS

Discarded computer keyboards roam contemporary urban streets like unwanted mutts in Latin America. Although bereft of cool wet noses or beseeching brown eyes, these electronic strays can nonetheless prove friendly companions. A computer keyboard consists of a switch matrix: instead of each key closing two discrete contacts (as with a typical SPST stand-along switch, as described in Chapter 16), it bridges specific lines in an X–Y grid, as shown in Figure 26.10.

Pressing Key 1 connects horizontal row 1 to vertical column A, Key 3 connects row 1 to column C, Key 4 connects row 2 to column A, etc. In a computer keyboard there are enough rows and columns to handle the full alphabet, plus numbers and all those extra function keys. A 10 by 8 matrix, for example, will handle the 80 keys. The computer "scans" the matrix to detect which key is depressed: it sends a pulse down each row and checks which column it comes out of (like a high-tech version of Splat-the-Rat).

If you open up the keyboard and scrutinize its circuit board, you should notice that the traces are arranged in a vague grid (see Figure 26.11). Solder a wire between the ground terminals of two female jacks; attach a clip lead to each of the hot terminals. Connect a sound source, such as an oscillator, to one jack, and connect the other jack to an amplifier.

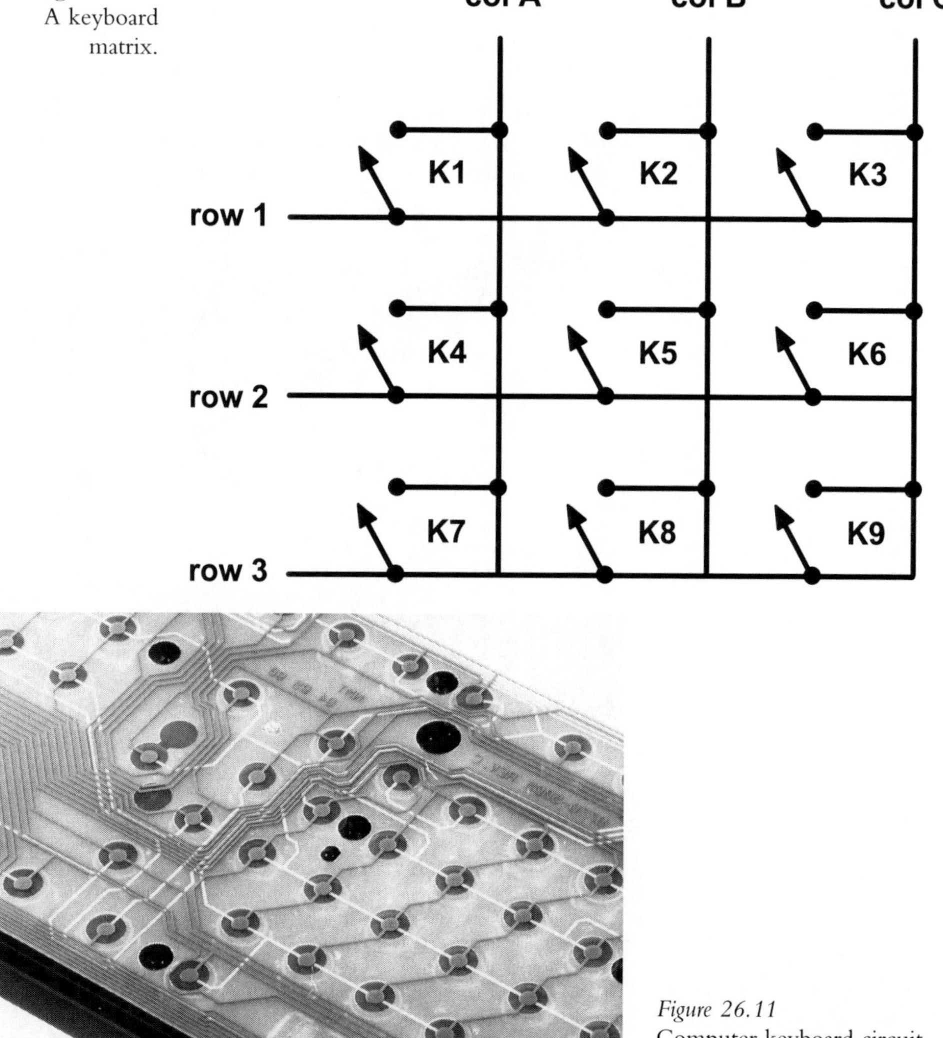

Figure 26.10 A keyboard matrix.

Figure 26.11 Computer keyboard circuit board showing matrix traces.

Use the clip lead to connect the input jack's signal/hot tab to one of the traces, which are probably routed to connectors at the edge of the circuit board (see Figure 26.12). Press down a key and touch the amplifier lead to other traces, one at a time, until you hear your sounds. Then release the key: if the sound shuts off, you've found a cross point in the matrix. Record this information somehow: use a sticker to mark the top of the key, or start a chart in which you keep track of the keycap legend that corresponds to a particular cross-point (i.e. "g" connects row 4 to column 3). If it doesn't turn off when you release the key it means you've just touched another point along the trace that the sound source is connected to, so keep testing other points until you find a cross point.

By repeating this admittedly arduous process you should be able to decode the matrix into rows and columns. Now solder each row and column to the hot terminal of an audio jack. Connect all the shields together. Use all the rows as inputs and all the columns as outputs, or vice versa. By pressing keys you can route any input to any output. You can use this device alone—as a signal router for spatial distribution, for example—or in conjunction with the matrix mixers described above to add switching to matrix-based signal processing.

If this decoding process sounds too daunting, you can find smaller, more easily mappable switch matrices in touch-tone telephones (3 columns × 4 rows), calculators, or various toys—all frequently discarded household items. "Raw" matrix keypads of various sizes can also be bought from a number of online retailers. Membrane switches,

Figure 26.12
Computer keyboard wired for testing audio routing.

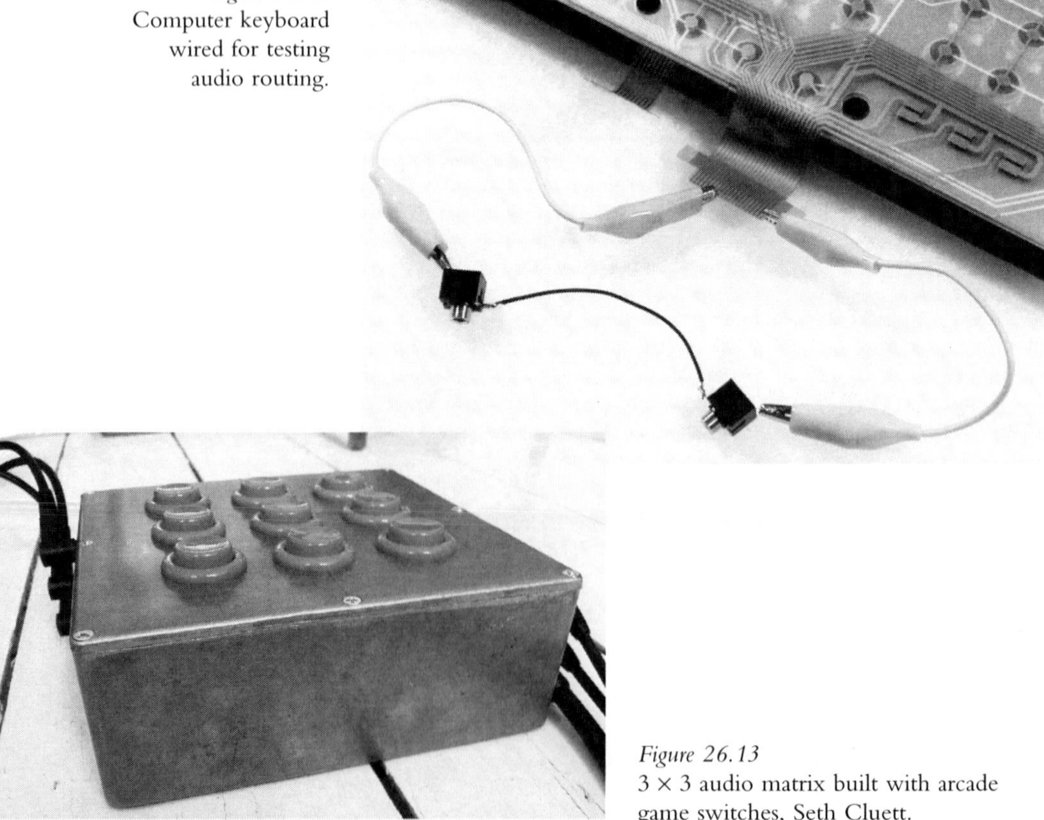

Figure 26.13
3 × 3 audio matrix built with arcade game switches, Seth Cluett.

commonly used in inexpensive keypads and inside many computer keyboards, have the advantage that, in addition to being able to close the switches by direct finger pressure, one can often activate them by "drawing" across the surface with a stylus of some sort, or rolling a billiard ball over it—nice gestural alternatives. For a more athletic matrix array, Seth Cluett wired up a bunch of arcade game switches (see Figure 26.13).

AUTOMATION

The basic photoresistor gating circuit described in Chapter 21 can be expanded into a matrix array as well (see Figure 26.14). The photoresistors can be activated by flashlights, ambient light and shadow, video or film projection, or oscillator-driven blinking LEDs, depending on the degree of control or indeterminacy you desire. For maximum dynamic range remember to cover the backs of the photoresistors with electrical tape, or place them in opaque lightpipes (like sections of soda straws or heat shrink).

The outputs of any of these various mixer designs can be patched directly to power amplifiers (guitar amps, stereos, etc.) or sent through DI boxes to a PA system. If you built a mixer to add inputs to a "normal" mixer (such as the ubiquitous Mackies or Behringers) you can patch the output of your homemade one into any unused inputs of the other—many mixer have "effect returns" that are perfect for this purpose.

These designs are cheap and simple enough that you might also consider building them into any of your "multi-voice circuits": you can add individual level controls to each of the six channels of a 74C14 hex oscillator (Chapter 18), or the various distortion and signal processing stages of the circuits we designed in Chapters 22 and 23.

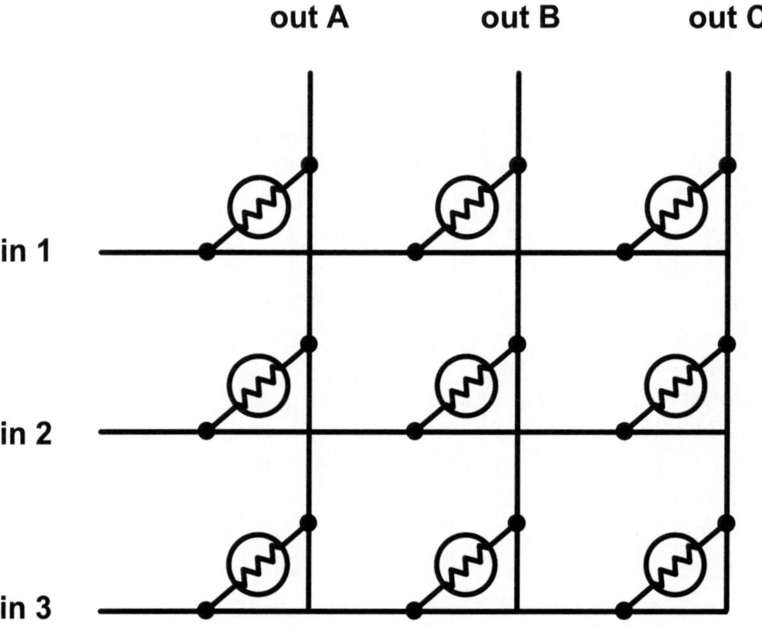

Figure 26.14 3 × 3 photoresistor matrix.

CHAPTER 27

A Little Power Amplifier: Cheap and Simple

You will need:

- Something to amplify: a guitar, an MP3 player, etc.
- A breadboard.
- Audio Power Amplifier chip, LM386.
- Assorted resistors, capacitors, and pots.
- Assorted jacks and plugs.
- A small speaker.
- Some solid hookup wire.
- A 9-volt battery and connector, or four AA batteries and a holder.
- Hand tools.
- An appropriately sized box to fit the circuit and speaker.

Whether in pursuit of a self-contained electronic instrument or some form of sound sculpture, one day you will tire of choosing between some putty-colored mini-amplifier and a bulkier, more expensive (and potentially more dangerous) PA system. There are a number of kits available from online retailers that include the essential integrated circuit, associated passive components (resistors, capacitors, etc.), and a printed circuit board (see Chapter 1, Figure 1.2). But if you want to save a few dollars and get some more design experience, consider soldering up your own using the venerable LM386 (see Figure 27.1). At less than $1 retail, this chip, combined with a few other components in a very simple configuration, makes a cheap but decent low-power audio amplifier. It is the heart of many mini-amps (including the Radio Shack 277–1008), and—once soldered up—can be substituted accordingly.

The basic configuration shown in Figure 27.2 gives a gain of 20 and is best for line-level signals such as CD players, computers, your oscillators, etc. By adding a 10 uF capacitor between pins 1 and 8 the gain rises to 200 (see Figure 27.3), which is more suitable for contact microphones, coils, and guitar pickups. You can add a switch to bring the capacitor in and out of circuit to select high or low gain for different input sources, as shown in Figure 27.4.

The "+" voltage from the battery connects to pin 6, and the "–"/ground connects to pin 4. You might want to include a power switch as well, or disconnect the battery from its clip when not in use, since this circuit drains more power than the others we've

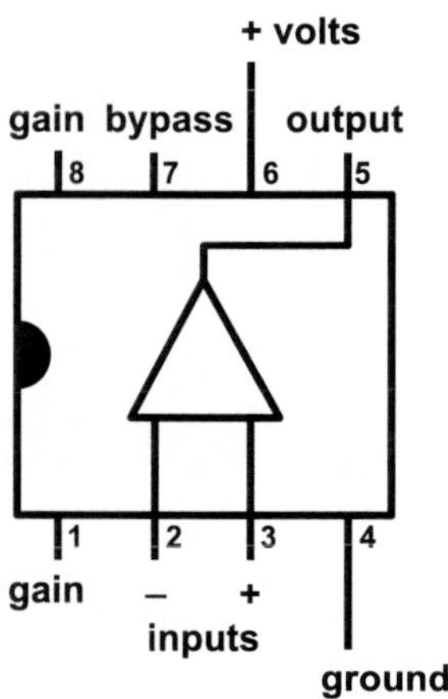

Figure 27.1
LM386 amplifier pinout.

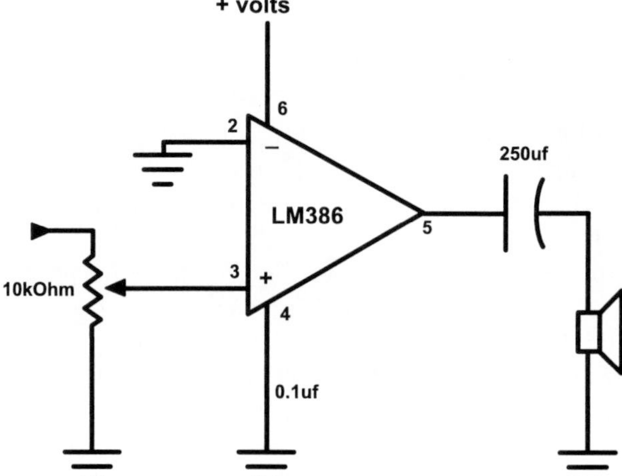

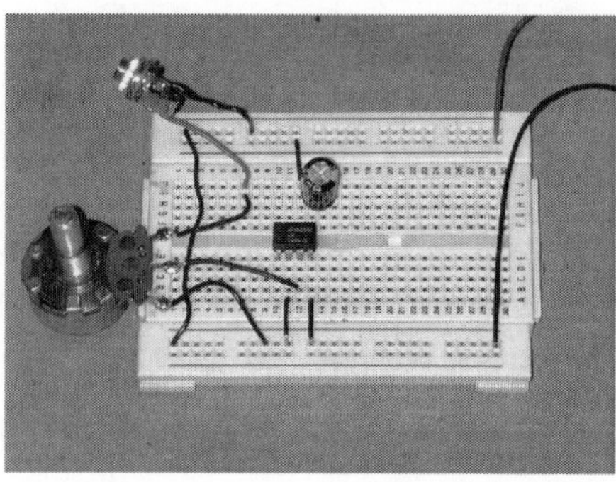

Figure 27.2
Amplifier with gain of 20.

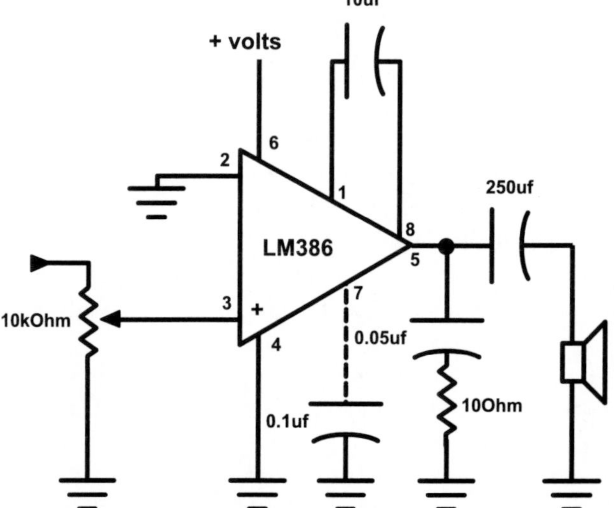

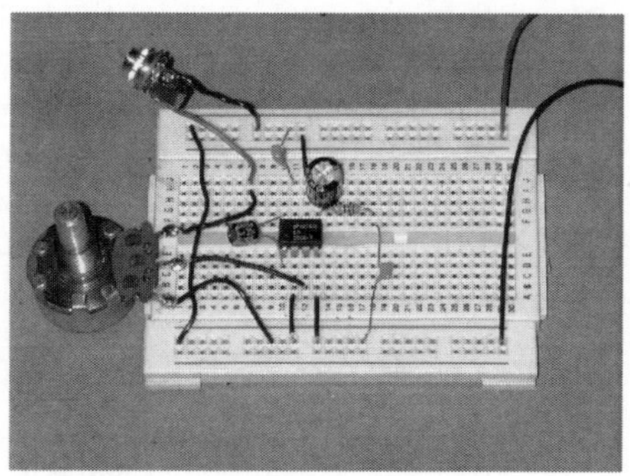

Figure 27.3
Amplifier with gain of 200.

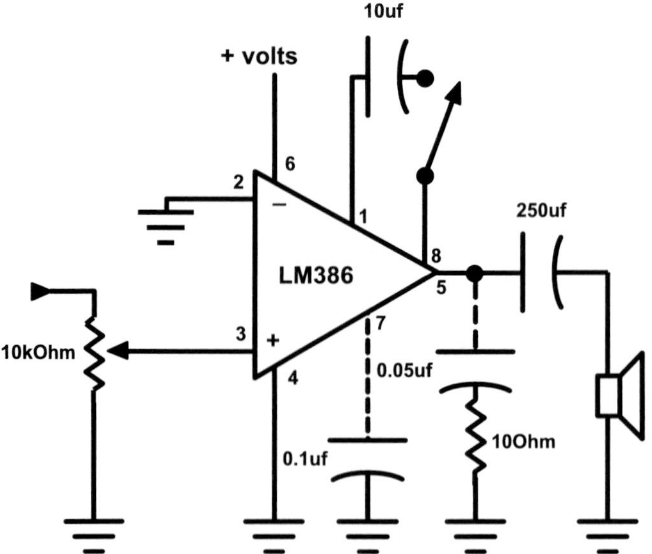

Figure 27.4 Amplifier with switch-selectable gain of 20 (open) or 200 (closed).

made. Pin 2 is also tied to ground, and the input signal goes to pin 3 after passing through a potentiometer used as a volume control. The 0.05 uF capacitor and 10 Ohm resistor shown at pin 5, and the 0.1 uF "bypass" capacitor at pin 7 are optional parts, to be added if the circuit oscillates and whines by itself.

One cautionary note: for all the apparent simplicity of this design, audio power amplifiers are notoriously finicky—worse, even, than the 4049 preamp we wrestled with back in Chapter 22. Especially in the high gain setting (200), this one is susceptible to oscillation. Guidelines for best performance are similar to those we followed with the 4049:

- Keep the jumper wires as short as possible—on the breadboard, and between the pots and jacks and the breadboard.
- Use shielded cable to connect to the circuit; speaker cable at the output.
- Install the various bypass capacitors shown on the schematics.

If nothing else works, consider buying a kit that includes a printed circuit board designed specifically for the amp. Such a board will do a much better job of suppressing spurious noises than a breadboard or generic circuit board such as we've used so far.

When you've finished soldering up you can mount it in a box with a speaker, like a self-contained mini guitar amp (Figure 27.5). Or leave the speaker out to keep the package smaller—you can then connect to various size speakers to suit the demands of the occasion. Figure 1.2 in Chapter 1 shows an amp in a small tin with jacks for input and output connections, but also on-board contact posts for quick hookups with clip leads.

This circuit puts out about 1/4 watt of audio power, and can be used to drive small speakers or headphones. It runs nicely off a 9-volt battery or a set of four AA batteries (the latter will last longer). Although 1/4 watt sounds pretty puny, it's enough to drive a piezo disk at pretty high sound levels using the backwards output transformer trick shown in Chapter 8. This amplifier can also drive directly a small motor (such as the vibrating motors from cell phones and pagers, also discussed in Chapter 8) or a low-power solenoid or relay. The EBow electromagnetic string driver described in Chapter 8 is basically a pickup coil connected to an LM386 wired to a driver coil extracted from a small speaker—all encapsulated in epoxy so we can't figure out exactly what else the designer threw in to make it sound so cool (but don't let that stop you from experimenting).

There are dozens of amplifier chips available these days, by the way, ranging in power from fractions of a watt like the LM386, suitable for battery power, to high-power ICs intended for use in home stereos, TVs or car sound systems. I've chosen the LM386 because of the extreme simplicity of the circuit design, but if you want more oomph you can do a bit of googling to find a louder chip.

An amplifier is the most basic and essential tool in electronic music. You can never have too many of them. So knit yourself one whenever you're feeling bored (I built a dozen of the little Altoid-tin amps shown in Figure 1.2 for use in workshops).

Figure 27.5
Mini-amplifier built from LM386.

CHAPTER 28

Analog to Digital Conversion, Really: Connecting Sensors to Computers Using Game Controllers

You will need:

- A computer running some kind of music software that accepts external control devices (joysticks, game pads, trackballs, etc.).
- A MIDI synthesizer, sampler, drum machine, etc., if the software on your computer does not generate sound directly.
- An expendable joystick, game pad, trackball, etc. compatible with your computer.
- Assorted pots, photoresistors, pressure sensors, etc.
- Some very light gauge hookup wire.
- Soldering iron, solder, a solder-sucker, clip leads and hand tools.

Now that the end of the book is in sight I will confess: I am not the raving anti-digital Luddite you might take me for after all these pages. I actually don't mind computers—in fact I've been making music with them since before the Apple was a twinkle in the collective Jobs and Wozniak eye. The issue for me has always been how to *play* them: sometimes an ASCII keyboard and a mouse don't give you enough control, or the right kind of control; sometimes you want to play something that feels less like a typewriter and more like a musical instrument (with apologies to Leroy Anderson). Of course you can use a MIDI interface to connect any commercial MIDI controller to your computer, but your choices are usually limited to piano-like keyboards, mis-triggering guitars, the odd pseudo-saxophone, and a handful of truly alternative controllers created by a few die-hard visionaries (see Art & Music 11 "The Luthiers").

Wouldn't it be nice to cobble together a computer instrument out of all the sensors we've been working with so far? Take a handful of switches, pressure sensors, pots, photoresistors, etc.; stick them onto a tennis racket, mailing tube or nerf ball; connect them to your computer; and play your favorite music software with a sort-of-guitar, a quasi-clarinet, or a stress-relief ball? Well, it's possible, even easy . . . .

There are a number of interfaces available that link sensors to a computer through the USB or Ethernet ports, or using a MIDI interface. The granddaddy of them all was STEIM's SensorLab (see Figure 28.4), which translated sensor data directly into MIDI,

THE LUTHIERS

The unsung heroes of electronic music are the engineers who build the instruments but don't necessarily make a career of playing them. While the name "Moog" is almost synonymous with "synthesizer," pioneering designers such as Donald Buchla, Serge Tcherepnin, Tom Oberheim (whose companies bore their names), Alan Pearlman of ARP, and David Cockerell of EMS are virtually unknown outside extremely geeky circles. Although the introduction of MIDI in the early 1980s resulted in explosive growth for the synthesizer industry, these new machines were purposefully generic, the musical equivalent of the putty-colored office PC. But on the sidelines of the industry, inspired and quirky engineers continue to flourish.

Today the late Robert Moog's company manufactures one of the best Theremins available (ironically, he got his start in business as a teenage entrepreneur selling Theremin kits of his own design), as well as a collection of sound processing devices based on his classic synthesizer modules. Likewise Donald Buchla continues to develop highly personal and expressive electronic musical instruments, such as Lightning, a controller that translates the position and movement of handheld wands into MIDI data. In 1975 Craig Anderton wrote what probably was the first book on musical circuits that could be understood by non-engineers, and he has been publishing how-to articles in musicians' magazines for years.

And in a world of off-the-rack musical tools, there still remains a special place for bespoke tailoring from the hands of the clever, personal luthier, such as Bob Bielecki, Bert Bongers, or Sukandar Kartadinata. Based in New York, Bielecki helped Laurie Anderson realize many of her idiosyncratic instruments, including the "Tape Bow Violin" (see Art & Music 6 "Tape," Chapter 9) and her first "masculating" pitch shifter. Later he built drone circuits for LaMonte Young (USA), precision sine wave oscillators for Alvin Lucier (USA), a light-to-sound interface for Stephen Vitiello (see Art & Music 10 "Visual Music," Chapter 24), and other personalized electronic devices for artists such as Charles Curtis (USA), Arnold Dreyblatt (USA), Annea Lockwood (USA), Linda Montana (USA), and Bill Viola (USA).

Bert Bongers (Netherlands) studied computer science and ergonomics before finding his calling as an instrument builder, initially for STEIM (Amsterdam) and The Institute of Sonology in The Hague. He later founded his own research labs in Barcelona and Maastricht, and now works at the University of Technology Sydney (Australia). He has built exquisite instruments for a number of artists, including Laetitia Sonami's (France/USA) "Lady's Glove," an evening wear take on virtual reality controllers (see Figure 28.1); Jonathan Impett's (UK) "Meta-Trumpet," a trumpet extended with ultra-sound position sensors, valve movement detectors, and additional switches; and Michel Waisvisz's (Netherlands) "Hands II" free-air gestural controller for computer music.

Sukandar Kartadinata (Germany) has been developing hardware and software tools for numerous artists since the early 1990s, including a hacked greeting card anklet for violinist Jane Henry (USA) (see Figure 28.2 and audio track 16), computer-controlled ratchet noise-makers for German artist Jens Brand (see also Art & Music 5 "Drivers," Chapter 8), and software/hardware hybrid instruments for Richard Barrett (UK/Germany), Annette Begerow (Germany), Steve Coleman (USA), Axel Dörner

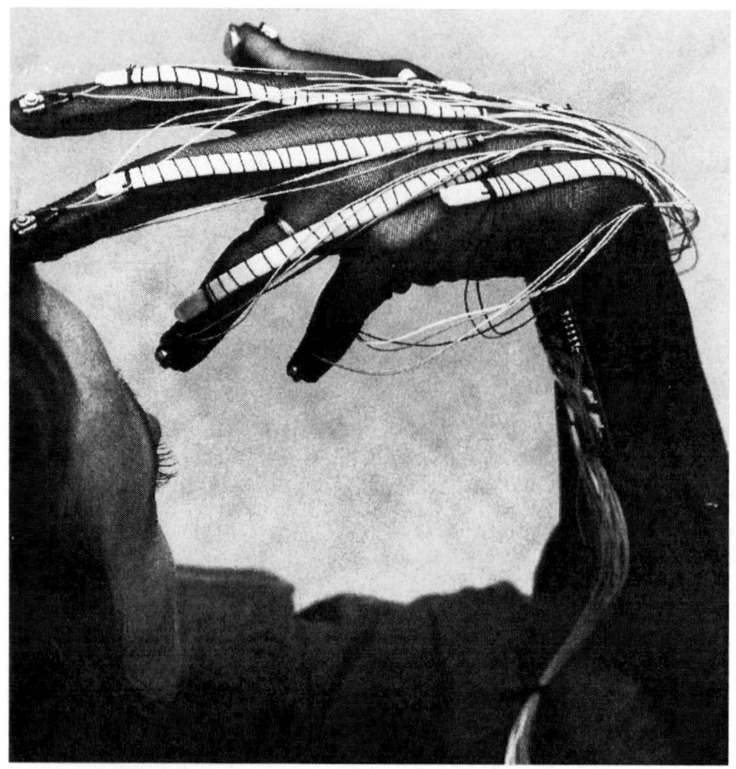

Figure 28.1 Laetitia Sonami, "Lady's Glove, v. 4," gestural computer music controller, built by Bert Bongers after a prototype by Laetitia Sonami.

(Germany), Walter Fabeck (UK), Sabine Schaefer (Germany). In 2004 he designed the "Gluion," a general purpose, configurable interface between sensors and computers (see Figure 28.6).

Bielecki, Bongers, and Kartadinata are virtuosos of "glue" technology—the seamless integration of hardware and software to solve problems that slip between the

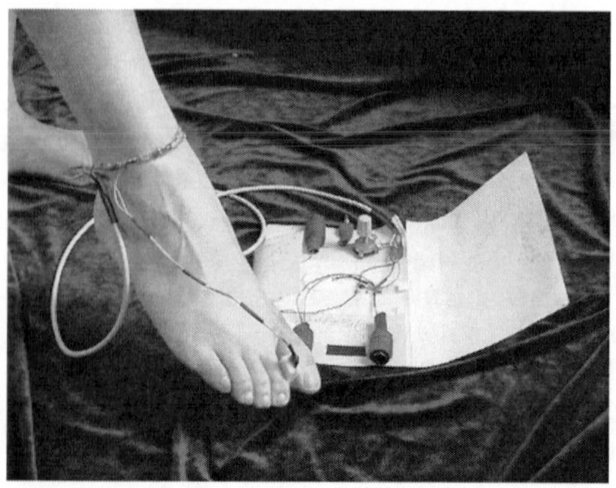

Figure 28.2 "Chipsaw," hacked musical greeting card anklet bracelet, designed for Jane Henry by Sukandar Kartadinata.

cracks of commercially available solutions. As with master printers, these designers are true collaborators, whose technical expertise and imagination result not only in workable solutions, but also in new inspirations and directions for the artists with whom they work.

Recently a new company, Machine Collective, has emerged to fill the niche between off-the-shelf electronic instruments and the one-of-a-kind personal devices designed and built by the luthiers I've mentioned. They produce a series of user-programmable modules that can be combined to form a wide range of unusual control surfaces, for use either with computers or for direct hardware control and circuit bending applications. Philadelphia-based Monome recently introduced an elegant keyboard and display matrix that can be interfaced to a variety of controllers and computers.

No list of resources for the befuddled artist would be complete without mention of STEIM—the Studio for Electro-Instrumental Music (or, in Dutch, Stichting voor Elektro-Instrumentale Muziek). Founded in 1968, STEIM (in Amsterdam) remains a spiritual retreat and think tank for musical innovation; through its residency program hundreds of artists have been helped to realize ambitious, heartfelt, and often impractical projects (see Figure 28.3)

Figure 28.3 "Mutantrumpet" by Ben Neill, with electronic extensions built by Jorgen Brinkman at STEIM and James Lo, New York City: Steim JunXion board, with eight momentary switches and eight continuous MIDI controllers (four pots, two joysticks next to the extra set of valves, and a fader).

and also let you program sophisticated interpretations of that data in a C-like language—a complete run-time computer music instrument, sadly no longer in production at the time of writing.

The I-Cube from Infusion Systems and Eric Singer's MidiTron are less expensive, computer-configurable, sensor-to-MIDI interfaces that can control MIDI devices directly or can connect to a computer for further processing of the raw sensor data. Various companies, including CH Products, manufacture inexpensive joystick development circuit boards that let you read multiple switches and analog sensors through your USB port (see Figure 28.5). STEIM has replaced the SensorLab with a versatile USB interface called the JunXion Board. Sukandar Kartadinata's "Gluion" is a high-speed Ethernet interface that can be configured for a wide range of sensors (see Figure 28.6). These

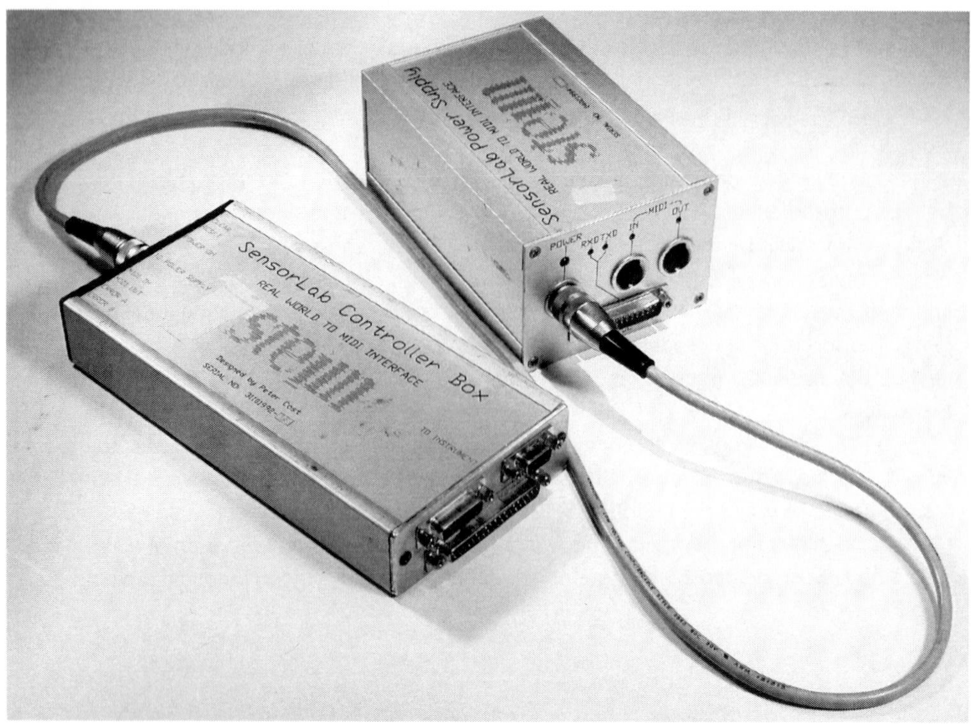

Figure 28.4 The STEIM SensorLab.

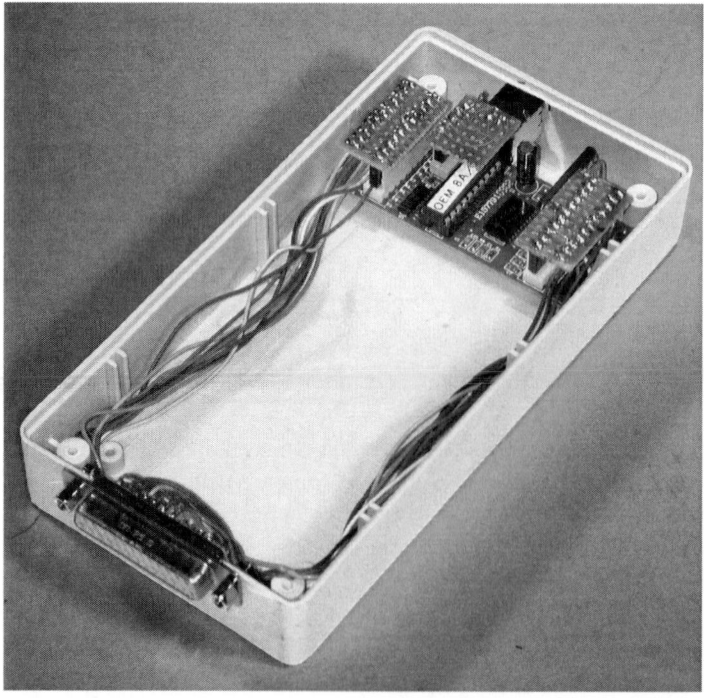

Figure 28.5 Sensor-to-USB interface built with CH Products interface board (open).

Figure 28.6 The Gluion.

days many hackers are programming Arduino boards to serve as sensor interfaces—the Arduino is very much the product of the Open Source generation, and its user base makes available an ever-expanding library of software routines and add-on products. Similar interface subsystems are available from a growing number of other sources. Prices range from under one hundred dollars to several thousand dollars. Some of these interfaces are bi-directional, allowing you to send control information through them to external relays, motors, etc., as well as receive data from various sensors. A few can transmit data wirelessly from sensors to your computer.

But there is a cheaper (if somewhat tackier) solution that should appeal to someone who's made it through this book: hack a game pad.

OH JOY!

Most controllers for computer games—joysticks, game pads, etc.—consist of a microprocessor that translates the state of various switches and analog sensors into data that is sent to the computer through its USB interface, sometimes via a wireless link. Some of these devices make excellent musical controllers in their own right—a good joystick gives several axes of motion that can be mapped to various aspects of your sound (pitch bend, modulation, filtering, etc.); this can be used by itself or in conjunction with a standard MIDI device (such as a keyboard) for very expressive musical control. A wireless game controller liberates you from sitting in front of your laptop like an accountant. And joysticks and game pads are cheap: they can be had for under $20 new (wireless ones cost a bit more).

All you need is software that accepts USB data from an external device. Programming languages such as Max, SuperCollider, and PD have objects or subroutines that receive USB data and configure it to control any aspect of your program. "JunXion," STEIM's elegant little utility, maps any USB data directly to MIDI; this MIDI output can then be used to control an external synthesizer or sampler directly, or can be patched within the computer to any music software that accepts MIDI syntax.

When you get tired of the fighter pilot imagery of joystick-controlled electronic music, it's time to get out the screwdriver and get down to work. As with the toys we hacked earlier in the book, it is the work of but a moment to remove the stock switches and joysticks and replace them with personalized variants. Sometimes simply rearranging the layout of switches and other controls changes the way you play them: replacing the cluster of trigger buttons with a line of switches can transform a Top Gun weapon into something more like a flute; substituting photoresistors or pressure pads for the pots in the joystick opens up new gestural options.

The first step is to boot up a program that can be controlled by the game pad and confirm that the device functions correctly *before* you start to hack it. Make sure all switches, joysticks, and other sensors are working, and familiarize yourself with what each one does. This way after you make component substitutions you have a reference for the correct functioning of the interface.

Now open the game pad (see Figure 28.7). The hacks themselves are pretty simple. Most switches on game pads are the kind you're probably familiar with from toys: conductive rubber "hats" that press down against gold interlocking traces on the circuit board (see Figure 28.8).

Figure 28.7
Open game pad showing rubberized switch pads.

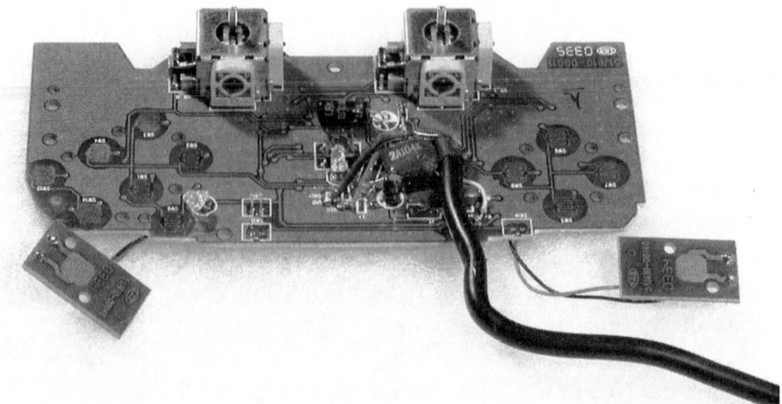

Figure 28.8
Switch pads removed to reveal switch traces on circuit board.

To substitute a different kind of switch simply solder one end of a thin, flexible insulated wire to each half of the trace network (see Figure 28.9), and solder the other ends to the contacts on your new switch (as we did with photoresistors in Chapter 21). Sometimes you can find a hole through a trace leading to the switch, somewhere nearby— it's easier to solder to one of these, if available, than directly down onto the switch pad traces themselves.

As we observed in Chapter 18, a joystick consists of two pots positioned so that when you move the stick front-to-back it rotates the shaft of one pot, while movement side-to-side rotates the other; complex movements are parsed into *x–y* data from the two pots interacting. If you look at the circuit board under the joystick, you should notice three solder pads in a line under each of the pots (see Figure 28.10).

The "ears" of the joystick pot (A and C) are soldered to the end pads, and the "nose" (B) to the center pad. Typically the pot is configured as a "voltage divider": one end pad is connected to ground, one to a positive voltage. As the pot rotates, the center tap puts out a voltage that varies between ground and the positive reference. This voltage is

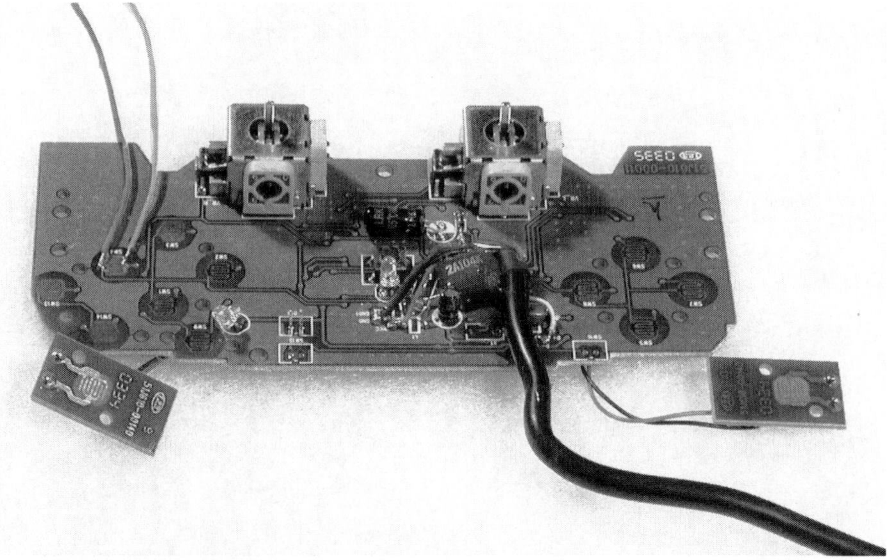

Figure 28.9
Switch pad traces with wires soldered on for remote switch.

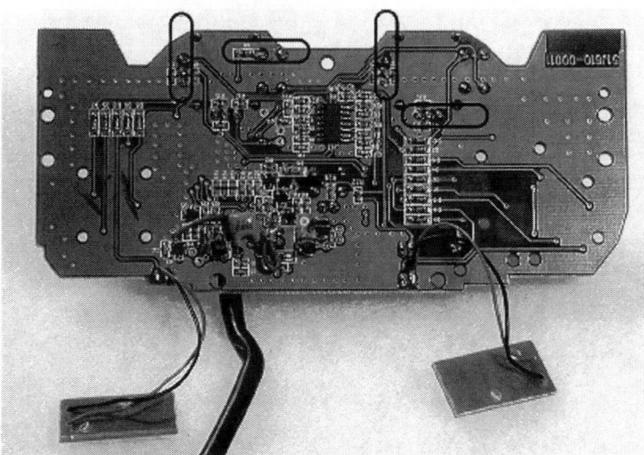

Figure 28.10
Joystick solder pads (outlined with black oval).

read by the microprocessor in the game controller and sent down the USB cable as a numeric value. To hook up an external pot or other analog sensor you first need to remove the joystick. A "solder sucker" is a useful tool here (see Figure 28.11): melt the solder at each pot pad with your soldering iron and use the vacuum of the solder sucker to extract as much of the solder as possible from around the connection (see Figure 28.12). You gotta move fast to slurp up the solder after you remove the iron but before it cools and hardens. There may be additional solder joints between the mechanical framework of the joystick and the circuit board; clear these out as well, as best as you can.

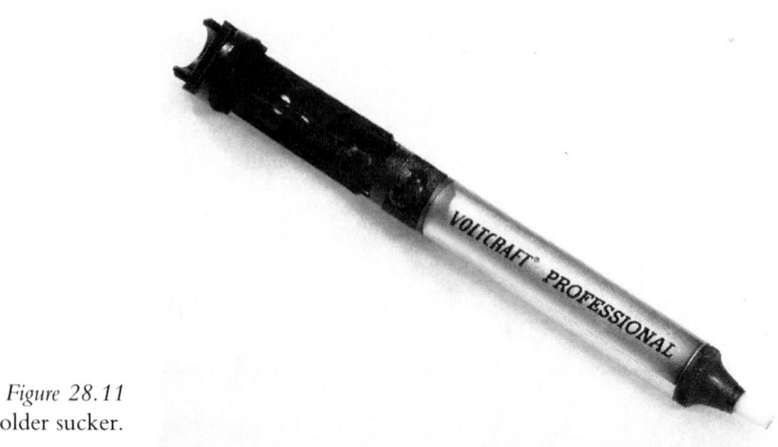

Figure 28.11
A solder sucker.

Figure 28.12 Sucking solder.

When all the pads have been cleared of solder gently pry the joystick off the board (see Figure 28.13). You may need to apply the tip of the soldering iron to various connections to melt last traces of solder as you wiggle the joystick free—this is not a pretty job. Once the joystick is removed, solder flexible insulated wire to each hole in the potentiometer triple-sets (see Figure 28.14).

These wires can now be connected to any electronic components that can be configured as a potentiometer-like "voltage divider." For example, Figure 28.15 shows a simple photoresistor-based voltage divider that you can swap into any place previously occupied by a pot. Here the photoresistor effectively replaces one-half of the joystick's pot (the section between one ear and the nose), while a new potentiometer fills in for the other half of the joystick's pot.

Solder the points labeled "A," "B," and "C" to the pads on the circuit board previously occupied by same terminals on the joystick pot you removed. The photoresistor changes resistance as light level fluctuates (you know this already). The pot in this configuration sets the operating range of the voltage divider. Start with a 1 megOhm

Figure 28.13
Joystick removed from right side of board.

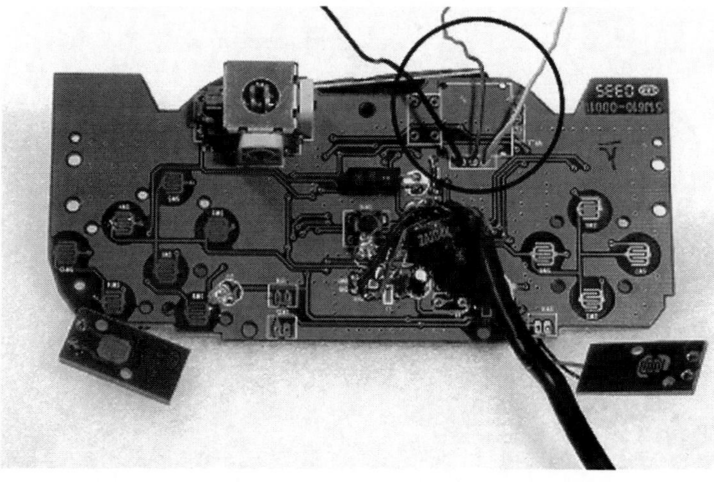

Figure 28.14
Wires attached to joystick solder pads.

pot. Run the test program you used to test the game pad earlier, and vary the light on the photoresistor. Adjust the pot until light variation produces a similar range of values in your program that the corresponding axis of the original joystick did. You may need to substitute a larger or smaller pot for the 1 megOhm one. When you find a setting of the pot that gives you a good range of response you can remove it from the circuit, measure the resistance between the two terminals used (with a multimeter), and substitute a fixed resistor. Alternatively, you can keep the pot in circuit, so you can trim the range to suit different light conditions; you can also buy a small, screwdriver-adjusted "trim pot" that gives you both adjustability and compactness. (For a nice example of photoresistors hacked on to a game controller see Kyle Evans' video on the DVD.)

To make a squeezable instrument substitute one of our homemade pressure sensors from Chapter 15 for the photoresistor in the same basic configuration (see Figure 28.16). You can also use this approach to interface electrodes for direct skin control (see Figure 28.17)—a Computer Cracklebox. As with the light sensor, adjust the pot to set the range; once again you can replace the pot with a fixed value resistor or trim pot (if desired) after determining its value experimentally.

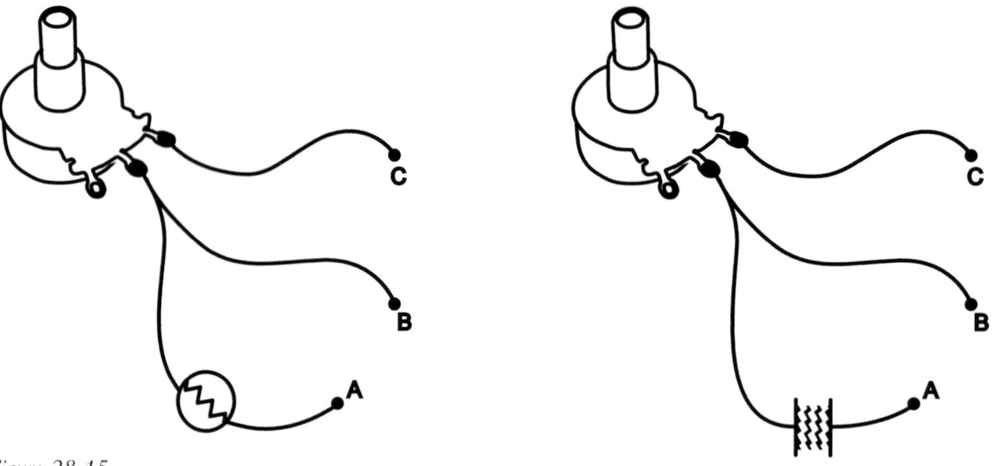

Figure 28.15
Photoresistor voltage divider.

Figure 28.16
Pressure-sensor voltage divider.

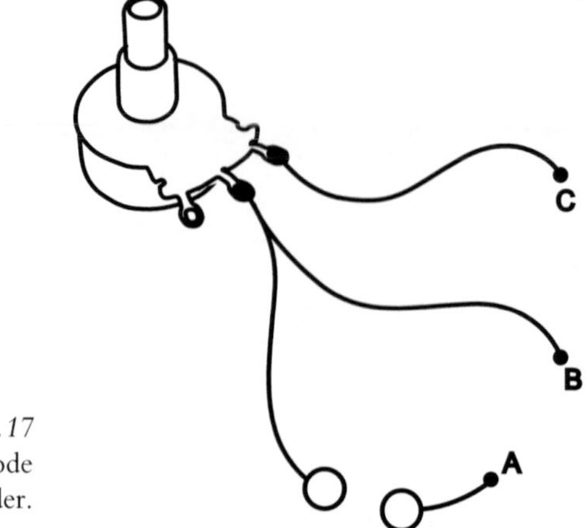

Figure 28.17
Electrode voltage divider.

Figure 28.18
A hacked wireless game pad, Nicolas Collins.

Once you've tried various sensor substitutes for the original joystick, the next task is to design an ergonometrically appropriate mechanical configuration for them—i.e. a physical instrument. If you've ever played air guitar on a tennis racket, now is your time to shine. On the other hand, if you're indecisive, slap a bunch of Velcro onto some switches, pots, pressure pads, photoresistors, and add chunks of Styrofoam or wood—voilá—musical Lego! Figure 28.18 shows a stretched wireless game pad, for example: the linear rearrangement of switches and joysticks makes it easier to play like a woodwind.

BEEP BEEP

Feeling boldly adventurous now? You can connect the center ("nose") of a joystick pad to the output of any sensor or circuit that generates its own output voltage, as long as it falls within the voltage range of the game pad: use a multimeter to measure the voltage between the ground pad and the "+" voltage present at the pad of the other ear of the removed pot—that voltage is the upper limit for your interface.

A Web search under "sensors" or "robotics" will turn up a dazzling array of gizmos capable of measuring temperature, humidity, distance, weight, air pressure, compass direction, acceleration, gas mixture—you name it. Air pressure sensors make great breath controls for digital wind instruments: the tea box in Figure 28.19 combines a breath control (blow into the lady's ear), a few slide pots, and 16 switches to make a sort of usb-melodica. Chikashi Miyama and Shingo Inao describe their "Qgo" as a kind of "invisible bandoneon," using accelerometers, distance sensors, and switches to translate hand motion into data for a Max/MSP program (see Figure 28.20).

Accelerometers can be used to map the movement of hands and bodies. Many pretty sophisticated sensors are surprisingly cheap. In the early days of homemade electronic instruments, all our components seemed to be guilt-laden byproducts of the dreaded military-industrial complex. Now, we lap up trickle-down from the automobile industry:

Figure 28.19 Wind controller built by Nicolas Collins from a USB interface, with four slide pots, sixteen switches, and an air pressure sensor, housed in a tea box.

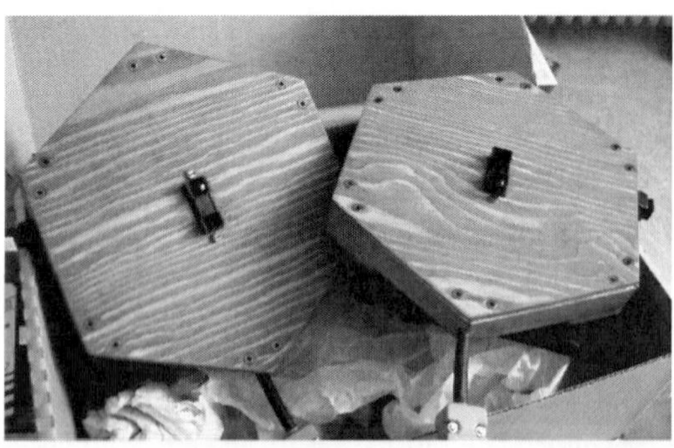

Figure 28.20 "Qgo" bandoneon controller by Chikashi Miyama and Shingo Inao, with one infrared sensor, one 2D accelerometer, and two buttons for each hand, using a Basic Stamp as the embedded controller.

pressure sensors control fuel injection systems, accelerometers trigger airbags, compasses keep us on the straight and narrow.

Look for sensors described as having a "voltage output" (as opposed to serial digital data or timed pulses) within the range we measured above. Download the documentation before you buy anything. You'll typically have to provide power (a battery will usually do, but some sensors are fussy about requiring 5 volts, which means a 9-volt battery will fry them), as well as a ground connection between the sensor and the game pad. Wire the sensor's output to the middle pad of one of the joystick's pots. With luck you'll be in the right range; if not, it's probably easier to trim the data values in software rather than fiddling with resistor values.

The cutting edge of gaming interface technology at the time of writing the second edition of this book is the Nintendo Wii. Highly hackable, laden with excellent sensors (especially accelerometers), and wireless (using Bluetooth), the Wii has generated a cult following amongst digital instrument designers, as a quick Google search will show. A lot of clever programming has already gone into the interpretation of sensor data within the controller to make it respond realistically when used as a tennis racket, guitar, or whatever, which gives you an excellent head start when designing a musical controller with gestural nuance (the realism and speed of response is also the reason the Wii has succeeded where so many other alternative game controllers, most notably the Mattel "Power Glove," have failed). That said, the lower cost and simpler circuitry of traditional game pads and joysticks favor them for the first experiments by the less skilled hacker, before you move on to the Wii.

One final oddball suggestion for a performance interface. An optical mouse contains a video camera and digital image processor integrated onto a custom chip. The camera is focused very close up, on your tabletop or mouse pad; the image processor compares sequential frames and calculates how far you have gone and in what direction; this *x-y* data is sent down the USB cable. If you mount a lens in a toilet paper tube and tape it to the bottom of the mouse you can shift the focus from desktop to across the room (you'll need to experiment with the lens strength and placement). Hang the mouse on the wall, walk and jump in front of it and your cursor should move around your screen. Now add a little USB-handling code and you've got yourself a $5- motion tracker.

And remember, you needn't use your new hybrid instruments exclusively to control sound: digital data is digital data, and can just as easily be used to manipulate video, control the speed of motors, or steer your RC car. It's only a question of software.

CHAPTER 29

Power Supplies: Carbon Footprints from AA to EEE

You will need:

- A DC "wall-wart" plug-in power supply.
- A diode, 1N4004 or equivalent.
- Some capacitors: 0.1, 100 uf, and as big as possible.
- A fixed-voltage positive regulator (LM7809 or LM7812).
- An LM317 variable voltage positive regulator.
- An 5 kOhm pot and assorted resistors.
- Hand tools, test meter, clip leads and soldering iron.
- A breadboard on which to test your circuit.
- A solderable circuit board on which to assemble your circuit.
- Some solid and stranded hookup wire.
- A few solar cells.
- Some fruits and vegetables and copper and galvanized nails.
- Spare change, paper towel, salt, water.
- A bicycle dynamo or small DC motor.

Oh dear, I feel like a father enrolling his son in a driver education class or explaining safe sex: I wish we could stop here, but one day you will leave home and must be prepared for the big world.

Although the Second Rule of Hacking barred you from touching an AC power cord, the time will come when batteries simply will not suffice. You will tire of the cost of replacing them, and the accompanying environmental guilt (although these concerns can be minimized by rechargeable batteries); or you will build a circuit that draws so much current that it drains the battery flat before you can say "Union Carbide." You will have to choose between living a long and virtuous (but possibly slightly constrained) life that adheres strictly to the 25-fold path of the Rules of Hacking, or enjoying the risky heretic pleasures of power supplies. As a stopgap solution to exposing your hands (and heart) to lethal voltages, you can advance to the ubiquitous "wall-wart" that powers so many domestic appliances these days (see Figure 29.1).

The power grid in North America delivers to your outlet a sine wave that fluctuates between 0 and 120 volts 60 times per second (in Europe, 240 volts at 50 cycles per second, in Japan 100 volts at 50 cycles). If you plugged a very strong speaker directly

Figure 29.1
A typical wall-wart power supply.

into the wall (not recommended, by the way), you would hear a loud, low pitch around 2 octaves below middle C. The wall-wart consists of a transformer encased in plastic and wired directly to an AC plug. The transformer takes the 100/120/240 volts of alternating current (AC) and steps it down to the non-lethal range suitable for powering electronic circuitry. The advantage of the wall-wart supply is that the dangerous voltages remain (in theory) within the plastic lump, and the ends of the wires present a mild, relatively safe, more battery-like voltage. The traditional power supply found inside your TV or guitar amplifier, on the other hand, brings the wall voltage right into the chassis, where it can easily be touched (ouch!) as you tinker. So, if you *must* use the electrical grid, let the wall-wart be your condom.

You can find wall-warts everywhere, often very cheap or even free (left over from old answering machines, for example). Since the physical con-struction of the thing is an indication of the quality of workmanship inside, you are advised to select one that looks pretty solid, with no holes, cracks or wobbly parts (just keep our French Letter analogy in mind—without smirking, please).

There are two basic types of wall-wart. An *AC* wall-wart consists solely of a step-down transformer; it puts out a low voltage 60 or 50 hz signal, which must be further conditioned to make it useful for powering circuitry. A *DC* wall-wart contains some additional circuitry (a few diodes and a big capacitor, to be specific) required to smooth out the fluctuating signal into a voltage that more closely resembles the steady DC output of a battery.

The wall-wart should be marked with the following information:

- The *primary voltage* that the wall-wart can be plugged into, i.e. 120 (North America) or 240 (Europe) volts AC (VAC). Some modern, "premium" wall-warts (such as those provided with most laptop computers) handle any input voltage between 100 and 240 volts. Choose the appropriate primary voltage for the country in which you are working.
- The *secondary voltage* that appears at the loose end of the long dangling wire, usually in the range of 3–24 volts. Some wall-warts include a switch for selecting amongst different output voltages. We need a secondary voltage between 5–15 volts.

- Whether it has an *AC* or *DC output*. We need DC.
- The amount of power the transformer can provide, usually measured in *watts* (W), *amps* (A or MA), or *volt-amps* (VA). We want a transformer that puts out a minimum of 100 milliamps (ma), which may be indicated as 0.1 amps or 5–10 watts.

For example, a wall-wart might be labeled "120 vac input, 12 vdc output, 200 ma."

Built into the plastic casing should be a power plug appropriate to the outlets in your country. In some instances the wall-wart is more like a rug-rat, with a power cord instead of a built-in plug; sometimes this cord is attached to the plastic lump, and sometimes it plugs into a socket of some sort. Two wires should emerge from the other side of the wart; they may be a parallel pair, like speaker cable or lightweight lamp cord, or may be shielded cable, with one conductor wrapped around the other. At the end of this cord will be one of a zillion different types of minimally distinguishable little plugs—usually some form of tortiglioni-like "coaxial power connector." Some cables terminate in a system of interchangeable connectors of different style. It's convenient if the plug matches some loose female jack you have in your parts collection, or one you can buy locally. It's risky to buy one part or the other sight unseen: never trust measurements on packaging, always test that the male and female connectors fit together snugly. On the other hand, matching connectors are not essential: you can always cut off the plug and connect the bare wires directly to your circuit, or solder on a new plug from a matching plug and jack set you acquire elsewhere.

When adapting a battery-powered circuit to be powered by a wall-wart you must observe two critical factors:

- The wall-wart's voltage must be within the two limits we described above (i.e. greater than 5 volts but less than 15 volts), but the current capability can be anything *higher* than the minimum need to power the circuit. For example, a circuit requiring 20 ma can be powered by a supply producing 20 ma, 100 ma, or 1,000 ma. The circuit will only draw as much current as it needs (sort of like having an oversized gas tank in the Subaru you only use to go shopping around the corner).
- Check the polarity of the secondary connector or wires: you must know which is "+" and which is "–" before connecting the wall-wart to your circuit, or catastrophic pyrotechnics may result.

Sometimes the wall-wart will be marked with information specifying which part of the connector is "+" and which is "–." Sometimes one of the cables will have a convenient white stripe that distinguishes it from its mate. But it is always safer to confirm the polarity with a multimeter.

Set the meter to measure "DC Voltage." One probe plugs into the meter's "ground" or "–" input, while the other connects to something probably marked "voltage" in red. Touch one probe to one part of the connector, and the other to the other; if you've already removed the connector measure between the ends of the stripped wires. If the meter reads out a voltage with no prefix (i.e. "13.6"), then the wire/connector touching the ground probe is the "–" output, and the other is the "+." But if the meter puts a "–" before the number (i.e. "– 13.6"), you know that the connections are reversed, that the wire touching the minus probe of the meter is actually the "+" output and the other is "–." Confused? Try this test on a nice familiar 9-volt battery. Then test the wall-wart again.

Once you figure out which wire is which, mark them carefully: scribble a drawing of the connector with appropriate + and – markings, or wrap some colorful tape around the + wire. If the output voltage of your wall-wart measures *less* than 15 volts DC you can connect it more-or-less directly to your circuit as shown in Figure 29.2. You can either find a matching connector to whatever plug is attached to the wall-wart's cord, or you can cut off the plug and solder the wires directly to the board. Check the polarity one more time before you plug it in! As a safety precaution against frying your circuit with a backwards power supply, you can connect a diode across the power supply as shown in the figure.

Cheap wall-warts will usually have some "AC ripple" in their voltage output—a sign of skimping on parts in the conversion from AC to DC (you usually don't get what you don't pay for, I'm afraid). A circuit powered with such a supply may hum—under its breath or quite loudly, depending on the quality of the wart. If you alternate between a battery and wall-wart powering the same circuit you should be able to hear the difference. Until you get good at designing or choosing better power supplies, a battery will usually sound cleaner. One easy fix that usually helps is to add a big capacitor, say between 100–10,000 uF, between "+" and "–" supply on your circuit board. As mentioned in Chapter 18, big electrolytic capacitors have polarity, like a battery, which is marked on the body. Make sure you connect "–" to the ground bus, "+" to "+"

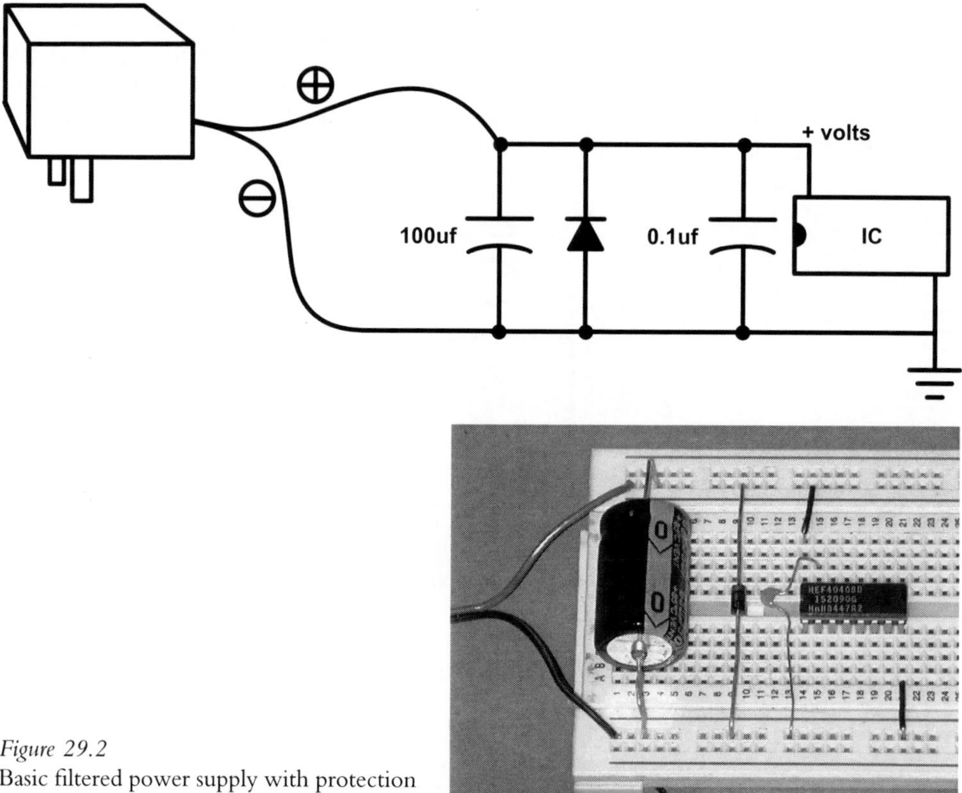

Figure 29.2
Basic filtered power supply with protection diode.

supply. Placing an additional 0.1 uF capacitor between the "+" and ground supply pins very near each chip also helps lower noise and reduce "crosstalk" between different parts of your circuit. Both these capacitors can be seen in Figure 29.2.

A quick measurement with a meter will show that even though a wall-wart might be marked "12 volts DC" in bright white letters, it could put out anything from 10 to 20 volts. The CMOS chips used in most of our circuits were chosen in part for their forgiving nature, but they have their limits—upper limits: they can run on power supplies from 3 volts to about 18 volts, but above 18 volts they can expire quite dramatically. A 9-volt battery sits comfortably between these two extremes. If you choose to use a wall-wart instead of a battery always measure its actual output voltage and polarity *before* connecting your circuit—don't rely on the markings on its case.

Rule #25: Never trust the writing on the wall-wart.

Alternatively, for a really quiet, reliable supply you can add a simple integrated circuit called a "voltage regulator" (see Figure 29.3). A regulator filters out the last of the AC ripple and sets the voltage to a precise level, which is specified by the last two digits of the chip's part number: 7812 = 12 volts, 7809 = 9 volts, 7805 = 5 volts, etc.

The input to the regulator must be at least 3 volts higher than it is expected to put out (i.e. 12 volts in for 9 volts out), and no higher than about 25 volts; so measure the output of your wall-wart with your meter to make sure it is within these limits. You can get regulators for a wide range of output voltages—the 7809 is a shoe-in for the 9-volt batteries we've been using, but the 7812 and 7805 are more commonly encountered at retailers, and our CMOS chips should run fine on any of these three. The basic design shown in Figure 29.4 works for any 78XX series regulator. It's normal for the regulator to get a bit hot—bolt the regulator's tab to a piece of metal to dissipate the heat (or you can buy yourself a fancy finned heat sink).

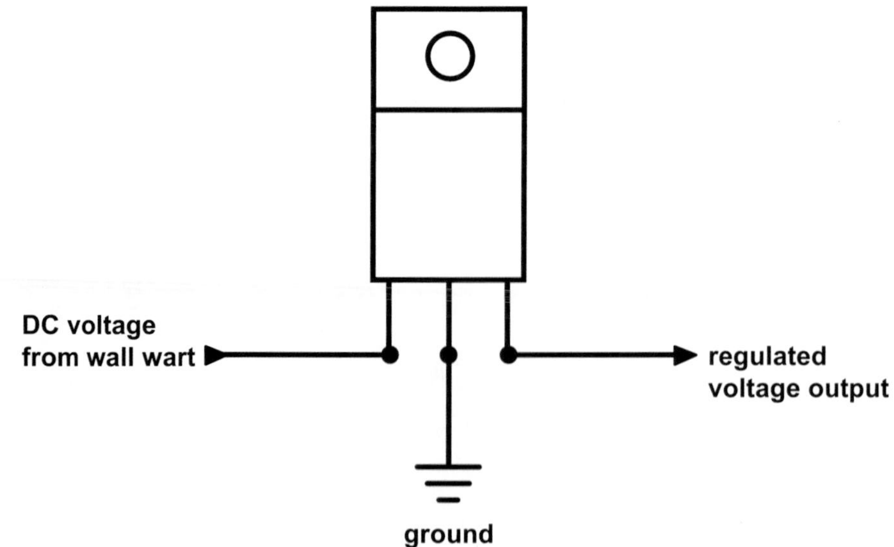

Figure 29.3 78XX-series voltage regulator pinout.

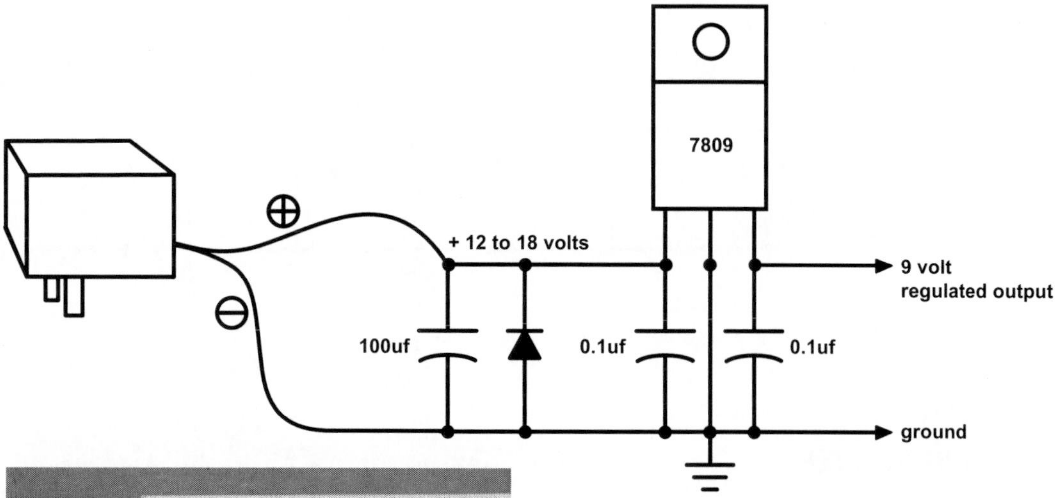

7809
+ 12 to 18 volts
9 volt
regulated output
100uf
0.1uf
0.1uf
ground

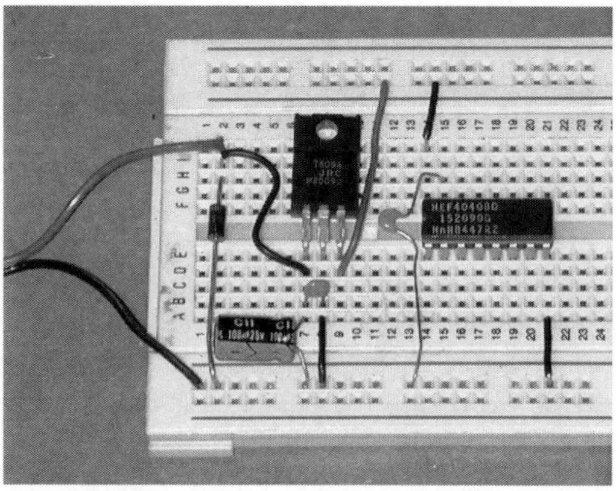

Figure 29.4
A simple regulated power supply.

If you have had success with the voltage starve circuit we described in Chapter 20 (Figure 20.14) you can implement it with a wall-wart (with or without regulator) feeding the pot just the same as a battery. Or you can build a power supply whose output voltage can be adjusted quite precisely, using a slightly different regulator: the LM317T (see pinout in Figure 29.5). The potentiometer in the circuit shown in Figure 29.6 lets you adjust the voltage anywhere between about 1.2 volts and almost the voltage coming out of the wall-wart. The LM317 looks similar to the 78XX regulators, with three legs, but the connections to the pins are different, so double-check your wiring before plugging in the wall-wart.

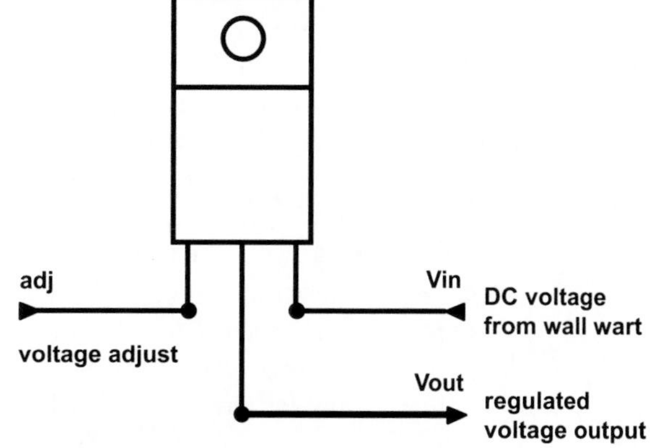

Figure 29.5
LM317 voltage regulator pinout.

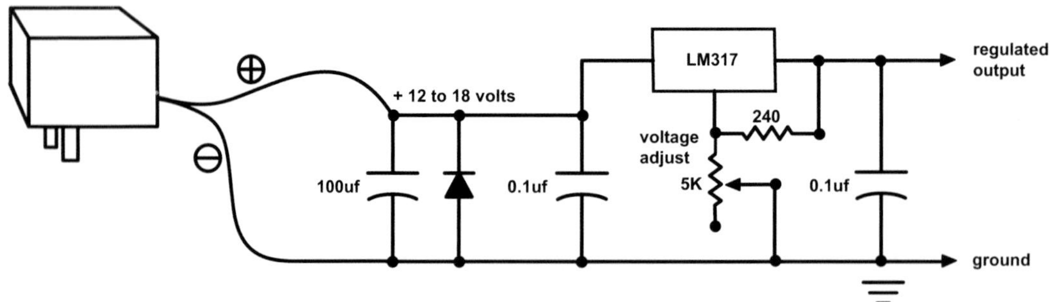

Figure 29.6 Variable voltage power supply with LM317.

GOING GREEN

If you feel like kicking your Duracell habit, but also want to stay off the grid, consider some alternative power suggestions that will reduce your carbon footprint to smaller than the business end of a stiletto heel. The obvious first step, as I suggested way back in Chapter 1, is to invest in some rechargeable batteries in lieu of the disposable kind. The newer NiMH (Nickel-Metal Hydride) batteries are much more efficient than any alkaline cell, packing more current (i.e. run time of your circuit) into a lighter package. They are available in many configurations, from AAA to D, as well as 9 volt. You can charge them quickly from any AC outlet, or invest in a solar charging unit.

Speaking of solar, there's no reason—aside from cloud coverage—that you can't power your circuits directly off old Sol. Solar power panels are pretty abundant these days, and they are decreasing in price as they increase in power. You can sometimes buy one that substitutes directly for a battery or wall-wart, providing anything from 5–28 volts at 500 milliamps or more of current (check out marine supply stores). But they can be a bit pricey. A more economical (and educational) alternative is to seek out—usually online—the individual cells that make up these panels. The cells are rated for voltage and current output. They have a pretty low output voltage—usually around 1.5 volts—but come in a wide range of current ratings, typically a function of the size (see Figure 29.7). Sometimes the cells are combined to add up to higher voltage. Pick a model whose *current* is sufficient for your circuit needs—100 milliamps (ma) is sufficient for most of our CMOS circuits, 500 ma if you're powering a small amplifier, such as the LM386 from Chapter 28, as well. Then buy enough to add up the voltage you need for your circuit. Bearing in mind that the CMOS circuits we've been powering on a 9-volt battery can run on voltages as low as 3 volts, you may be able to get away with only a handful of cells (see Figure 29.8).

A critical component in Figure 29.8 is the capacitor. Remember how, in Chapter 18, I compared the capacitor to a bucket? In this circuit the capacitor is your canteen in the desert, tiding you over between oases (or a cistern collecting rainwater so you can continue to shower after the storm has passed). The capacitor acts a little bit like a rechargeable battery whose strength is proportional to its size in uf. Without it the circuit responds directly to the amount of light present: when the sun shines the circuit runs; when a cloud passes the circuit shuts off. If we insert a medium-size capacitor, say

Figure 29.7 Assorted solar panels (center unit with rechargeable batteries attached).

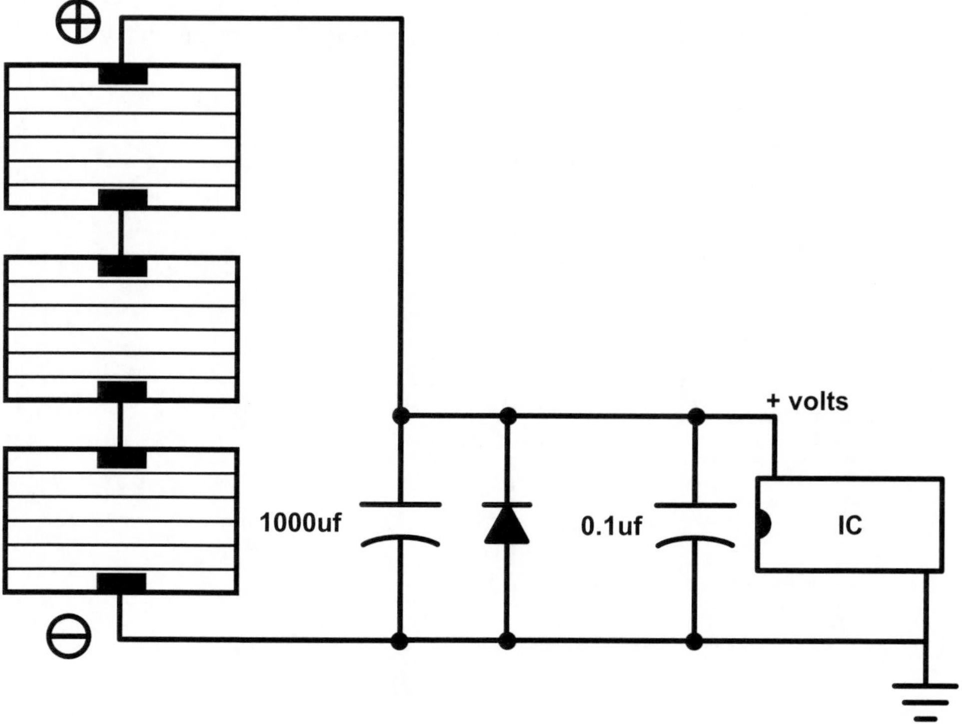

Figure 29.8 Direct solar power of a circuit by adding voltage output from multiple solar cells.

1,000 uf, it will retain enough energy to bridge the gap caused by a scudding cloud, but not enough to keep it running overnight. If we get an insanely large capacitor, say around 1 farad (not micro-farad), such as are used in some memory backup applications, it holds the charge much longer, acting much more like a rechargeable battery. Finally, we can substitute actual rechargeable batteries (usually a set of 1.5-volt cells stacked to add up to your desired voltage) for the capacitor, and modify our design slightly, as shown in Figure 29.9. At this point whatever we hang off this circuit is really being powered by the battery, not the sun, and the battery is in turn being "trickle charged" by the sun, weather permitting. Be sure that the series voltage of your photocells is *greater* than the battery voltage, or the batteries won't charge, and *less* than the maximum voltage for the circuit (i.e. 15 volts), or the chip might fry.

As on a few other occasions in this book, I should explain that Figure 29.9 is a simplified design. Proper battery chargers can get quite complex—these days many incorporate microprocessors that regulate the charging current and monitor the cell's condition to minimize the charge time and maximize the battery's lifetime. If you are serious about the long-term survival and performance of your project you should consider either springing for a commercially available solar-powered battery charger, or type those key words into Google and spend a few minutes downloading schematics by people who know more about this arcania than I do.

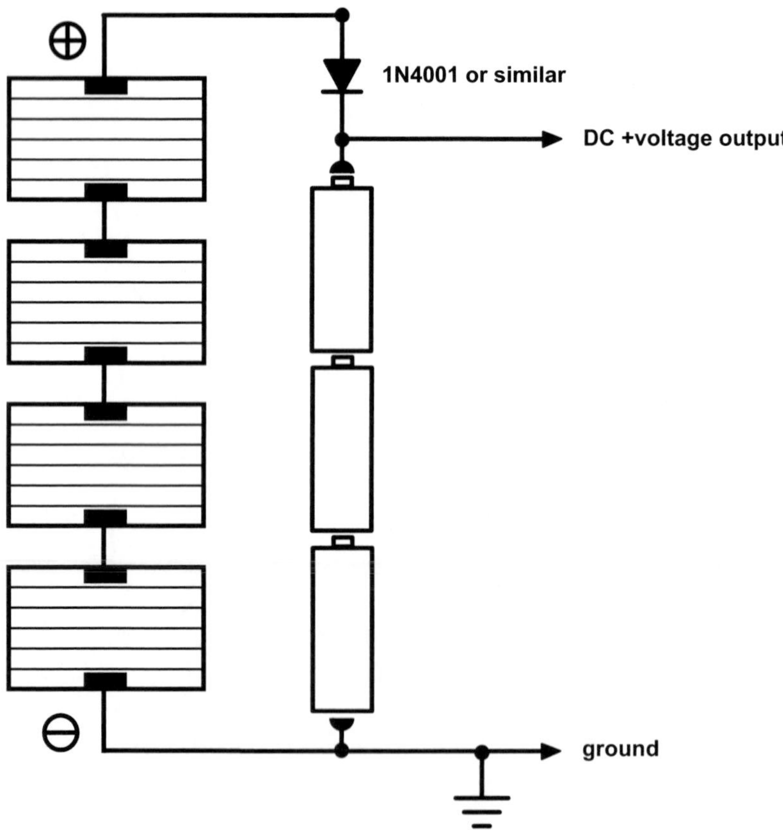

Figure 29.9 Solar-powered battery charging circuit.

MY FIRST BIO-FUEL

Those of you with dim memories of grade school science fair projects with potatoes and nails and light bulbs might be wondering about the possibility of building your own battery with the contents of your crisper bin. Not only can it be done, but with proper lobbying you might convince Congress to subsidize your efforts to free American electronic music from the shackles of foreign oil.

The potato battery functions on the same basic principle as a traditional battery: when two strips of different metals (typically copper and zinc) are inserted into an acid solution (in this case the moisture inside the potato) an electrochemical reaction takes place, which generates a potential difference between the metal strips. A pair of electrodes (as the metal strips are called) inserted into a potato will generate around 1 volt at a very small current. Individual bio-cells can then be added in series and parallel to produce higher voltages and more current, respectively.

A copper penny or strip of copper wire can serve as your positive electrode, and a galvanized nail for the negative one. Insert them into a potato, lemon, or apple, and measure the voltage with a multimeter. Put several electrode pairs in series as shown in Figure 29.10—you can use a separate vegetable or fruit for each pair, or insert multiple electrodes in a single large tuber. Make sure none of the electrodes touch each other. A quick check with the meter should confirm that your power station is putting out around one volt per potato. You can connect an LED and, if the orientation is right, it should light up. Hook up your new battery to a simple oscillator circuit and see if it runs. If it doesn't you might need to add a few more potatoes in *series* to increase the *voltage*, or in *parallel* if your circuit requires more *current*.

The voltage output of this kind of battery is a function of the metals used for the electrodes, rather than your choice of produce: copper and zinc (present in galvanized hardware) yield about 1.1 volts per cell. Substituting magnesium for the zinc increases this to about 1.6 volts (I'm not sure what the most convenient source of magnesium

Figure 29.10 Adding multiple potatoes in series to increase voltage.

scrap would be). The lifetime of your battery is primarily a function of how long it retains its moisture, so you might experiment with more succulent options, such as lemons, grapefruit, etc. And do bear in mind that these earthy batteries put out a very small amount of current—if you're lucky our wonderfully low-power CMOS oscillators will whistle comfortably on potato juice, but don't expect to run an amplifier or motor this way.

You may remember that earlier, in Chapters 15 and 18, we discussed using bits of vegetables and fruits as alternative resistors in toy clocks and oscillators (see also Figure 30.27). The resistive character of these materials is primarily a function of their moisture content, although we can speculate now that their electrolytic interaction with the wires we used to make the connections might also influence the behavior of our circuits: two wires of the same metal (i.e. copper) should not induce voltage, but different metals can. You might experiment with the contents of your garden for both a power supply and a timing component in your circuits. Too bad we can't grow our own chips . . . . (David Tudor once told me a story of a Brazilian electronic-music composer who, in the early 1960s, was making his own transistors by baking crystals in his oven but, sadly, I didn't think to follow up this lead before Tudor passed away.)

One last observation: the voltage output of these bio-batteries is not as steady as your basic Duracell. In fact, the chemical process at play generates a fair amount of noise—a capacitor added to our power supply circuit should smooth this out if it's problematical. On the other hand, if instead of connecting the electrodes to a circuit or LED you wire them to the tip and sleeve of your headphones, you can *listen* directly to the crackly sound of a singing potato. Not loud enough? Connect the electrodes to the input of your amp or mixer, and turn it up.

KITCHEN SINK

As long as you're rummaging about in your refrigerator for alternatives to Union Carbide, you might scan your kitchen for some paper towel and salt. As Cy Tymony has shown in his wonderful book, *Sneaky Uses For Everyday Things*, you can build potato-like batteries from nothing more than spare change, wet paper towel, and table salt. Each cell requires one penny and one nickel, for the copper and zinc electrodes respectively (sorry, Euro-toting hackers, you're out of luck). You can solder a wire to each. Fold a piece of paper towel into a coin-sized pad, dampen it, sprinkle it with salt, and sandwich it between your coins. Measure the voltage between the two electrodes and it should be in the potato range. The clever thing about this style of homemade battery is that you can just stack your change in a folded strip of paper towel to build up whatever voltage you want, rather than digging up a spud per volt. Wire cells in parallel to increase the current capacity.

HAND JOBS

For the ultimate in self-sufficiency consider combining alternative energy resources with a physical fitness regimen. A *generator* is a device to convert repetitive mechanical motion into electricity. Most function electromagnetically, and bear the same relationship to an electric motor as a microphone has to a speaker: spinning a shaft rotates a coil in a field of fixed magnets, inducing a current flow. Traditional power plants use coal-, oil-, or gas-powered turbines to spin these generators; hydroelectric dams channel water past turbine blades; wind farms are popping up all over. Our Dutch hackers will immediately think of the whirring little dynamo on the front wheel that powers the bicycle headlight—these can easily be adapted to stationary exercise bikes, skateboards, egg beaters, spinning wheels, fishing reels, water wheels, windmills, steam engines, etc. Whether a response to soaring energy prices, or merely a green affectation, generators are showing up in a number of common consumer products: flashlights that light when you squeeze or shake them, radios and laptops with hand-cranks, etc. The generator element can be removed from any of these devices and re-wired to power your circuit.

As you might infer from the interchangeability of speakers and microphones, any DC motor can be used as a generator: connect the motor terminals to an LED, spin the shaft, and see the light (see Figure 29.11).

As with our solar circuits, the critical factor is smoothing the erratic output of the generator so that the circuit doesn't shut down when you park the bike or the wind dies down. The solutions are basically the same as those shown elsewhere in this chapter: add a big capacitor, or configure the generator as a charger for a rechargeable battery. Then again, if you had interesting results from the voltage starve circuit you may find that a relatively unfiltered generator makes a good "performable power supply." Ithai Benjamin's and Alejandro Abreu's "Synthinetic" (see Figure 29.12 and their video on the DVD) and Phil Archer's "Music Boxes" (see Figure 30.6 in the next chapter and his video on the DVD) are lovely examples of this type of power supply in action.

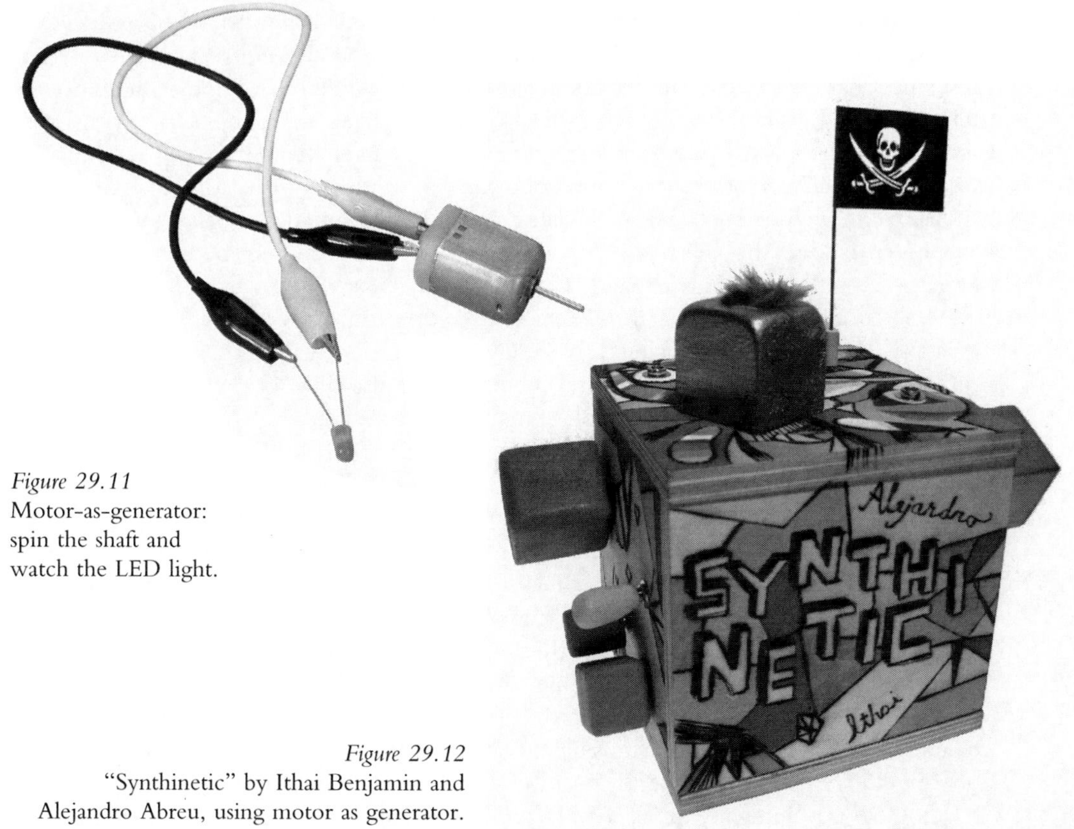

Figure 29.11
Motor-as-generator: spin the shaft and watch the LED light.

Figure 29.12
"Synthinetic" by Ithai Benjamin and Alejandro Abreu, using motor as generator.

CHAPTER 30

The Future is Now

In this, the second edition of *Handmade Electronic Music*, I have attempted to correct errors and oversights; to satisfy the desire, among readers of the first edition, for further projects; and to respond to how the world has changed since 2006. When I began developing the materials for this book back in 2002, most younger artists working with technology were focused on computers and the multitude of programs available at the time. The contrarian remainder was under the thrall of Circuit Bending. Accordingly, I took basic Bending as the center of my workshops and text, and expanded it backwards and forwards, so to speak. I started with alternate ways of *listening*: making and using contact mikes, coils, tape heads, electret elements, and other alternative microphones, and playing radio circuit boards with naked flesh as a way to sensitize our hands to circuitry. Then, when it seemed that my students, like children at Christmas, would burst if they couldn't open up their toys RIGHT NOW, we moved on to hacking clocks, shorting circuits, and indulging in the other heady, pseudo-random practices that fell under the rubric of Circuit Bending. When boredom set in or students tired of subsidizing toy stores and Goodwill shops, we moved on to our first oscillator. From there our silicon future unfolded like a gnomen, and as the years passed I was pressed to keep adding to the collection of circuits—as reflected in the expanded content of this text.

Subsequent to the publication of the first edition in 2006 I came into contact with a vast sea of benders and hackers. Every few weeks I received a request for assistance on a recalcitrant project, or someone would send me a link to a YouTube clip or a Web site demonstrating some cool thing whose connection to my book they felt compelled to point out. My correspondents all share a "whatever it takes" attitude toward technology. This new wave of hackers is obsessed with neither technology nor methodology, but moves with grace from hardware to software, from soldering to sawing, from audio to video, from stage to gallery, from adaptive re-use to spontaneous creation, from happenstance to intent, from idiocy to genius. If there is any identifying trait, it is the desire to disrupt technology's seemingly perfect inviolability.

As I said years and pages ago, in the Introduction to the first edition of this book, the sticker claiming "no user serviceable parts inside"—whether affixed to a toy, a TV, a piece of software, a computer, a motor, or even a single integrated circuit—should be taken as a challenge. And for that challenge we must all—hacker and non-hacker alike—be grateful: today's "breakers" become tomorrow's "makers" (to use my son's terminology). Behind any goofy YouTube video might lurk the next Steve Jobs, Pierre

Omidyar, or Sergey Brin, as likely as the next John Cage, Jasper Johns, or Nam June Paik. Like that of snails or finches, technology's evolution seems to follow the pattern of "punctured equilibrium," and our hackers are holding the pointed sticks. And while we wait for the NEXT BIG THING we can enjoy the delicious din of all the now little things.

The DVD includes a gallery of 87 very now things, each squeezed into a 60-second package. Taken together they provide a reasonable overview of what is happening at the moment of this writing. The diversity of activity taking place makes it hard to categorize in a meaningful way, but seven threads seem to weave through the disk: Beyond Bending—the state of Circuit Bending today; Feedback—the guitarist's friend grows up; Off The Grid—abandoning batteries; I'm With The Band—making music together; Sound and Vision—an art school style; Mechanics—getting very physical; Swashbucklers—some virtuosic iconoclasts.

BEYOND BENDING

Circuit Bending has changed since Reed Ghazala coined the term. One factor has been toy technology's shift toward greater integration of functions onto a single chip. At the end of the last century, control of a toy's various functions (making sound, blinking lights, reading switches, defining the clock speed, etc.) was typically distributed amongst several different integrated circuits and associated components, and benders delighted in messing around with the myriad connections between those components. Now integration has reached the point that everything is controlled by a single malevolent-looking black blob. There are no exposed connections to rearrange. And with more and more on-chip clocks, which are not dependent on external resistors, not even the most basic changing-the-clock-speed-bend is possible. Today's bender is condemned to wander the aisles of thrift shops, garage sales, and eBay, continually confronted with the same stock of cast-off toys—while the new models, with their seductively novel sounds, evade corruption.

The frustration this causes is often compounded by ennui. Bending celebrates the first rule of hacking (ignorance is bliss). It shuns theory (Ghazala's Web site is aptly named "anti-theory.com"), and encourages instead the sharing of empirical observations: "insert a jumper between these points and this will happen, don't worry about why; or try anything else and let me know if it works." In contrast to the laborious analytical work that had previously accompanied most electronic engineering, even in hobbyist and musical circles, this philosophy is tremendously liberating for the first-time hacker. But after the thrill of "how" wears off, some of us ask "why?" Accordingly, many younger artists gain access to circuitry through classic bending activities, but then move on to diversify their electronic portfolio: interconnecting toys, combining handmade circuitry with bent toys, hacking other found technology (effect pedals, video circuits, mechanical devices), writing software, etc.

Rather than bending toys, for example, Neal Spowage (UK) built his "Wands," a responsive performance instrument, from metal detectors and security wands (see Figure 30.1). Kaspar König's (Netherlands) "Musguitear" uses hacked mosquito killers. Chris Powers (USA) has modified guitar effect pedals, bringing touch-sensitive contact points from the circuit board out to a set of electrodes he plays with his bare toes

Figure 30.1 "Electro Magnetic Wands," Neal Spowage.

(see Figure 15.14 in Chapter 15). Charles McGhee Hassrick (USA) takes a defiantly ecological approach, and creates audio-visual installation projects using only the discarded trash of the museums, galleries, and other venues where he exhibits.

Many artists extend their bends with computers or scratch-built circuitry. For his "After Math" Chester Udell (USA) hacked a TI "Speak and Math"—one of the most popular bending toys—so that it could be controlled by flight simulator controllers through a computer and MIDI (see Figure 30.2). Ian Baxter (UK) fabricated external body contacts to control a child's voice-changer toy for a performance instrument he calls "The Masher." Japanese artist Haco modified an electronic organ kit to be controlled by a combination of pencil lines on paper (as explained in the section on alternate resistors in Chapter 15—see Figure 15.10) and electrodes for body contact.

Several musicians have "bent" electric guitars and other acoustic instruments by embedding circuitry. Jeffrey Byron (USA) installed bent toys in his electric guitar. Zach Lewis (USA) replaced the neck pickup in his Fender Toronado with photoresistor-controlled oscillators, played with shadows cast by the pick-wielding hand (see Figure 30.3). Peter Blasser's (USA) "Radiothizer" resembles a koto, but contains Theremin-esque circuits that transform the sounds of the plucked strings in response to movement of the arms in the spatial field of the instrument. Ben Neill (USA) added switches and pots to his trumpet in order to control a computer music system via MIDI (see Art & Music 11, "The Luthiers" in Chapter 28, and Figure 28.3).

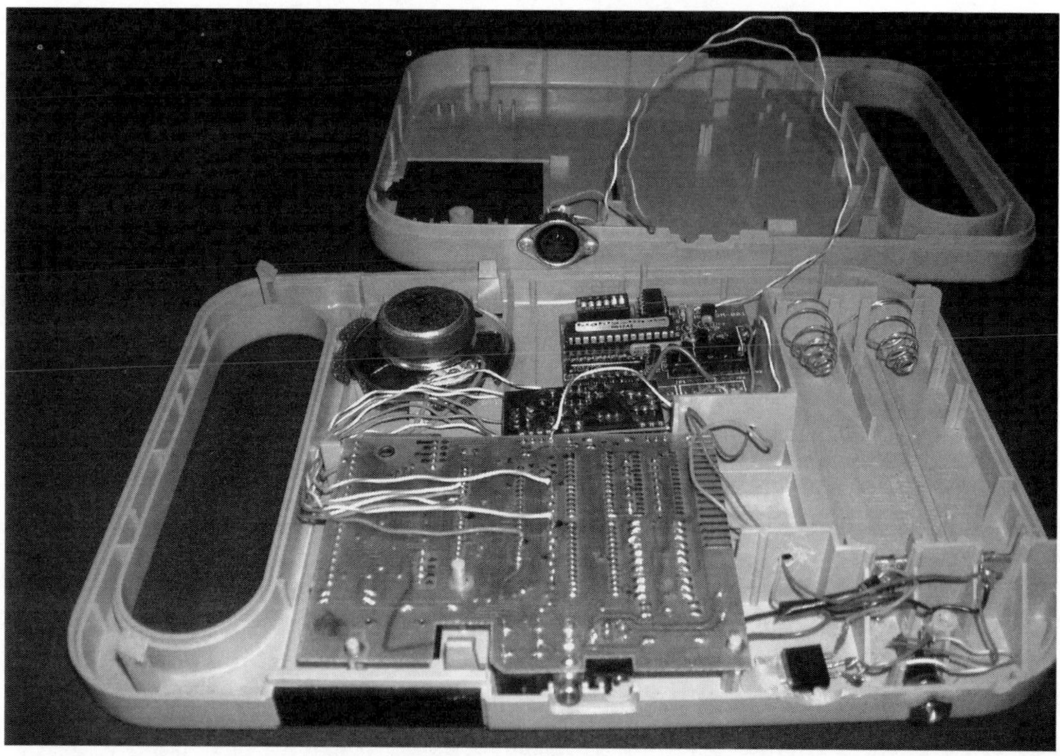

Figure 30.2 "After Math" hacked TI "Speak and Math" showing added MIDI interface board (between speaker and battery terminals), Chester Udell.

Figure 30.3
Oscillators embedded in electric guitar, Zach Lewis.

FEEDBACK

Interest in live performance and interactivity has led many to experiment with one of the most basic electronic sound resources, feedback. For her "Speaker Synth" Lesley Flanigan (USA) places contact mikes inside the cones of five open speakers and controls their feedback (see Figure 30.4). Minoru Sato (Japan) embeds microphones and small speakers at the opposite ends of glass tubes and adjusts the gain to create a feedback organ. For their performance work, "Crude Awakening," Chris Black and Christine White (New Zealand) use an array of speakers and contact mikes to resonate common objects (metal pans, grills, coils of wire, etc), which are physically twisted to affect their feedback pitch (see Figure 8.14 in Art & Music 5 "Drivers" in Chapter 8). Frederick Brummer (Canada) filters feedback through drum heads. Gert-Jan Prins (Netherlands) continues to do amazing things with radio-frequency feedback between homemade transmitters and receivers (see Figure 3.4 in Art & Music 1, "Mortal Coils," in Chapter 3).

Matrix feedback amongst multiple circuits, as pioneered by David Tudor (discussed in Chapter 26), came back into popularity in the 1990s through the work of Toshimaru Nakamura (Japan) and other advocates of "no-input mixing," in which mixers and arrays of circuits are transformed into oscillators by patching cables from outputs to inputs. Vic Rawlings (USA) configures naked circuits and numerous small speakers in complex feedback loops, which he manipulates by rubbing wire brushes across the circuit boards (see Figure 15.19 in Chapter 15).

OFF THE GRID

Well before gasoline topped $4/gallon hackers began experimenting with alternate sources of energy. Both Daniel Schorno (CH/NL) (see Figure 30.5) and Fred Lonberg-Holm (USA) harness solar power for performance instruments and installation projects. Phil Archer's (UK) "Music Boxes" (see Figure 30.6) and Ithai Benjamin's and Alejandro

Figure 30.4
"Speaker Synth,"
Lesley Flanigan.

Figure 30.5 "Desert Scorpio," Daniel Schorno—installation, on solar chargeable battery, Erg Chebbi Desert, Morocco, 2007.

Figure 30.6 "Music Boxes," with hand-cranked dynamo power, Phil Archer.

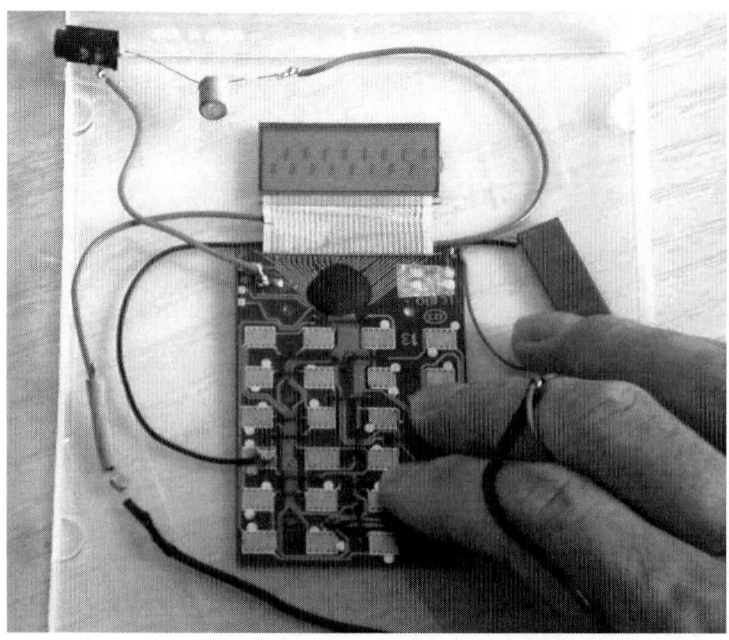

Figure 30.7
Solar power calculator touch synthesizer, Emir Bijukic.

Abreu's "Synthinetic" (see Chapter 29, Figure 29.12) incorporate hand-cranked generators as power sources whose instability becomes a defining characteristic of the circuit's performance. Emir Bijukic (Italy) listens to various points on the circuit board of a solar-powered calculator as he varies the light level and brushes his fingers across the exposed traces (see Figure 30.7). Lorin Edwin Parker's steam-driven synthesizer was described in Chapter 4 (see Figure 4.2). And in a tour de force of engineering, M. R. Duffey (USA) is designing "Solar Thermal Automata" based on plans and descriptions dating back to classical Greece and Rome, which utilize heat engines to generate organ-like sound directly from solar heat, without the need for any electronic circuitry.

I'M WITH THE BAND

Among the many benders and hackers that have formed bands are NotTheSameColor, an Austrian duo (Billy Roisz and Dieter Kovacic) that combines live manipulation of sound and video (see Art & Music 10 "Visual Music" in Chapter 24, and Figure 24.1). Japan's (e)-bombers are a bending big-band: six guys with bent toys and homemade circuitry, each wearing his own PVC-encased speaker system. Oscillatorial Binnage (Toby Clarkson, Chris Weaver, and Dan Wilson), Owl Project (Simon Blackmore and Antony Hall) and P. Sing Cho (Knut Aufermann, Moshi Honen, Sarah Washington, Chris Weaver, and Dan Wilson) are all electronic ensembles active in the UK today. (The Owl Project also developed the "m-Log"; a USB controller interface embedded in a hollowed out log, which can be purchased through their web site.) Grace and Delete (James Dunn and Chris Cundy) mix hacked electronics (including bent keyboards and, appropriately enough, a tinnitus analyzer) with bass clarinet. Stewart Collinson and Duncan Chapman (UK) recently formed the Bent Radio Orchestra, which—building on the grand tradition of the Scratch Orchestra and Portsmouth Sinfonia—invites the public to bring, open, and lay hands upon radios (as demonstrated in Chapter 11) for mass performances. The RGB~Toysband, formed after a workshop I gave in Brussels in 2005, takes electronic music into the streets, train stations, and other public spaces; its members have published a manifesto of sorts encouraging others to create satellites of the original group (similar Hacking bands have sprung up out of workshops in Cuneo and Padova, Italy, and Zurich, Switzerland.)

Finally, several new ensembles make a point of building circuits in front of their audience, rather than bringing completed instruments to the stage. The Swiss Mechatronic Art Society gives regular performances that feature live soldering—once, with all members of the group linked to a central master clock powered by dripping water. (This group also conducts workshops and publishes circuit designs online.) The members of New York-based Loud Objects (Tristan Perich, Kunal Gupta, Katie Shima) program microcontroller chips to produce and process sound, then solder them together on top of an overhead projector in front of the audience, who can watch the connections proliferate on a big screen (see Figure 30.8). My own workshops always end with a public event in which the audience wanders amongst the busy hackers as they assemble and test their last contact mikes, bent toys, and noisy circuits—a sort of factory-floor concert. In my recent piece "Salvage—Guiyu Blues," seven players use test probes to make a dozen connections between a simple circuit of my design and contact points on a dead circuit board (from a computer, cell phone, fax machine, mixer, etc.), reanimating

Figure 30.8
Live soldering performance by Loud Objects, showing overhead projection of work surface.

Figure 30.9 "Salvage—Guiyu Blues," Nicolas Collins—control circuitry, probes, and dead French Telecom circuit waiting to be re-animated.

the dead circuit and transforming it into a complex oscillator (see Figure 30.9). At the NIME Festival in Brooklyn in June 2007 I battled my young British *dopplegänger*, Nick Collins, for the Nic(k) Collins Cup (commissioned from British Potter Nic Collins) in a concert that pitted live SuperCollider programming (by Nick) against live circuit building (by Nic).

SOUND AND VISION

To state the obvious: sound is more than music. This has always been true, but the traditional distinction between "found" sound (produced in the world but living its life outside human intent) and "constructed" music (the imposition of human genius upon selected, designed sounds) is increasingly moot. Not only is music often "found," sound is often "designed." This is true in the rarified realms of art, the globalized realms of commerce, and the erudite realms of science.

Perhaps I'm attuned by virtue of having taught in an art school for the better part of a decade, but my ear tells me museums, galleries, and artists' studios just keep getting noisier: it's not that there is so much more "sound art" now than ten years ago, but rather that so much more art has sound. There are myriad reasons for this, from the high-falutin' and theoretical to the incidental and pragmatic. For many artists the digital camcorder has become the new sketchbook, and it is so difficult to defeat the camera's built-in microphone that most video footage has sound by default. And, just as a camera often redirects the artist's eye, so the constant presence of a soundtrack, whether intentional or not, draws attention to sound. My students shoot video with the lens cap on for the sole purpose of gathering sound, and they play back audio from camera tapes, without bothering to hook up a video monitor. When it comes time to edit, video and sound are cut and pasted using the same keystrokes and mouse clicks, with neither media privileged over the other.

As a result, music (or "sound art," if you prefer) is emerging from art schools, utilizing many of the same materials and tools found in music schools, but often with a very different set of skills, aesthetic concerns, and historical baggage. Some of this work takes the form of "pure sound"—CD tracks or MP3 files—but frequently it is characterized by a heightened sensitivity to visual implications of the technology of its production. That technology can be pretty diverse, since artists today tend to be "multi-instrumentalists": the majority of my students don't identify with a specific medium—they flit with ease between paper, film, videotape, wood, metal, computers, canvas, and circuits.

As materials go, electronic toys with their lurid plastic casings, and homemade circuits packaged in cigar boxes and Pringles tubes, are more varied and seductive to the eye than laptops and software. Moreover, for those who want to work with sound but lack traditional musical training, the learning curve is easier than that of most "normal" musical instruments.

Brett Balogh, Adrian Bredescu, Kyle Evans, Alex Inglizian, James Murray, Chris Powers, and Aaron Zarzutzki—all former students of The School of the Art Institute of Chicago—are typical of this new generation of electronic artists. Balogh works extensively with radio technology, adapted as a creative medium, rather than a source of news or existing music. Bradescu and Evans have built rich extensions of the circuits featured in this book. Inglizian is very active on the Chicago bending scene, conducting workshops as well as performing regularly, bending toys and building his own circuits (see Figure 30.10). Murray draws on his background as a DJ to design homemade circuits, developing instruments for cutting and mixing (see Figure 26.8 in Chapter 26), as well as multiple touch-radio synthesizers (such as those in Figure 11.5 in Chapter 11). A typical Chris Powers' hack to a guitar effect box is shown in Figure 15.18 in

Figure 30.10 Various homemade circuits by Alex Inglizian.

Chapter 15. Zarzutzki built a wonderful instrument from crystal oscillators intended for computer clocks: the crystals oscillate at frequencies in the megahertz range, well above human hearing, but the difference tones that result from non-linear diode mixing (like we used in Chapter 18, Figure 18.19) yields unstable tone clusters that we can hear (see Figure 30.11).

This trend is not limited to trained artists, of course. A visit to YouTube makes it very clear that video cameras (especially those built into cell phones) have supplanted not only the sketchbook, but the Brownie, Instamatic, Polaroid and Super-8 as well—the documentary tools of previous generations of amateurs. And the content on the YouTube site demonstrates the international popularity of bending and hacking as musical forms.

Some of the most interesting work to come out of the new audio-visual sensibility involves circuitry that either generates simultaneous audio and video signals, or allows the two to interact. Americans Jon Satrom (see Art & Music 10 "Visual Music" in Chapter 24 and Figure AM10.6) and J. D. Kramer (see Figure 30.12) have independently hacked children's video paintboxes to generate fractured digital video and glitch-laden sound. The English trio of Phil Archer, Luke Abbott, and Dan Thomas, as well as the American Jordan Bartee, use hacked Sega game consoles for live video and sound. Phillip Stearns' (USA) "Pixel Maelstrom" is a video synthesizer constructed by radically hacking an old TI99/4a microcomputer; Stearns has also created a series of pieces using an array of analog electronics to process sound and video through feedback loops.

In the spirit of Yasunao Tone's "Molecular Music" (see Art & Music 10 "Visual Music" in Chapter 24 and Figure 24.6), Jeffrey Byron and Jay Trautman (USA), Joe

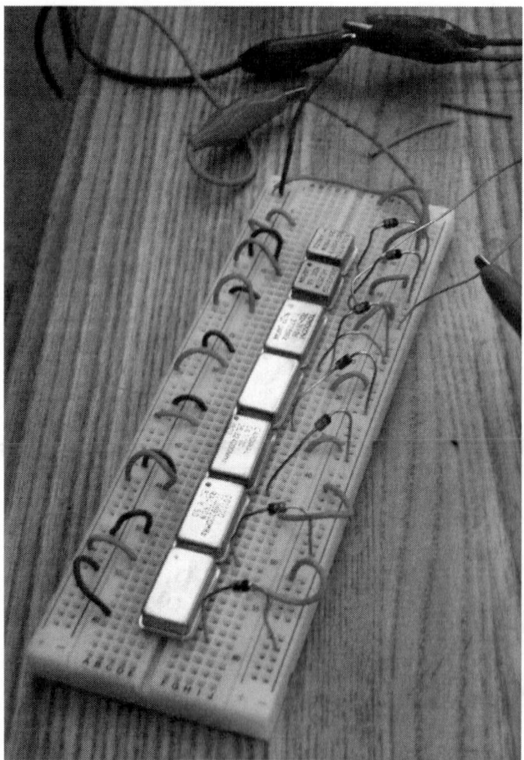

Figure 30.11
Multi-crystal beat-frequency oscillator, Aaron Zarzutzki.

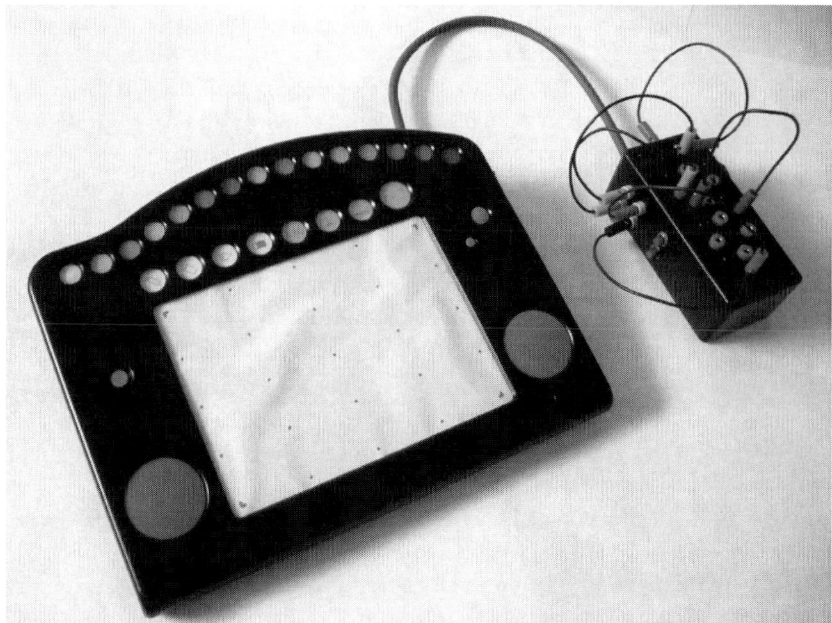

Figure 30.12 "Permutator" video synthesizer, J. D. Kramer.

Grimm (USA), and Infusion (USA) have built oscillators and signal processors that respond to film and light projection or sensors taped to a video screen. Michał Dudek (Germany/Poland) has connected oscillators and a bent keyboard to LCD TVs to disrupt the video sync and thereby modulate the broadcast images. As described in the box in Chapter 24, several artists—including BMBCon (NL) and myself (USA)—have built simple video projectors using LCD screens from small TVs and handheld games.

At the more sophisticated end of the technological spectrum, the New York-based duo LoVid (Tali Hinkis and Kyle Lapidus) have designed the "Syncarmonica," a system of circuit modules, set into a table top, that can be patched together like an old-fashioned analog synthesizer to generate sound and video (see Figure 24.7 in Art & Music 10 "Visual Music" in Chapter 24). LoVid has also embedded video and sound circuitry into soft sculpture, such as their "Ghoti" ("fish," in the variant orthography suggested by William Ollier in 1855 as a comment on the irregularity of English spelling, but often attributed to George Bernard Shaw). Ilias Anagnostopoulis has built an analog video synthesizer that harkens back to the Sandin Image Processor and Paik-Abe video synthesizer of the 1960–70s.

MECHANICS

Alvin Lucier once said that circuitry didn't interest him because it was "flat," while sound in space was three-dimensional. Ron Kuivila and I, his students at the time, bristled, but 33 years later I'm inclined to accept his analysis as a simple, non-judgmental fact. Several contemporary artists have reacted to the flat nature of circuits and computers by incorporating mechanical devices in their hacks. In Leah Crews Castleman's installation "Compose, Construct, Control," visitors press organ pedals with their feet to activate a beautifully crafted array of motors, cams, levers, and cables that animates an automated percussion ensemble, composed largely of trash (see Figure 30.13). Martin Riches' elegant "Motor Mouth" (1998–99) is an acoustic speech synthesizer, a mechanical version of the human mouth with moving lips, teeth, tongue, and larynx, with a blower which serves as the lungs; its controlling computer is programmed with a few sentences and can also speak individual vowels, semi-vowels, nasals, labials, and fricatives (see Figure 30.14).

A vaguely erotic physicality drives the work of Catalan musician Ferran Fages, who presses cones of styrofoam against the spinning platter of a cheap record player; the result sounds electronic but is produced purely acoustically, through friction, in a nod to the record player's pre-electronic, gramophone roots. Similarly, Chris DeLaurenti's (USA) "Flap-o-phone" is a manually activated, acoustic turntable: folded cardboard, a needle, and a stick exhume sound from 78 rpm records; this is a variant of the CardTalk record player devised in the 1950s by Christian missionary Joy Ridderhof for playing back Bible recordings in unwired locations.

The historicism implicit in mechanical devices is knowingly exploited by Guillermo Galindo, who describes his "MAIZ" as a "cybertotemic instrument": an assemblage of trash (wine crate, street sweeper parts, credit card, cigar box) and instrument parts (guitar neck and three strings), all whacked by a powerful motor controlled by a computer in

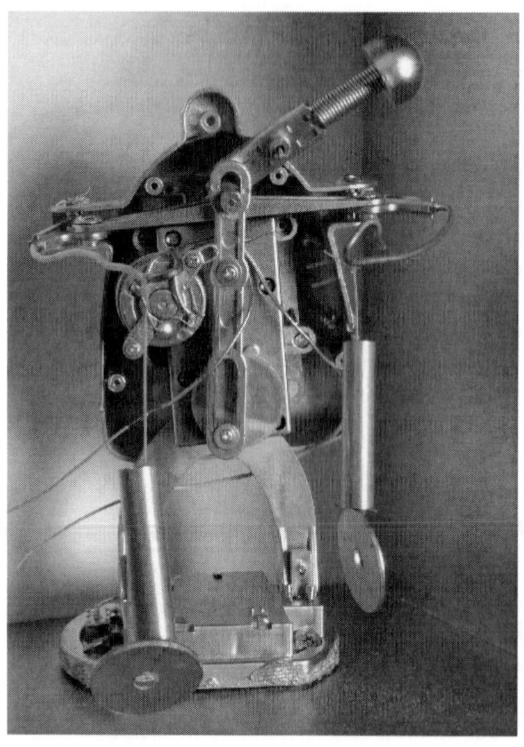

Figure 30.13
Detail from "Compose, Construct, Control," Leah Crews Castleman.

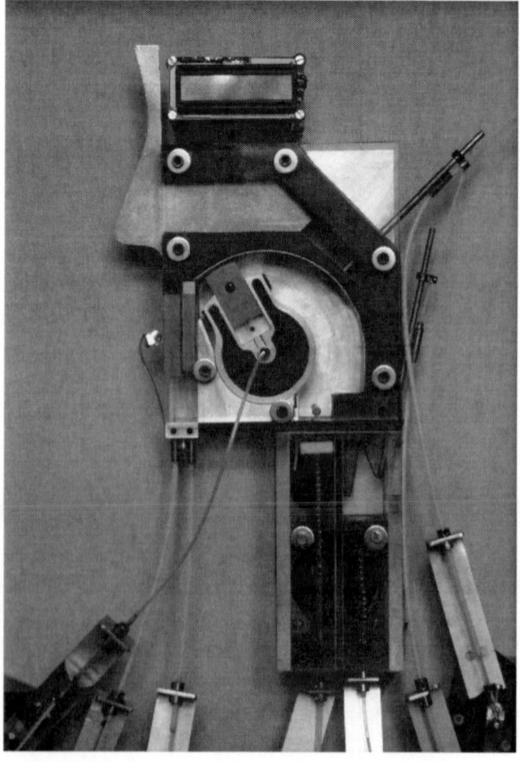

Figure 30.14
"Motor Mouth" (1998-99), Martin Riches.

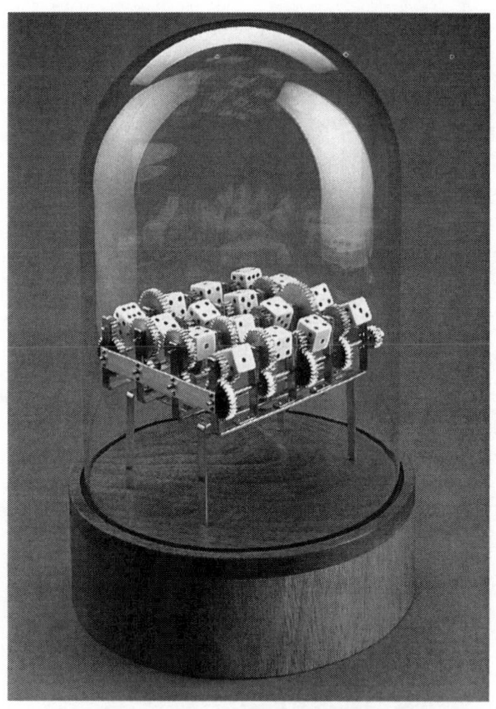

Figure 30.15
"Mandala #2" (2000), mechanized dice thrower, Marc Berghaus.

response to various sensors, the MAIZ merges pre-Colombian imagery, hurdy-gurdy mechanics, and digital interactivity. Marc Berghaus chose gears over computers, and uses Newtonian technology to create "chance machines" for generating haikus, throws of dice, and permutational music on tiny acoustic pianos (see Figure 30.15).

SWASHBUCKLERS

Driven by the desire to hear a world just over the horizon, more and more artists are building circuits from scratch, either to aid and abet the transformation of found technology, or in its stead. Sebastian Tomczak and Christian Haines in Australia, Tuomo Tammenpää in Finland (see Figure 30.16), Alejandra Perez Nuñez in Chile, and Douglas Ferguson and Steve Marsh in the USA have all developed beautiful extensions of the sort of circuitry featured in this book. In addition to his beautiful spring-and-piezo gamelans, as shown in Art & Music 3 "Piezo Music" in Chapter 7 (Figure 7.15), Adachi Tomomi (Japan) has been building and selling quirky Tupperware-encased music circuits since 1994 (see Figure 30.17). Florian Kaufmann (Switzerland) and Osamu Hoshuyama (Japan) are both serious fans of

Figure 30.16
"NAND Can," oscillator group, Tuomo Tammenpää.

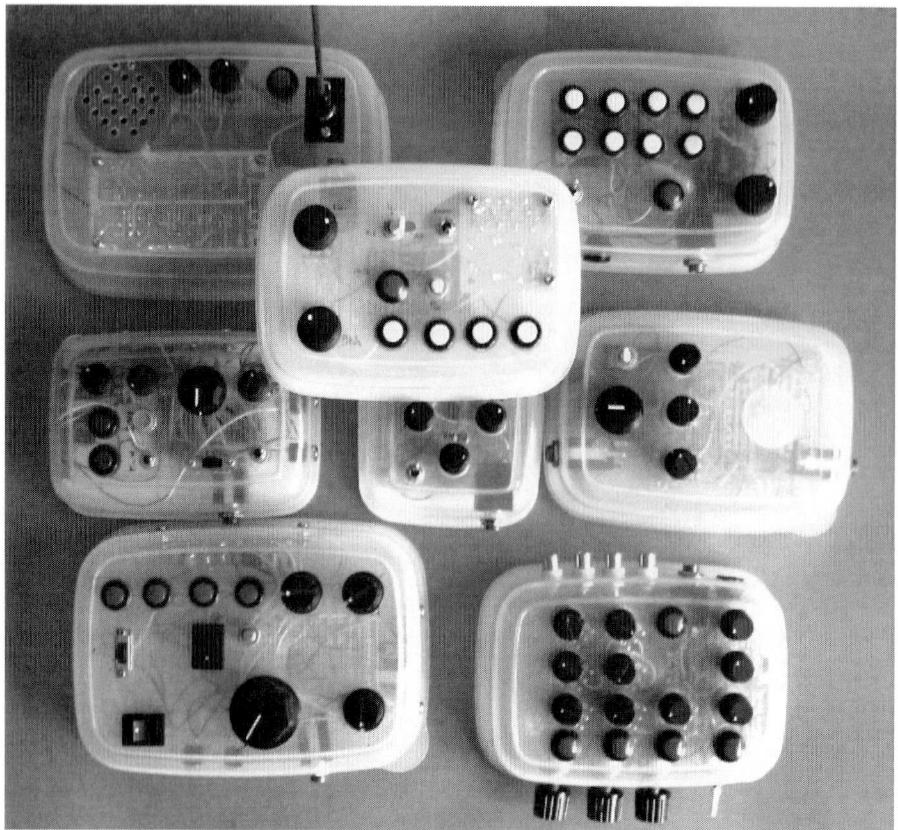

Figure 30.17 Assorted Tupperware-encased instruments, Adachi Tomomi.

Figure 30.18 "Where's The Party At," hacker-friendly audio sampler kit, Todd Bailey.

Figure 30.19 "Beavis Board" (left) and parts kit (right), Beavis Audio Research.

CMOS audio, and have set up Web sites with designs that go well beyond those I discuss.

Todd Bailey (USA) incorporated numerous contact points and an open area for extra components in his audio sampler kit, "Where's The Party At," to encourage customization by the user (see Figure 30.18). Beavis Audio Research has produced a guitar effect kit for those who fear solder: preamps, distortion circuits, filters, tremolos, etc. can be plugged together on a breadboard that connects to jacks and pots on a sturdy metal base (see Figure 30.19)—"learning to solder sucks," states the Beavis website.

Hans w. koch (Germany) has created a series of pieces based on the physical idiosyncracies of specific computers: in "Bandoneon Book" he opens and closes the lid of an Apple Powerbook, with accordion-like gestures, to affect feedback between the computer's built-in mikes and speakers; the feedback is further processed by software running on the computer, controlled from its keyboard. He bows the case of another laptop in "Electroviola"; the sounds of rosin on plastic are picked up by the internal mikes and, once again, transformed by koch's code. And in the melancholy tradition of Mahler's "Kindertotenlieder," in "Core-sound" we listen to the death rattle of an old PC as koch drips water onto its motherboard.

Phil Archer's (UK) early work with classic keyboard bending, percussion activated by a dot matrix printer, and a CD player-driven Hawaiian guitar were described in the "Bending" sidebar in Chapter 15 (Figures 15.11 and 15.12). Gutting another pair of CD players, Archer connected the sled of each one to the circuit board of the other, causing the lasers to chase each other's tails and play back a torrent of glitches. His series of hand-generator powered music boxes are described earlier in this chapter. Like hans koch, Archer has experimented with the heretic interaction of moisture and electricity, dripping water onto a Yamaha keyboard in his blithely-titled installation, "What's the Worst That Could Happen?" He has also incorporated growing plants in his circuits as components that react to human touch and presence in a distinctly spooky fashion, reminiscent of Tom Zahuranec's experiments in the 1970s. Archer's compatriot Dan Wilson has also used a discarded dot matrix printer as the core for an instrument: he amplifies the printer's internal springs, motors, and coils, as well as some added strings, in his "Printar" (see Figure 30.20). Wilson has built instruments out of floor sweepers, used the principles of the Victorian Oscillator (Chapter 5) to pluck strings and springs, employed worms to play Cracklebox-style electrode-controlled circuits, resonated objects

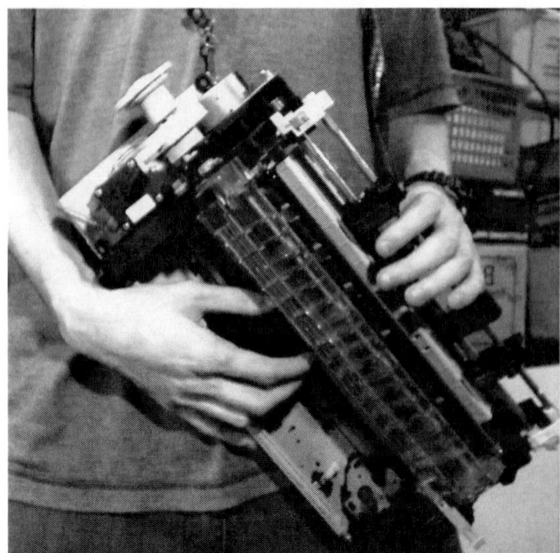

Figure 30.20
"Printar," Dan Wilson.

Figure 30.21
Thumb piano with electromagnetic drivers, Dan Wilson.

large (lampposts) and small (mbiras) with electromagnetic feedback (see Figure 30.21), and, very briefly, persuaded a pair of hedgehogs to play a zither in his back garden.

Alex Baker (UK) has created a number of pieces that explore the interaction of sound and mechanical forces. His "Wind Powered Record Player" (see Figure 30.22) is an acoustic gramophone whose platter is turned by the wind. In "Catch" a ping-pong ball is tossed into the air by a speaker cone pulsed with sound. Transducers attached to the heads of his "Autonomous Drum Kit" transform the skins into reversible microphone/speakers: first they are used to record sticks striking the drums; then they are reversed to play back the recorded sound through the heads, evoking a ghostly, phantom drummer.

Douglas Repetto (USA) was instrumental in setting up Dorkbot ("for people doing strange things with electricity to get together to talk about their work") in New York in 2000; subsequently, satellite groups have sprung up all around the world. For years

Figure 30.22 "Wind Powered Record Player," Alex Baker.

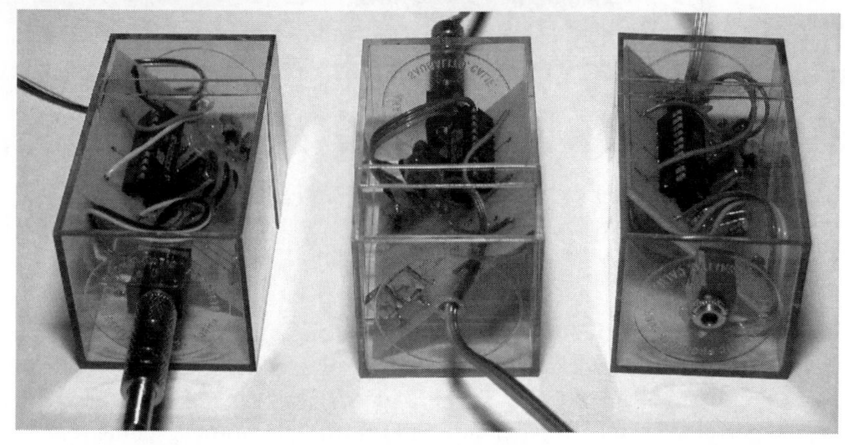

Figure 30.23 Three of 42 circuits from "Crash and Bloom," Douglas Irving Repetto.

he has been at the radical fringe of hacking culture as an educator (he is director of research at Columbia University's Computer Music Center), organizer, inventor, and prolific artist. In "Crash and Bloom" 42 identical small circuits emulate the cycle of growth and collapse of certain biological systems (see Figure 30.23). For "Fuseboxes" Repetto built miniature noisemaking circuits into 20 tiny tin boxes from fuses. "Slowscan Soundwave" attempts to make sound waves visible by translating air pressure patterns in a room into the movement of suspended plastic sheets.

Like Repetto, Phillip Stearns (USA) evokes the biological world in his impossibly complex "AANN" (Analog Artificial Neural Network). Stearns soldered up 50 identical neuron-simulation circuits, interconnected them, and ended up with a "squid baby" that responds to sound and light by "lighting up like a Christmas Tree" and "shrieking like so many dying seagulls" (see Figure 30.24). For his "Burlap" series Stearns wove circuitry into fabric for exhibition on a gallery wall (see Figure 30.25). Stearns' video work, using both digital and analog technology, was described earlier in this chapter.

Electronic components, as small as they are these days, are not quantum—a change of scale and the mysteries of what had previously been the lowest level of operation

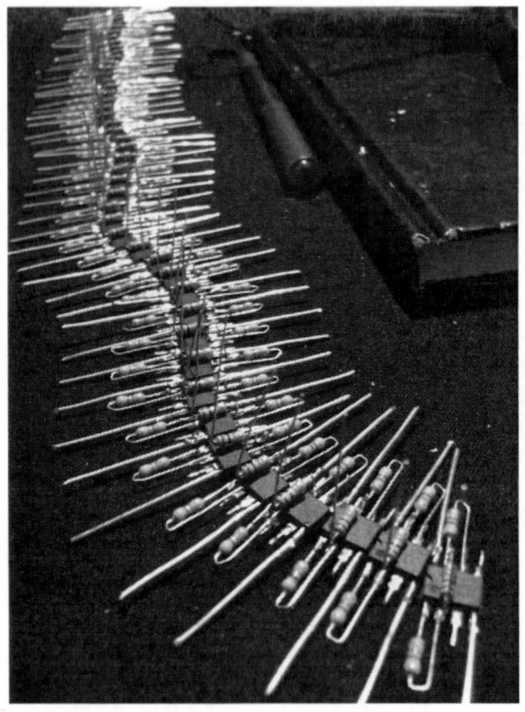

Figure 30.24
Series of neuron subassemblies from "AANN," Phillip Stearns.

Figure 30.25
Detail from "Burlap-II," Phillip Stearns.

become tangible, touchable, hackable. Some artists experiment at this component level. Patrick McCarthy (USA) makes his own potentiometers with cardboard, safety pins, and pencil lines. Nyle Steiner's (USA) website, "Spark, Bang Buzz," guides viewers through making their own diodes from zinc, using a flame as an amplifier, listening to a drop of salt water, building TV picture tubes and lasers from scratch, and other esoteric projects. Substituting cozily familiar materials for the arcane can have a humanizing effect: Peter Blasser (USA) builds circuits on sheets of paper, rather than fiberglass circuit boards. Grégoire Lauvin has created several gallery installations that use live plants as capacitors in oscillators, causing the pitch world to change as the plants grow (see Figure 30.26); he's also built a beautiful "Potato Organ" whose notes are tuned by impaling vegetables and fruits on nails emerging from the instrument's housing (see Figure 30.27).

THERE IS NO CONCLUSION

But there are anecdotes. Here's one: a few years ago I was conducting a hacking workshop at a very high-tech European music research institute. The group was made up of composers, computer programmers, acousticians, and electronic engineers. Real smart people. One student got very confused as soon as we started breadboarding our first oscillator. I asked if he was by any chance either dyslexic or a conservatory-trained composer (two population groups for whom matrix topology seems unusually vexing).

Figure 30.26
"Bio Oscillator,"
Grégoire Lauvin.

Figure 30.27
"Potatoes Organ,"
Grégoire Lauvin.

"No," he replied cheerfully, "I'm an electronic engineer. I designed a complete Digital Signal Processor for my senior thesis, but I did it all with software on a computer—I never actually touched a chip before today."

My engineer is not alone. Sociologist Richard Sennett observed that after computers became ubiquitous in the 1980s "we tended to forget the importance of physical senses." Or, as Bill Burnett, Executive Director of the Product Design program at Stanford University, put it: "a lot of people got lost in the world of computer simulation." Burnett goes on to add the all-important truth that "you can't simulate everything." That simple and obvious idea—that there is a world before and beyond simulation—has become ever more important as simulation has snaked its way into more and more of our waking hours.

The material world is now making a comeback. *Make* magazine, the de facto journal-of-record for hackers of all stripes, boasts a paid circulation of 100,000 and 2.5 million visits per month to its web site. Bug Labs, a recent New York-based startup, sells Lego-like components that let the user snap together GPS receivers, cameras, LCD displays, motion detectors, and other sub-modules to design and build their own digital products; the company "envisions a future where CE stands for Community Electronics, the term 'mashup' applies equally to hardware as it does to Web services." A West Coast activist who goes by the name of Mr. Jalopy has drafted a "Maker's Bill of Rights," advocating that "meaningful and specific parts lists shall be included," "batteries shall be replaceable," etc. "I want companies to start thinking about shared innovation," says Jalopy, "to realize that they're not selling to customers, but to collaborators." The software giant Adobe invited Gever Tulley, who normally teaches children at his Tinkering School in Montar, CA, to get their designers to put down their mice, pick up their screwdrivers and make *things*. "The physical act of making things helps the whole person," says Tulley.

The first edition of *Handmade Electronic Music* was intended as an invitation to reach out and collaborate with commodities, and as an introduction to those pioneers who had been doing so since the dawn of the silicon age. All puns intended, it seems to have touched a chord, or at least to have bumped one string of an ongoing arpeggio. Today there's more to listen to, more to look at, more to learn, more pages in this book.

There you have it: 30 chapters of circuits, suggestions, and glimpses of the work of some of the most interesting artists ever to confront a clip lead or soldering iron. The rest is up to you. Run inside and play.

APPENDIX A

Resources

THE WEB

Several years ago I walked into the office of a technically minded colleague at my school to ask for a reference manual in which I could look up the pinout and schematic of an unfamiliar chip. Clamping a large, Chicago-style hand to my shoulder, Ed replied, "Nic, I *could* loan you the book, but let me ask you this: give a man a fish and he's fed for one day, teach him how to fish and he . . . ?" ". . . wastes all his time fishing when he should be helping out around the house?" I continued. "No," sighed a disappointed Ed. "Type the part number into Google and you will find the data sheet in the first hit," he muttered as he closed the door. The point I missed in his parable: it's never been easier to hack.

In the early days of homemade electronic music, schematics and suggestions were exchanged by word of mouth and sleight of hand, like cures for colicky babies. Then a few dumbed-down circuits crawled out of engineering journals into magazines for electronic hobbyists and aspiring electric guitarists; one or two books appeared, written in something vaguely like English rather than Technese. Finally Tim Berners-Lee birthed the World Wide Web and a hundred fuzztones flowered.

Anything you want to know is out there; all you need to do is find it and understand it. Finding it is easy—understanding it may take some work. You will have to teach yourself a bit more of the vocabulary of electronics than was demanded by this book—you'll need to get comfotable reading schematics. As Ed suggested, typing a part number or name of a component will usually retrieve a PDF of a manufacturer's data sheet—here you'll find all the basic information you need to start working with it: what pins are connected to power, which are inputs, which are outputs, etc. Enter a descriptive phrase instead (" 'Phase Shifter' + schematic") and you'll be directed to any number of wacky Web sites hosted by people who seem to have nothing better to do than compile vast collections of circuit diagrams and provide links to like-minded fanatics. All you need to do is figure out how to translate the schematic onto the breadboard. A little trial and error, persistence, patience, and an occasional glance back at this book should get you there.

BOOKS

PDF data sheets can be downloaded as you need them, but thick data books are still available from the major chip manufacturers—they're expensive, but worth it if you get in deep. Often they include schematics of suggested basic circuits for specific chips, or more detailed "application notes."

There are several books that can help fill in theoretical gaps between the one you are reading now, and the more engineer-oriented data you will find on the Web or in the data books. Some of them date from the 1970s but are still relevant. Don Lancaster's *CMOS Cookbook* (Indianapolis, IN: SAMS Publications, 1977) provides a thorough introduction to the chip family we have misused throughout my book. Walter Jung's *OpAmp Cookbook* (Indianapolis, IN: SAMS Publications, 1974) will introduce you to the component out of which most audio circuits are built, but which I have avoided completely in our designs. Craig Anderton, the grandfather of electronic hacking for musicians, published *Electronic Projects for Musicians* back in 1975 (New York: Amsco Publications) and it's still an excellent guide to basic musical circuits and general principles of design and construction, written in large, reassuring, musician-friendly letters. For super low-tech, "foxhole technology," *Sneaky Uses for Everyday Things* by Cy Tymony (Kansas City, KS: Andrews McMeel Publishing, 2003) is a wonderful source of circuit designs built from little more than stationary supplies, salt, spare change, and wet paper towel. Reed Ghazala, the patron saint of Circuit Bending, put out a book that is an excellent companion to the one you are reading now: Reed Ghazala, *Circuit Bending: Build Your Own Alien Instruments* (New York: Wiley Publications, 2005). Brian Wampler's *How to Modify Effect Pedals for Guitar and Bass* (CreateSpace, 2007) provides detailed instructions on hacks to numerous stomp boxes. And it's always useful keeping an eye on the most popular journal of hacking culture, *Make* magazine: http://makezine.com/.

STUFF

What holds true for information also goes for material resources: although the Big Apple's Canal Street no longer teems with the warrens of weird electronic and mechanical surplus shops that enthralled me throughout my childhood, the Web has become a virtual medina of the misplaced and unwanted. Add "+price" to the search field after anything you desire—plugs, piezo disks, tape heads, tilt-switches—and you'll soon find a place to buy it. Since you're not manufacturing missiles or airbags in quantity, you'll need to find a source that will sell to the common man or woman. For ICs, resistors, capacitors, and other small components a straightforward electronic retailer is probably the best bet. As of the time of writing some good sources that stock a wide range of parts include Digikey (www.digikey.com), Jameco (www.jameco.com), and the delightfully-named Mouser Electronics (www.mouser.com).

For hackable gizmos, used equipment, pots, jacks, boxes, and general inspiration, however, you really must try the "surplus" outlets. Here are a few reputable sources that have been around for a while selling cool stuff:

All Electronics: www.allelectronics.com
Marlin P. Jones: www.mpja.com

Electronic Goldmine: www.goldmine-elec.com
B.G. Micro: www.bgmicro.com
Surplus Shed: www.surplusshed.com
Unicorn Electronics: www.unicornelex.com

Kits are a great way to ease the transition from circuit bending to free-form electronic design. The project-specific printed circuit boards minimize the chances for mis-wiring, and they often come with suggestions for variations. Todd Bailey's "Where's the Party At" sampler kit is specifically designed for hacking (see Chapter 30 and www.narrat1ve.com/). Paia is a long-standing manufacturer of kits for building everything from modular synthesizers to electronic wind chimes: www.paia.com. Velleman (www.velleman.be) and a few other companies (mostly European) make kits for a wide range of circuits, from amplifiers to zappers-of-bugs. A convenient source in North America is Quality Kits in Kingston, Ontario: www.qkits.com. Other useful kit outlets are Ramsey Electronics (www.ramseyelectronics.com) and Electronics 1 2 3 (www.electronics123.com). The Beavis Board, described in Chapter 30, provides a terrific solder-free route to numerous cool audio circuits (www.beavisaudio.com/bboard/).

The Web knows no national boundaries, but sadly the US Post Office does. The above U.S. sources will charge a premium for shipping abroad. In England Maplin (www.maplin.co.uk) carries a wide range of components. In Germany, Conrad Electronics is a good bet (and they have retails shops in Berlin and several other cities): www.conrad.de. RS Components is a full-range dealer of new parts that delivers across Europe: http://rswww.com.

For impulse shopping in Yourtown, USA there's always Radio Shack—overpriced but over here. They also have an online presence: www.radioshack.com.

Finally, I am pleased to report that the authentic, old-fashioned "electronic junkyard" is not entirely extinct. As I write this I have just returned from a delirious crawl through Apex Surplus at 8909 San Fernando Road in Sun Valley, California, under the guidance of Lorin Parker. Their overstuffed shelves and murky corridors abound with 50 years of electronic technology and just plain weird stuff. The Web address is www.apexelectronic.com but this is no time for a virtual experience—having read this book you owe it to yourself to visit Apex once before you (or it) die, even if you never get to Mecca, Jerusalem, or St. Peters. Weird Stuff Warehouse (www.weirdstuff.com) in Sunnyvale, California, is the elephant graveyard of obsolete computers and computer accessories—just the place to pick up a fully functional first-generation Mac. American Science and Surplus (www.sciplus.com) and Triode Electronics (http://store.triodestore.com) in Chicago have provided my students with a wide range of excellent materials. The Ax Man (www.ax-man.com), with several locations in the Minneapolis/St. Paul area, includes a fair amount of electronic parts amongst a wide range of general surplus material (such as East German crossing guard blinking braces). There are many other wonderful stores around the country, I've been told, but these are ones I have visited and can personally vouch for.

CULTURAL ARTIFACTS

The music made by the artists mentioned in this book is, with few exceptions, not only unavailable on major labels, but unlikely to be found on most file sharing sites. Individual artist Web sites are a good start, and I've included as many as possible in Appendices B and E. An excellent, non-Amazon source for CDs, books, and periodicals of fringe music is CDemusic (www.cdemusic.org).

Two good general histories of electronic music, featuring many of the artists discussed in this book, are:

Joel Chadabe, *Electric Sound—The Past and Promise of Electronic Music*. New York: Prentice Hall, 1997.
Thom Holmes, *Electronic And Experimental Music* (third edition). New York: Routledge, 2008.

Recording the Beatles—The Studio Equipment and Techniques Used to Create Their Classic Albums, by Kevin Ryan and Brian Kehew (Houston, TX: Curvebender Publishing, 2006), provides an extraordinarily detailed look at the evolution of the recording studio as a musical instrument. Detailing dozens of cool vintage gizmos, this is tech-porn at its most erotic. The best $100 you'll ever spend.

TapeOp magazine (www.tapeop.com) publishes interviews with innovative audio designers, features the occasional article on building or modifying gear, and is a great way to keep track of what people with modest budgets are doing with music technology.

Leonardo Music Journal (http://leonardo.info/lmj) has for many years published writings by experimental musicians and sound artists, and is a rare source of first person accounts of struggles with circuits.

This book is very much about *stuff*, and the single best book about how stuff is designed and consumed today is: Rich Gold, *The Plenitude*. Cambridge: MIT Press, 2007.

APPENDIX B

Tools and Materials Needed

This is a list of the essential supplies needed to do the projects in this book. Most can be obtained from a variety of sources. I have listed online outlets that stocked the parts at reasonable prices as of September 2008. Be advised that the stock of "surplus" retailers can fluctuate wildly; when the specified part vanishes, however, there's often an acceptable alternative available from the same vendor. Most parts are cheaper from these surplus sources, when available, but for chips I suggest going to a non-surplus source, since the prices are usually pretty similar (especially for small quantities) and the quality of surplus chips can be a bit dodgy. Many of the parts can be purchased at your local Radio Shack, although inevitably at a ludicrously inflated price.

Most of the designs are forgiving of a wide range of component variation and substitution; those few items that are critical and should not be substituted without due care and attention have been noted in the list. You should be able to pick up all the needed parts for under $50; tools might set you back another $25–$50.

I have not specified in this list all the scrounged objects, such as tape heads and loudspeakers, as these are best hunted locally, following the advice in specific chapters.

SOURCE KEY

The major sources in North America for the materials needed are indicated as follows:

AE = www.allelectronics.com (All Electronics)
J = www.jameco.com (Jameco)
RS = www.radioshack.com (Radio Shack)
EG = www.goldmine-elec.com/ (Electronics Goldmine)
MPJA = www.mpja.com (Marlin P. Jones)
UE = www.unicornelex.com (Unicorn Electronics)
QK = http://store.qkits.com/ (Quality Kits)

Within Europe the major sources are:

Conrad Electronics (Germany): www.conrad.com
Maplin Electronics (UK): www.maplin.co.uk/
RS Components: http://rswww.com

TOOLS

The tools needed to do the projects in this book are:

Good soldering iron, with as fine a tip as possible; 15–60 watts.
Roll of "rosin-core" electrical solder (not "acid core" solder).
Battery-powered amplifier:
    RS: Mini Audio Amplifier, 277–1008, $18.19;
    Fender "Mini Twin Amp" is very good, *c.* $39.00 from www.samash.com.
Or an amplifier kit, with a scrounged speaker added:
    QK: 1 watt amp (LM386), QK17, $8.95;
    QK: 2 watt amp (TBA820M), FK602, $5.95.
Assorted patchcords to connect to amplifier from various jacks.
Prototyping breadboard:
    AE: PB-400, $4.00;
    EG: G16567, $4.49.
Inexpensive digital multimeter (voltage, resistance, current).
Small diagonal wire cutters, suitable for light gauge wire.
Simple wire strippers, suitable for light gauge wire.
Set of small screwdrivers (sometimes called "jeweler's screwdrivers"), flat and Phillips tips, suitable for opening electronic toys, portable radios, etc.
A solder-sucker.
"Sharpie"-style fine-tipped permanent marker.
Roll of insulating electrical tape.
Flashlight.
Small saw for plastic and metal.
Double-stick tape.
Files.
Electric drill and bits.
Small spring clamps or clothespins (non-conductive: plastic or wood, not metal).
Scissors.
Utility knife.
Swiss Army knife.

PARTS AND SUPPLIES

Additional parts and supplies needed to work on projects in this book are listed below. Just to be on the safe side, always pick up a spare or two especially if the part is cheap. The prices are accurate as of September 2008 and are listed in US dollars.

Pcs = pieces (each number of pieces needed will be specified). Where two prices are separated by a "/" the first number equals the cost for 1–10 pieces; the second equals the cost for 10–100 pieces (i.e. $0.35/0.30)—sometimes you can save a lot by buying in bulk, if you're in for the long haul or order for all your friends as well.

Insulated wire, 22–24 gauge, stranded, approximately 20 feet.
Insulated wire, 22–24 gauge, solid, approximately 20 feet.
Shielded audio cable, 1 conductor + shield, as thin as possible, approximately 20 feet.

Can of "Plasti-Dip" tool-handle insulation paint; $13.00.
Plastic terminal barrier strip, 1 piece:
AE: TB-20, $2.25.
Sheet of antistatic foam (used for packaging integrated circuits):
J: 13864, $9.75 (1pc 24″ × 12″ × 1/4″).
5: 1/8″ (3.5mm) male plugs (if you are using an amplifier with 1/8″ input and output jacks—otherwise select appropriate jacks for your amplifier):
AE: PMP, $0.40/$0.35.
5: 1/8″ (3.5mm) female jacks (or 1/4″ female jacks if you are working with that standard):
AE: MJW, $0.45/$0.40.
2: piezo disks:
AE: PE-53 (15 mm diameter), $0.75/$0.65;
EG: G14378 (0.78″ diameter), $0.99/3 pcs.
4: 9-volt battery connector clips:
AE: BST-3, $0.25/$0.15;
EG: G2210A, $1.00/5 pcs.
10: test leads with alligator clips at each end:
AE: MTL-10, $2.50/10 pcs;
EG: G1498, $1.99/10 pcs.
1: telephone pickup coil
AE: TPX-1, $2.00.
1: Inductor (for making coil pickup):
EG: G16604, $0.50.
1: audio output transformer
RS: 273–1380, $2.99 (this part is difficult to find from any other vendor).
2: Electret microphone elements:
J: 160979, $0.75/0.65.
6: 1.0 megOhm linear taper potentiometers:
J: 29066, $1.09/$0.85.
4: audio taper potentiometers:
J: 10 kOhm, 255426, $1.35/$1.09.
10: 10 kOhm linear potentiometers
J: 29082, $1.09/$0.95.
6: photocells:
J: 120310, $1.45/$1.20 (this is a very good, wide range part, preferable to most others I've found);
EG: G15175, $1.00/2 pcs (closest surplus part to more expensive Jameco part).
Resistor Assortment, common values (most important values are 1 k, 2.2 k, 10 k, 100 k, and 1 meg—if you wish to buy something less than a full set get 10 of each of these):
AE: RES-61, $11.50 (610 pcs);
RS: 271–308, $6.49 (100 pcs).
Capacitor assortment; most values are not critical. Any general purpose capacitor assortment that covers this range will do:
monolithic ceramic caps
10: 10 pf; J: 15333, $0.07/$0.06;

10: 100 pf; J: 15341, \$0.05/\$0.035;
10: 0.01 uf; J: 25507, \$0.09/\$0.06;
10: 0.1 uf; J: 25523, \$0.07/\$0.06.

electrolytic caps

10: 1.0 uf; J: 330431, \$0.045/\$0.039;
10: 2.2 uf; J: 93731, \$0.047/\$0.041;
10: 4.7 uf; J: 31002, \$0.042/\$0.036;
10: 10.0 uf; J: 94220, \$0.083/\$0.057;
10: 47 uf; J: 31116, \$0.067/\$0.056.

6: LEDs:
J: 152864, \$0.29/\$0.24.

6: small signal diodes (1N914 or equiv):
J: 655269, 0.025/0.015.

1: phototransistor:
J: 112169, \$0.36/\$0.30.

3: 74C14, CD4585, or CD40106 Hex Schmitt Trigger Integrated Circuit (all integrated circuits should be "dual in-line package," not "surface mount"). These MUST *NOT* BE 74AC14 or 74HC14!
J: 44257, \$0.49/\$0.46 or 13768, \$0.39/\$0.35.

3: CD4093 Quad NAND Schmitt Trigger Integrated Circuit:
J: 13400, \$0.35/\$0.30.

2: CD4040 Binary Divider Integrated Circuit:
J: 12950, \$0.31/\$0.29.

2: CD4049 Hex inverter Integrated Circuit:
J: 13055, \$0.35/\$0.29.

2: CD4046 Phase Locked Loop Integrated Circuit:
J: 13013, \$0.45/0.39.

2: CD4017 Decade Counter:
J: 12749, \$0.39/0.35.

1: LM386N-3 Power Amplifier Integrated Circuit:
J: 24133, \$0.69/\$0.59.

1: Fixed Positive Voltage Regulator:
9 volt: J: 876352, \$0.56/0.47;
12 volt: J: 876395, \$0.44/37.5.

1: LM317T Adjustable Voltage Positive Regulator:
J: 898800, \$1.01/0.94.

2: Big Electrolytic Capacitor:
4,700 uf, J: 93675, \$1.69/1.52.

2: 4001 Diode:
J: 35975, \$0.04/0.028.

2: Solar Cell:
AE: SPL-61, \$3.75.

1: 8 pin DIP IC socket:
J: 112206, \$0.09/\$0.08.

4: 14 pin DIP IC socket:
J: 112214, \$0.10/\$0.075.

4: 16 pin DIP IC socket:
    J: 112221EF, \$0.13/\$0.115.
1: PC board:
    UE: 32–3008, \$1.89;
    R: 276–0170, \$3.49.
2: 9-volt batteries for circuit breadboarding.
A tape head (see Chapter 9).
A portable radio (see Chapter 11).
Several electronic toys (see Chapter 12).
Batteries for radio and toys, as needed.
Wall-wart (see Chapter 29).

APPENDIX C

The Rules of Hacking

Rule #1: Fear not (Chapter 2)!

Rule #2: Don't take apart anything that plugs directly into the wall (Chapter 2).

Rule #3: It is easier to take something apart than put it back together (Chapter 2).

Rule #4: Make notes of what you are doing as you go along, not after (Chapter 2).

Rule #5: Avoid connecting the battery backwards (Chapter 2).

Rule #6: Many hacks are like butterflies: beautiful but short-lived (Chapter 2).

Rule #7: In general try to avoid short circuits (Chapter 2).

Rule #8: In electronics some things are reversible with interesting results, but some things are reversible only with irreversible results (Chapter 4).

Rule #9: Use shielded cable to make all audio connections longer than 8 inches, unless they go between an amplifier and a speaker (Chapter 7).

Rule #10: Every audio connection consists of two parts: the signal and a ground reference (Chapter 7).

Rule #11: Don't drink and solder (Chapter 7).

Rule #12: After a hacked circuit crashes you may need to disconnect and reconnect the batteries before it will run again (Chapter 13).

Rule #13: The net value of two resistors connected in parallel is a little bit less than the smaller of the two resistors; the net value of two resistors connected in series is the sum of the two resistors (Ohm's Law for Dummies) (Chapter 14).

Rule #14: Kick me off if I stick (Zummo's rule) (Chapter 17).

Rule #15: You can always substitute a larger 1.5-volt battery for a smaller one, just make sure you use the same number of batteries, in the same configuration (Chapter 17).

Rule #16: It's always safer to use separate batteries for separate circuits (Chapter 17).

Rule #17: If it sounds good and doesn't smoke, don't worry if you don't understand it (Chapter 18).

Rule #18: Start simple and confirm that the circuit still works after every addition you make (Chapter 18).

Rule #19: Always leave your original breadboard design intact and functional until you can prove that the soldered-up version works (Chapter 19).

Rule #20: All chips may look alike on the outside without being the same on the inside—read the fine print (Chapter 20)!

Rule #21: All chips expect "+" and "–" power connections to their designated power supply pins, even if these voltages are *also* connected to other pins for other reasons—withhold them at your own risk (or entertainment) (Chapter 20).

Rule #22: Always use a resistor when powering an LED, otherwise the circuit and/or LED might blow out (Chapter 21).
Rule #23: Distortion is Truth (Poss's Law) (Chapter 22).
Rule #24: It is easier to drill round holes than slots (Chapter 26).
Rule #25: Never trust the writing on the wall-wart (Chapter 29).

THE LAWS OF THE AVANT-GARDE

Law #1: Do it backwards (Chapter 4).
Law #2: Make it louder, a lot (Chapter 7).
Law #3: Slow it down, a lot (Chapter 13).

APPENDIX D

Notes for the DVD

The DVD can be viewed on any DVD player or computer.

The disk is divided into three sections:

1. Project Tutorials: video demonstrations of projects from the book.
2. Gallery of short video clips of hacking and bending projects by artists.
3. Audio files of musical examples included on the CD accompanying the first edition of this book.

NOTE: the video is formatted for a 4:3 image mode, not 16:9 "wide screen"/High Definition. When viewing on a television please make sure the aspect ratio is set to 4:3 or the image will appear distorted.

PROJECT TUTORIALS

1. Circuit Sniffing (Chapter 3)
2. In/Out—Electromagnetism Explained (Chapter 4)
3. Jumping Speakers (Chapter 5)
4. How to Solder (Chapter 6)
5. How to Make a Contact Mike (Chapter 7)
6. Drivers (Chapter 8)
7. Tape Heads (Chapter 9)
8. Electret Microphones (Chapter 10)
9. Laying of Hands (Chapter 11)
10. Hack the Clock—Basic Circuit Bending (Chapters 12–15)
11. Your First Oscillator (Chapter 18)
12. From Breadboard to Circuit Board (Chapter 19)
13. On/Off—Gating Audio with Light (Chapter 21)

GALLERY OF ARTISTS' WORK

Most of the artists featured in this section are discussed in the book—please consult the index. The urls listed below indicate personal or primary Web sites for more information.

1. Luke Abbott, Phil Archer and Dan Tombs: *Bent New York 2006 (extract)*
2. Adachi Tomomi (www.adachitomomi.com): *Tomoring II*
3. Adachi Tomomi: *Instruments in Tupperware Boxes*
4. Vasco Alvo (www.vascoalvo.com): *FMkbrd*
5. Ilias Anagnostopoulis (http://iknewthem.tripod.com): *The Bob System Analog Modular Audio/Visual Synthesizer*
6. Eric Archer (http://ericarcher.net/): *Big Box o' Techno*
7. Phil Archer (www.philarcher.net/): *Plants*
8. Phil Archer: *What's the worst that could happen?*
9. Todd Bailey (www.narrat1ve.com/): *Where's The Party At?*
10. Alex Baker (www.alexbaker.co.uk/): *Autonomous Drum Kit*
11. Brett Ian Balogh (www.brettbalogh.com): *Random Access Radio*
12. Jordan Bartee: *Bent Gear*
13. Ian Baxter (www.ianbaxter.net): *The Masher*
14. Ithai Benjamin (www.ithaibenjamin.com) and Alejandro Abreu (www.alexabreu.com/): *Synthinetic*
15. Marc Berghaus (www.marcberghaus.com/): *Proof #1 Stone*
16. Emir Bijukic (www.bijukic.org/): *Solar Cell Calculator Directly Wired to Amp*
17. Chris Black and Christine White: *Crude Awakening*
18. Peter Blasser (www.ciat-lonbarde.net/rzither/index.html): *Radio Zither*
19. BMBCon (www.bmbcon.demon.nl/con/): *LCD Video Projector*
20. Frederick Brummer (http://overmindproductions.com): *Speaker/Drum*
21. Michael T. Bullock (www.finenoiseandlight.net/): *Open-backed Transistor Radio*
22. Jeffrey Byron and Jay Trautman: *Light-governed Self-oscillation of 555 Timers*
23. Lauren Carter and Joe Grimm (www.thewindupbird.com/): *Benjamin Franklin Forever*
24. Lauren Carter and Joe Grimm: *Picnic Party*
25. Leah Crews Castleman: *Compose, Construct, Control*
26. Duncan Chapman and Stewart Collinson: *The Bent Radio Orchestra*
27. Seth Cluett (www.onelonelypixel.org/): *Open-backed Cassette Walkman*
28. Stewart Collinson: *Hacked "Cyberhead"*
29. Christopher DeLaurenti (www.delaurenti.net): *Christopher DeLaurenti Plays the Flap-o-Phone*
30. DIY Kamikazee Group (www.mechatronicart.ch/): *Live Solder Performance*
31. Michał Dudek: *Modulating an LCD TV*
32. (e)-Bombers (http://e-bombers.jugem.jp): *Excerpt from Live Performance at Tama Art University, Tokyo*
33. Arthur Elsenaar and Remko Scha: *Face Shift—Algorithmic Facial Choreography*
34. Kyle Evans: *Hacked PC Game Controller*
35. Ferran Fages (www.ferranfages.net): *Acoustic Turntable with Styrofoam and Wood*
36. Robyn Farah: *Performance Video with Bent Toy*
37. Douglas Ferguson (www.douglasferguson.us/): *10-Step Sequencer w/Poly-Rhythmic Clock*
38. David First (www.davidfirst.com/): *The Radiotron*
39. Lesley Flanigan (www.lesleyflanigan.com/): *Speaker Synth*
40. Joe Grimm (www.thewindupbird.com/): *Epiphenomenal Boogie*
41. Andy Guhl (http://andy.guhl.net/): *The Instrument: LED-Painting 1*

42. Andy Guhl: *The Instrument: Double Frequencies*
43. Alex Inglizian (http://cliplead.com): [*dreamPhone*]
44. Alex Inglizian: [*optoSynth*]
45. Sawako Kato (www.troncolon.com/): *Love Love Crystal Radio* ("ishi ~ listening stone" was commissioned by Roulette with funds provided by the Jerome Foundation.)
46. Florian Kaufmann (www.floka.com): *Typosynth*
47. Florian Kaufmann: *FLooper*
48. hans w. koch (www.hans-w-koch.net/): *Computer Music I: Exploded View*
49. Kaspar König (www.kasparkoenig.com/): *Musguitear*
50. J. D. Kramer (www.fauxaudio.net/): *The Permutator*
51. Grégoire Lauvin (www.gregorth.net/): *Potatoes Organ*
52. Eric Leonardson (http://ericleonardson.org/): *The Springboard*, as featured on Gearwire.com (www.gearwire.com/)
53. Zach Lewis (www.disposablethumbs.com/): *Electric Guitarscillator*
54. Loud Objects (www.loudobjects.com/): *Live Soldering Performance*
55. LoVid (www.lovid.org): *Venus Mapped*
56. Steve Marsh (www.myspace.com/stevemarshtx): *Circuit Bent Stix Drum Machine*
57. Antony Maubert (www.antonymaubert.ift.cx): *Extending Korg M1*
58. Patrick McCarthy (www.rothmobot.com/): *How to Hack Together a Safety Pin Potentiometer*
59. James Murray (www.collateralwork.blogspot.com): *Spanish Radio*
60. Joker Nies (www.omnichord.de/Omnibend_E.html): *Circuit Bending the Omnichord*
61. Alejandra Perez Nuñez (www.elpueblodechina.org/): *elpueblodechina: A Performance of (Dis)Integrated Circuits of High Energy Electronics and People*
62. Brendan O'Connell (www.futureeelectrode.blogspot.com): *Osc-a-lot*
63. Oscillatorial Binnage: *Realisation of David Tudor's "Rainforest"*
64. Ivan Palacky (www.palacky.org): *Dopleta 180 amplified knitting machine*
65. Lorin Edwin Parker (www.electricwestern.com): *Steam Powered Synthesizer*
66. Chris Powers (www.myspace.com/cpowersbeats): *Mouth & Hips*
67. Vic Rawlings (www.vicrawlings.com): *Radio Shack Electronic Reverb and Roland Enhancer EH-2*
68. Tom Richards: *The Charming of Beautiful Car*
69. Martin Riches: *MotorMouth*
70. Jon Satrom (http://jonsatrom.com/): *The Vitch*
71. Janek Schaefer (www.audioh.com/): *The Tri-Phonic Turntable*
72. Jesse Seay (http://jesseseay.com): *Untitled (Resonant Objects)*
73. Neal Spowage: *Electro-magnetic Wands*
74. Phillip Stearns (www.art-rash.com/pixelform/): *AANN: Artificial Analog Neural Network*
75. Phillip Stearns: *Pixel Maelstrom*
76. Swiss Mechatronic Art Society (www.mechatronicart.ch/): *8 Step Sequencer*
77. Tuomo Tammenpää (http://misusage.org): *NandSynth*
78. Sebastian Tomczak (http://little-scale.blogspot.com): *"Mr. Bleeper" 3-Voice CMOS Sequencer*
79. Chester Udell (http://grove.ufl.edu/~cudell/): *Schizophrenic Zoot*

80. Valve/Membrance (Keisuke Oki, ASUNA, Minoru Sato) (www.archive.org/details/Transmediale08_ValveMembrance): *Excerpt From Performance*
81. David Watson: *Reap The Wind*
82. Dan Wilson: *Worm Cracklebox*
83. Dan Wilson: *Resonated Grille & Pitchfork Assemblies*
84. Dan Wilson: *Electromagnetic Staccato Resonator*
85. Dan Wilcox (www.soundmetal.robotcowboy.com): *soundMetal*
86. Aaron Zarzutzki (www.myspace.com/aaronzarzutzki): *Cymbal and Speaker*
87. Aaron Zarzutzki: *Crystal Oscillators*

AUDIO TRACKS

In assembling the CD that accompanied the first edition of this book, I chose existing works by musicians and sound artists that connect to the topics, techniques, and aesthetics discussed in the text. The intent was to place the technical details of the book in a musical context, and to preserve for the experimenter the surprise and satisfaction of hearing a circuit for the first time without a preconceived notion of what it should sound like. Most of the tracks are excerpted from longer pieces; a few were produced especially for the CD. They have been re-formatted for the second edition to play back on any DVD player or computer.

AUDIO TRACK NOTES

Circuit Sniffing

1. Andy Keep: "My Laptop Colony—Colony in My Laptop," 3′11″
2. Nicolas Collins: "El Loop" (excerpt) 2′17″
3. David First: "Tell Tale 2.1", 1′06″

The three tracks listed above all use electromagnetic signals as their primary sound material (see Chapter 3 and Art & Music 1 "Mortal Coils" in Chapter 3). For "My Laptop Colony—Colony in My Laptop," Andy Keep placed a telephone pickup (see Figure 3.1) on his laptop and recorded while he booted the computer and began running his concert software. "El Loop" is an excerpt from a ride on a Chicago Transit Authority elevated train, recorded through stereo telephone pickups. The sounds in "Tell Tale 2.1" result from the Theremin-like interference between two portable radios placed next to each other.

Jumping Speakers

4. John Bowers: "Study One for Victorian Synthesizer," 4′34″

John Bowers created the sounds in this piece using only batteries, clip leads, and scrap metal (see Chapter 5).

PIEZO MUSIC

5. Collin Olan: "rec01" (excerpt), 3′10″
6. Peter Cusack: "Baikal Ice" (excerpt), 2′39″
7. Richard Lerman: "Changing States 6" (excerpt), 3′38″

These tracks reveal the range of micro-sounds produced by melting and heating that can be heard through inexpensive contact mikes (see Chapter 7). Collin Olan froze two Plasti Dip-encased piezos in a block of ice and recorded as it melted. Peter Cusack used similar homemade hydrophones to record the breakup of ice flows during the spring thaw in Lake Baikal. In "Changing States 6" Richard Lerman uses a butane torch to "bow" metal whiskers solder to the edge of piezo disks (see Art & Music 3 "Piezo Music" in Chapter 7).

Drivers

8. Ute Wassermann: "Improvisation" (excerpt), 3′03″
9. Nicolas Collins: "It Was A Dark And Stormy Night" (excerpt), 1′10″

These two tracks demonstrate the ethereal effects of resonating materials with sound (see Chapter 8 and Art & Music 4 "David Tudor and 'Rainforest'" and "Drivers," both in Chapter 8). Ute Wassermann sings through a "corked" speaker (see Chapter 8, Figure 8.3 and Art & Music 5 "Drivers") held against a gong, in an improvised trio with Vladimir Tarasov and Manos Tsangaris on percussion. Nicolas Collins' voice drives electromagnetic coils placed above guitar strings (see "Drivers") and triggers percussion samples, accompanied by Guy Klucevsek on accordion. In both tracks the voices are filtered, resonated, and reverberated by tuned metal to produce sounds akin to an old plate-reverb, or shouting into a piano with the sustain pedal down.

Laying of Hands

10. Josh Winters: "Radio Study #4," 2′14″

An AM radio is played with damp fingers (see Chapter 11).

Composers Inside Electronics

11. David Behrman: "Players With Circuits," 6′00″
12. Christian Terstegge: "Ohrenbrennen" ("Ear-burn") 13′38″

In "Players with Circuits," David Behrman and Gordon Mumma, pioneers in the field of hardware hacking, mix oscillators (see Chapter 18) and acoustic piano into homemade ring modulators to produce unstable, sideband-intensive sound explosions. In "Ohrenbrennen" ("Ear-burn") four oscillators are controlled by photocells inside small altar-like boxes containing candles; the pitches of the oscillators rise in imperfect unison, punctuated by swoops that trace the sputtering of the candles as they burn down (see Art & Music 8 "Composing Inside Electronics" in Chapter 14).

Circuit Bending

13. Phil Archer: "Yamaha" PSS-380, 4′37″
14. P. Sing Cho: "Long Nosed," 5′10″
15. Maestros: "Electricity and its Double," 2′54″
16. Jane Henry: "The Chip is Down: Jerry Goes to Glory" (excerpt), part 2 of "In Jerry's Wake," 2′01″

Here come the Benders! All four of these tracks were made with bent electronic toys and other simple hacked circuits (see Chapters 12–17, and Art & Music 9 "Circuit Bending" in Chapter 15). Phil Archer makes arbitrary connections between components on the circuit board of a Yamaha keyboard with a piece of wire, triggering notes, bursts of noise, and warped "autoaccompaniment" sequences (see Figure 15.11). "Long Nosed" is a live performance by the all-bending quintet, P. Sing Cho; their sounds are born of tables strewn with electronic ephemera. In "Electricity and its Double" the Maestros (David Novak and James Fei) pair a bassoon with a toy that plays barnyard sounds. "The Chip is Down" is Jane Henry's duet for violin and hacked greeting card, excerpted from a larger work dedicated to the maverick Texas composer, Jerry Hunt (see Art & Music 11 "The Luthiers" in Chapter 28).

Distortion

17. Robert Poss: "Dicer," 2′35″

Robert Poss' guitar is chopped and channeled by two "Super Fuzzy Dicers" (see Chapter 23, Figure 23.2).

Optical Music

18. Stephen Vitiello: "World Trade Center Recordings," 2′05″
19. Norbert Möslang: "Blinking Lights," 4′00″
20. Yasunao Tone: "Imperfection Theorem of Silence," 2′05″

These three pieces link light and sound (see Chapter 24 and Art & Music 10 "Visual Music" in Chapter 24). Stephen Vitiello placed a photocell on the eyepiece of a telescope (see Figure 24.4) and focused it on nighttime traffic far below his studio on the ninety-first floor of the World Trade Center. Norbert Möslang's similar circuits extract rhythmic patterns from blinking bicycle lights. Yasunao Tone uses Scotch tape to disturb the laser reading a CD, resulting in a plethora of glitches.

BIOGRAPHIES

Andy Keep is a performer/composer, who researches behavioral models and timbrel languages of audio feedback set-ups through responsive strategies in improvisation and composition. He works in Bath, UK.

Nicolas Collins, the author, has been hacking and designing musical circuits since he was eighteen years old. He was also one of the first artists to use microcomputers in live performance (1978). New York born and raised, he currently lives in Chicago. Additional information on his work is available at www.NicolasCollins.com.

David First, a New Yorker, has long been interested in the secret worlds of minimal gestures, hyper-sensual tuning relationships, and ritual phenomena. Additional information on his work is available at www.davidfirst.com.

John Bowers is a composer, performer, computer scientist, and researcher of early electrical instruments. He currently works in Norwich, UK. Additional information on his work is available at www.onoma.co.uk/jmbowers/victorian.html.

Collin Olan is a New York-based artist and performer working with computers and recording technology to create alternate visions of everyday phenomena.

Peter Cusack's work combines his long-standing interests in improvisation, live electronic performance, and environmental recording. London-based, he travels extensively in pursuit of sound.

Richard Lerman has been working with piezo disks since the late 1970s, integrating them into performance instruments, audio installations, and recordings. He lives in Phoenix, Arizona. Additional information on his work is available at www.west.asu.edu/rlerman.

Ute Wassermann is a singer, sound artist, and instrument designer based in Berlin. She has collaborated with numerous composers and improvisers. Her multiphonic singing brings the various resonant spaces of the body into vibration, and generates resonant phenomena in space.

Josh Winters is a visual artist and performer working in Chicago and Urbana, Illinois.

David Behrman, a New Yorker, was one of the first composers to build his own electronic instruments in the 1960s. He has also been at the forefront of live computer music since the advent of the Kim 1 microcomputer (see Foreword).

Christian Terstegge, the Hamburg-based sound artist, has specialized in elegant, minimal installations incorporating homemade circuitry since the mid-1980s.

Phil Archer's work revolves largely around the creative modification of consumer electronic devices, altering their sound production capabilities to generate previously impossible results. He is based in Norwich, England. Additional information on his work is available at www.philarcher.net/.

P. Sing Cho is an all circuit-bending band of five London artists: Knut Aufermann, Moshi Honen, Sarah Washington, Chris Weaver, and Dan Wilson.

The Maestros (James Fei and David Novak) perform live with acoustic instruments, hacked toys, contact mikes, and vintage analog synthesizer modules. Additional information on their work is available at www.jamesfei.com.

Jane Henry performs free improvisations and original compositions as a solo violinist, and sometimes with low-tech electronic devices, and/or human partners. She is currently active in presenting international new music improvisers in Texas venues. Additional information on her work is available at www.janehenry.com.

Robert Poss was a founding member of Band Of Susans. He has performed and recorded with Nicolas Collins, Rhys Chatham, Ben Neill, Phill Niblock, David Dramm, and Bruce Gilbert and recently began working with choreographers Alexandra Beller and Sally Gross. A New Yorker, he continues to perform his innovative guitar and electronics pieces in the United States and Europe. Additional information on his work is available at www.distortionistruth.com.

Stephen Vitiello is an electronic musician and media artist. His sound installations, photographs, and sculptures have been exhibited internationally. He currently resides in Richmond, Virginia where he is in the faculty of the Department of Kinetic Imaging at Virginia Commonwealth University. Additional information on his work is available at www.stephenvitiello.com.

Norbert Möslang co-founded the legendary Swiss electronic duo Voice Crack in 1972. He currently performs solo and in collaboration with many musicians around the world. He is a master of "cracking everyday electronics" and a virtuoso of circuit-based performance. Additional information on his work is available at www.moeslang.com.

Yasunao Tone was raised and educated in Japan, but has long been a fixture on the New York avant-garde music scene. He is the first person to have made music out of intentionally-induced CD malfunctions.

CREDITS

Artists retain full copyright on their submitted tracks.

1. Andy Keep, "My Laptop Colony—Colony in My Laptop" was previously released on *A Call For Silence*, Sonic Arts Network, 2004. Used by permission of Andy Keep.
2. Nicolas Collins, "El Loop" was commissioned by Chris Cutler for broadcast on Resonance FM, London, 2002. Used by permission of Nicolas Collins.
3. David First, "Tell Tale 2.1" was previously released on *A Call For Silence*, Sonic Arts Network, 2004. Used by permission of David First.
4. John Bowers, "Study One for Victorian Synthesizer" was recorded expressly for this CD. Used by permission of John Bowers.
5. Collin Olan, "rec01" was released in full on *rec01*, Apestaartje CD, 2002. Used by permission of Colin Olan and Apestaartje.
6. Peter Cusack, "Baikal Ice" was released in full on *Baikal Ice (Spring 2003)*, ReR CD. Used by permission of Peter Cusack.
7. Richard Lerman, "Changing States 6" was recorded in concert at the Audio Arts Festival in Krakow, Poland, 1999. Used by permission of Richard Lerman.
8. Ute Wassermann, "Improvisation" was recorded in concert at Akademie Schloss Solitude in 1995, and was released in full on *Improvised Music From Solitude in Eleven Parts* (Vladimir Tarasov, artistic director,) Akademie Schloss Solitude CD, 1995. Used by permission of Jean-Baptiste Joly, artistic director, Akademie Schloss Solitude, Manos Tsangaris, Vladimir Tarasov, and Ute Wassermann.

9. Nicolas Collins, "It Was A Dark And Stormy Night" was released in full on *It Was A Dark And Stormy Night*, Trace Elements CD, 1992. Used by permission of Nicolas Collins.
10. Josh Winters, "Radio Study #4" is used by permission of Josh Winters.
11. David Behrman, "Players With Circuits" was recorded in concert at Lincoln Center Library, New York City, in 1966, and was previously released on *Wave Train*, Alga Marghen CD, 1999. Used by permission of David Behrman.
12. Christian Terstegge, "Ohrenbrennen" was recorded in concert at Künstlerhaus Hamburg, Hamburg, Germany, 1988. Used by permission of Christian Terstegge.
13. Phil Archer, "Yamaha PSS-380" is used by permission of Phil Archer.
14. P. Sing Cho, "Long Nosed" was recorded live at the 12 Bar Club, London, England, 2005. Used by permission of Sarah Washington.
15. Maestros, "Electricity and its Double" was previously released on *Precision Electro-Acoustics*, Organized Sound Recordings, 2001. Used by permission of James Fei.
16. Jane Henry, "The Chip is Down: Jerry Goes to Glory" was recorded in concert at Roulette, New York City, by Jim Staley, 1996. Used by permission of Jane Henry.
17. Robert Poss, "Dicer" was recorded expressly for this CD. Used by permission of Robert Poss.
18. Stephen Vitiello, "World Trade Center Recordings" (1999) was previously released on *A Call For Silence*, Sonic Arts Network, 2004. Used by permission of Stephen Vitiello.
19. Norbert Möslang, "Blinking Lights" was recorded expressly for this CD. Used by permission of Norbert Möslang.
20. Yasunao Tone, "Imperfection Theorem of Silence" was previously released on *A Call For Silence*, Sonic Arts Network, 2004. Used by permission of Yasunao Tone.

DVD PRODUCTION CREDITS

Chapter project videos shot and edited by James Murray
Audio Mastering of CD material: Robert Poss
DVD Design and Mastering: James Murray

Notes and References

The following urls, books, periodicals, and recordings are relevant to specific chapters. For many of the people and things referenced in the text Web sites are the fastest, most economical first step towards learning more. URLs are valid as of September, 2008.

Foreword to the First Edition

David Behrman: www.lovely.com/bios/behrman.html.
Henry Cowell: www.schirmer.com/composers/cowell_bio.html.
Conlon Nancarrow: www.kylegann.com/index2.html.
Harry Partch: www.corporeal.com/.
David Tudor: see Art & Music 4 "David Tudor and 'Rainforest'," Chapter 8.
Gordon Mumma: www.brainwashed.com/mumma/.
www.lovely.com/bios/mumma.html.
Max/MSP: www.cycling74.com/.

Introduction

John Cage: see Art & Music 2 "John Cage," Chapter 7.
Don Buchla: see Art & Music 11 "The Luthiers," Chapter 28.
STEIM: see Art & Music 7 "The Cracklebox," Chapter 11, and Art & Music 11 "The Luthiers," Chapter 28.

1: Getting Started

Weller soldering irons are the choice of the serious hacker: www.cooperhandtools.com/brands/weller/.

3: Circuit Sniffing

Irdial records (www.irdial.com) has been involved in some of the more arcane aspects of radio culture—the "Conet Project" CD set documents the bizarre phenomenon of shortwave "number readers."

Art & Music 1: Mortal Coils

Leon Theremin: there are hundreds of Web sites devoted to Theremin and his instruments, but this one, by his niece, is a good place to start: www.lydiakavina.com/therem.html.

Another, with a good bibliography: www.thereminworld.com/pubs.asp.

And Steven Martin's film is wonderful and available on video: "Theremin: An Electronic Odyssey" (1994).

Gert-Jan Prins: www.gjp.info/.

Alvin Lucier: http://alucier.web.wesleyan.edu/.

A recording of "Sferics" can be found on *Sferics*, Lovely Music, Ltd. VR 1017, 1988.

A collection of interviews, writings, and scores by Alvin Lucier: Alvin Lucier, *Reflections. Interviews, Scores, Writings (Reflexionen, Interviews, Notationen, Texte)*, English–German Edition, MusikTexte, Köln, 1995.

Lauren Carter and Joe Grimm: www.thewindupbird.com/.

Karlheinz Stockhausen: www.stockhausen.org/.

Disinformation: www.discogs.com/artist/Disinformation. www.ashinternational.com/.

Joyce Hinterding: www.sunvalleyresearch.com/Luminoska/index2.htm.

Haco: www.japanimprov.com/haco/.

Haco. *Stereo Bugscope 00*. Improvised Music from Japan, IMJ-523, 2004.

David First: www.davidfirst.com/.

"Tell Tale 2.1," *A Call for Silence*. London: Sonic Arts Network CD/book, 2004.

Andy Keep. "My Laptop Colony, Colony In My Laptop" *A Call for Silence*. London: Sonic Arts Network CD/book, 2004.

Jérôme Noetinger: www.metamkine.com/.

Rob Mullender: www.ear-rational.com/detail.php?id=8341&searchparm=Mullender&order=&offset=.

Christina Kubisch: www.christinakubisch.de/.

Information on her inductive transmissions and electrical walks can be found at www.christina kubisch.de/english/klangundlicht_frs.htm.

Sawako Kato: www.troncolon.com/.

Information on her crystal radio work: www.troncolon.com/0CR/.

Vasco Alvo "FM kbrd": www.vascoalvo.com/ProjectsBox/fmkbrd.html.

4: In/Out

For a good history of the Telharmonium see: Joel Chadabe, *Electric Sound—The Past and Promise of Electronic Music*. New York: Prentice Hall, 1997, pp 3–8.

Lorin Parker www.electricwestern.com.

Information on the steam-powered synth: www.electricwestern.com/stynth.html. and the DVD.

5: The Celebrated Jumping Speaker of Bowers County

John Bowers: www.onoma.co.uk/jmbowers/victorian.html.

http://suborderly.blogspot.com/2007/03/suborderly-music-victorian-synthesizer.html.

John Bowers and Vanessa Yaremchuk, "The Priority of the Component, or in Praise of Capricious Circuitry." *Leonardo Music Journal*, Vol. 17 (2007), p. 39.

Aaron Zarzutzki: www.myspace.com/aaronzarzutzki.

Alex Baker: www.alexbaker.co.uk/.

For an example of the Jumping Speaker principle applied to resonating guitar strings see Dan Wilson's video "Electromagnetic Staccato Resonator" on the DVD.

6: How to Solder

For information on clip lead fanaticism see Phil Archer, "Clip Art." *Leonardo Music Journal*, Vol. 17 (2007), pp. 29–30.

7: How to Make a Contact Mike

A good history of piezoelectricity: www.piezo.com/tech4history.html.
Information on Plastidip: www.plastidip.com/index.php.
Measurement Specialties: www.meas-spec.com/myMeas/sensors/piezo.asp.
Their piezo film transducers can be bought retail from DigiKey (www.digikey.com): http://search.digikey.com/scripts/DkSearch/dksus.dll?Cat=1966393;keywords=Measurement%20Specialties.

Art & Music 2: John Cage—The Father of Invention

Some good Cage Web sites: www.music.princeton.edu/~jwp/texts/worklist.html www.johncage.info/.
Read this: John Cage, *Silence: Lectures and Writing*, Wesleyan University Press, CT, 1961.
Lou Harrison: www.harrisondocumentary.com/.

Art & Music 3: Piezo Music

Hugh Davies, *Sounds Heard*, Soundword, Chelmsford, UK, 2002.
Richard Lerman: www.west.asu.edu/rlerman/.
Eric Leonardson: www.ericleonardson.org/.
Eric Leonardson. "The Springboard: The Joy of Piezo Disk Pickups for Amplified Coil Springs." *Leonardo Music Journal,* Vol. 17 (2007), pp. 17–20.
Adachi Tomomi: www.adachitomomi.com/a/sminst.html.
Ivan Palacky: http://carpetscurtains.fiume.cz/about_ip.html.
Otomo Yoshihide: www.japanimprov.com/yotomo/.
Janek Schaefer: www.audioh.com/.
Vasco Alvo "Stylus Pen": www.vascoalvo.com/ProjectsBox/stylus.html.
Michael Graeve: www.michaelgraeve.com/.
Alan Lamb: www.rainerlinz.net/NMA/22CAC/lamb.html.
Fence music: www.abc.net.au/arts/adlib/stories/s873159.htm.
www.jonroseweb.com/f_projects_great_fences.html.
Peter Cusack, *Baikal Ice*, ReR CD, 2004.
Collin Olan, *Rec01, Listen 001*, Apestraartjje CD, 2002.

8: Turn Your Tiny Wall Into a Speaker

Regarding stocking policy of Radio Shack see: www.theonion.com/content/news/even_ceo_cant_figure_out_how.
An informative history of the spring reverb: www.accutronicsreverb.com/history.htm.
A good tube amp resource page, with links to sources for beefier output transformers: www.diyguitaramp.com/parts.html.
Rolen-Star Audio Transducer: www.rolen-star.com/.
Bass Shakers: www.aurasound.com/.
I-Beam: www.feeltheaudio.com/.
ButtKicker: www.thebuttkicker.com/.
Soundbug: www.feonic.com/fr.asp?section=products.
Sonic Impact Technologies: www.si5.com/.
A new version of the speaker: www.si5.com/products.php?pID=4022.

Zelco "Outi": http://zelcocom.nationprotect.net/Merchant2/merchant.mvc?Screen=PROD&Product_Code=03312&Category_Code=NewProducts.
EBow: www.ebow.com/.
Fernandes Sustainer: www.fernandesguitars.com/sustainer.html.

Art & Music 4: David Tudor and "Rainforest"

The David Tudor quote is from "From Piano to Electronics." *Music and Musicians*, Vol. 20, August (1972), interview with Victor Schonfeld.
Tudor Web sites: www.emf.org/tudor/
www.getty.edu/research/conducting_research/digitized_collections/davidtudor/.
"Composers Inside Electronics—Music After David Tudor." *Leonardo Music Journal*, Vol. 14 (2004).

Art & Music 5: Drivers

For information on Alvin Lucier see Art & Music 1 "Mortal Coils" in Chapter 3.
Alvin Lucier, *Music For Solo Performer*, Lovely Music, Ltd., VR 1014, 1982.
Ute Wassermann with Windy Gong: *Improvised Music From Solitude in Eleven Parts*, CD, animato acd 6008–3, Germany, 1994.
Jens Brand: www.jensbrand.com/.
Nicolas Collins: www.NicolasCollins.com.
Collins' backwards electric guitar can be heard on:
Sound Without Picture, Periplum Records, 1999.
It Was A Dark And Stormy Night, Trace Elements Records, 1992.
Felix Hess, *Light As Air*. Heidelberg, Germany: Stadgalerie Saarbrücken, Kehrer Verlag, 2001.
Jesse Seay, "Untitled": http://jesseseay.com/section/18322.html.
Toshiya Tsunoda, "Listening to the Reflection of Points":
www.westspace.org.au/program/toshiya-tsunoda.html.
www.helenscarsdale.com/haynes/words/crossplatformtsunoda.htm.

9: Tape Heads

Art & Music 6: Tape

Alvin Lucier, *I am Sitting in a Room*, Lovely Music, Ltd., 1981/1990.
Steve Reich, "Come Out," on *Steve Reich: EarlyWorks*, Nonesuch Records, 1987.
Pauline Oliveros, "I of IV" on *New Sounds in Electronic Music: works of Oliveros, Reich, Maxfield*, Odyssey Records, 1967.
Terry Riley, *A Rainbow in Curved Air*, CBS Masterworks, 1969/1990.
Nam June Paik, Random Access: www.nydigitalsalon.org/10/artwork.php?nav=artists&artwork=13.
Laurie Anderson: www.laurieanderson.com/.
The Tape Bow Violin can be heard on:
"Three Walking Songs (for Tape Bow Violin)" from *United States Pt. 1*.
"Sax Solo (for Tape Bow Violin)" from *United States Pt. 1*.
"I Dreamed I Had To Take a Test . . ." from *United States Pt. 3*.
The quote is from: Laurie Anderson, *The Record of the Time*. Lyon: Musée Art Contemporain, Lyon, 2002.
Mark Trayle: http://music.calarts.edu/~met/.
Mark Trayle. "Free Enterprise: Virtual Capital and Counterfeit Music at the End of the Century." *Leonardo Music Journal*, Vol. 9 (1999), pp 19–22.

10: A Simple Air Mike

A good site for source material on the theory of microphone operation (including electret capsules): www.educypedia.be/electronics/microphonestypes.htm.

Electret theory: www.nationmaster.com/encyclopedia/Electret-microphone.

Good mike builder site: http://tech.groups.yahoo.com/group/micbuilders/.

Rob Danielson has done extensive experiments with electret microphone elements. His Web site lists many resources for mike builders: www.uwm.edu/~type/audio-art-tech-gallery/.

An excellent guide to amplifying *any* instrument (on a small budget) is:

Bart Hopkin, *Getting a Bigger Sound: Pickups and Microphones for Your Musical Instrument*. Tucson, AZ: Sharp Press, 2003. Hopkin covers making and using coil pickups, contact mikes, and air mikes, and includes a very neat trick for rewiring electret elements to improve their ability to record very loud sounds without distortion.

A good explanation of Plug-In-Power: www.telinga.com/pipwp.htm.

A great general resource for budget-bound audio experimenters is *Tape Op—The Creative Music Recording Magazine* (www.tapeop.com). They review gear, publish how-to articles, interview independent music engineers and artists, and provide an excellent useful resource for alternative recording practice.

David Watson: www.youtube.com/watch?v=yBWdLAagPgA.

Felix Hess's barometric recordings are included on a CD in *Light As Air* (see notes to "Drivers," Chapter 8).

11: Laying of Hands

Duncan Chapman and Stewart Collinson have recently started a "Bent Radio Orchestra" that gave its premiere performance at Expo Brighton in July 2008—see Chapter 30, http://expofestival.org/?page_id=23, and http://elists.resynthesize.com/mulch-discuss/2008/05/1948215/-Bent-Radio-Orchestra-EXPO-2008.html.

Mike Bullock: www.finenoiseandlight.net/.

Seth Cluett: www.onelonelypixel.org/.

Susan Stenger: www.susanstenger.co.uk/.

Art & Music 7: The Cracklebox

For a detailed history of the Cracklebox go to: www.crackle.org/CrackleBox.htm

To purchase a Cracklebox go to: www.steim.org/steim/cracklebox.php.

14: Ohm's Law For Dummies

Art & Music 8: Composing Inside Electronics

An excellent resource on the history of the San Francisco Bay area music scene can be found at: http://o-art.org/history/index.html.

Center for Contemporary Music at Mills College: www.mills.edu/academics/undergraduate/mus/center_contemporary_music.php.

Robert Ashley: www.lovely.com/bios/ashley.html.

"Music With Roots in the Aether" (1976), Ashley's video portrait of seven American composers, provides an excellent insight into the Zeitgeist of the post-Cagean culture of "seat of the pants" electronic music.

Kenneth Atchley: www.katch.com/.

John Bischoff : www.johnbischoff.com/.

Douglas Kahn. "A Musical Technography of John Bischoff." *Leonardo Music Journal*, Vol. 14 (2004), pp 75–79.

Chris Brown: www.cbmuse.com/.
Laetitia de Compiegne Sonami: www.sonami.net/.
Scott Gresham-Lancaster: www.o-art.org/Scot/.
Frankie Mann: www.lovely.com/bios/mann.html.
Tim Perkis: www.perkis.com.
Mark Trayle: see Art & Music 6 "Tape," Chapter 9.
Nicolas Collins: see Art & Music 5 "Drivers," Chapter 8.
Ron Kuivila: www.lovely.com/bios/kuivila.html.
http://framework.v2.nl/archive/archive/node/actor/default.xslt/nodenr-65793.
Serge Tcherepnin: www.serge-fans.com/history.htm.
Paul De Marinis: www.well.com/~demarini/.
Voice Crack: www.moeslang.com.
http://andy.guhl.net/.
Jim Horton: http://mitpress2.mit.edu/e-journals/Leonardo/lmj/horton.html.
The Kim 1: www.kim-1.com/.
An insider's history of the emergence of early microcomputer music from the Bay area homemade circuitry scene can be found at: http://crossfade.walkerart.org/brownbischoff/introduction_main.html.

15: Beyond the Pot

T. Escobedo's "Synthstick": www.geocities.com/tpe123/folkurban/synthstick/synthstick.html. Lots of other cool instruments are featured on his "FolkUrban Music" Web site: www.geocities.com/tpe123/folkurban/index.html.
Mike Challis: www.mikechallis.com/.

Art & Music 9: Circuit Bending

Reed Ghazala: www.anti-theory.com.
Reed Ghazala, *Circuit Bending: Build Your Own Alien Instruments*. New York: Wiley Publications, 2005.
Experimental Musical Instruments: www.windworld.com/.
The Bent Festivals: www.bentfestival.org/.
Joker Nies: www.klangbureau.de/.
Jon Satrom: see notes to Art & Music 10 "Visual Music," Chapter 24.
Speak and Spell: http://en.wikipedia.org/wiki/Speak_&_Spell_(game).
The Phil Archer quotation is from personal correspondence.
Phil Archer: www.philarcher.net/.
John Bowers: see notes to Chapter 5.
Bowers, J., and Archer, P. "Not Hyper, Not Meta, Not Cyber but Infra-Instruments." In Proceedings of NIME, 05 (New Interfaces for Musical Expression), May 26–28, 2005, Vancouver, BC, Canada. Downloadable from http://hct.ece.ubc.ca/nime/2005.
Rawlings, Vic. "The Boss GE-7 E.Q. and Flexible Speaker Array as Tonal Filters." *Leonardo Music Journal*, Vol. 17 (2007), pp. 37–38.
The Sarah Washington quotation is from personal correspondence, May 24, 2005.
A good Web site for information on the British Circuit Bending scene is: www.lektrolab.com/.
Leonardo Music Journal, Vol. 17 (2007), *My Favorite Things—The Joy of the Gizmo*, features articles and artists' statements by many Circuit Benders, as well as a CD, curated by Sarah Washington, with 17 tracks by various artists.

17: Jack, Batt, and Pack

Peter Zummo: www.kalvos.org/zummope.html.
Alex Inglizian: http://cliplead.com/.

18: World's Simplest Circuit

For general information on the CMOS family of integrated circuits, suggested applications, and a hands-on primer in binary logic get a copy of:
Donald Lancaster, *CMOS Cookbook*. Indianapolis, IN: SAMS Publications, 1977.
Dan Wilcox: www.soundmetal.robotcowboy.com.
Grégoire Lauvin: www.gregorth.net/.
For a great example of early photoresistor-controlled oscillators, listen to: David Behrman, "Runthrough," on *Wave Train*, Alga Marghen CD, 1998.

19: From Breadboard to Circuit Board

A strong argument against my comment on the fragility of breadboards can be found in the fabulous effect-pedal kit known as "The Beavis Board": www.beavisaudio.com/bboard/index.htm. See Chapter 30, Figure 30.19.
And: Phil Archer, "Clip Art." *Leonardo Music Journal*, Vol. 17 (2007), pp. 29–30.

20: Getting Messy

Passive filters designed by Robert Moog: www.electro-music.com/forum/viewtopic.php?highlight=passive&t=26643.
Online calculator for selecting capacitor values for specific filter cutoff frequencies: www.muzique.com/schem/filter.htm.
A neat little Voltage Starve stand-alone module can be bought from Beavis Audio: www.beavisaudio.com/Projects/DBS/.

21: On/Off

Numerous discussions on the virtues of optical gates, compressors, and the like can be found in *Tape Op Magazine* (see references for Chapter 10).
Lowell Cross: see Art & Music 10 "Visual Music," Chapter 24.
Frederic Rzewski: http://en.wikipedia.org/wiki/Frederic_Rzewski.
"Bruit Secret": http://arthist.binghamton.edu/duchamp/Hidden%20Noise.html.

22: Amplification and Distortion

Craig Anderton: see Art & Music 11 "The Luthiers," Chapter 28, and Appendix A.
Robert Poss: www.distortionistruth.com/.
Robert Poss. "Distortion is Truth." *Leonardo Music Journal*, Vol. 8 (1998), pp. 45–48.

23: Analog to Digital Conversion, Sort Of

"Pools of Phase Locked Loops" is the title of a 1972 composition by David Behrman. It can be heard on: Behrman, David. *My Dear Siegfried*. XI Records CD. 2005.
For examples of advanced audio application of CMOS chips see Tom Bug's wonderful "Bug Brand" web site: http://bugbrand.co.uk/pages/electronics.htm.

24: Video Music/Music Video

Soundhack: www.soundhack.com/.
Big Eye: www.steim.org/steim/bigeye.html.
Jitter: www.cycling74.com/products/jitter.html.

Art & Music 10: Visual Music

Nam June Paik, "Magnet TV": www.nydigitalsalon.org/10/artwork.php?artwork=27.
Bill Viola: www.billviola.com/.
Billy Roisz: http://gnu.klingt.org/.
Watts/Behrman/Diamond, "Cloud Music":
www.vasulka.org/Kitchen/PDF_Eigenwelt/pdf/152-154.pdf.
Yasunao Tone: www.lovely.com/bios/tone.html.
Yasunao Tone (Asphodel Ltd., Asphodel 2011), 2003. *Solo for Wounded CD* (Tzadik, TZ-7212), 1997.
Lowell Cross: www.LowellCross.com www.8ung.at/fzmw/2001/2001T1.htm.
Lowell Cross, "'Reunion': John Cage, Marcel Duchamp, Electronic Music and Chess." *Leonardo Music Journal*, Vol. 9 (1999), pp. 35–42.
Stephen Vitiello: www.stephenvitiello.com/.
www.ubu.com/sound/vitiello.html.
For Norbert Möslang and Andy Guhl see Art & Music 8 "Composing Inside Electronics," Chapter 14.
BMBCon: www.bmbcon.demon.nl/con/.
Jon Satrom: http://jonsatrom.com/.
LoVid: www.lovid.org.
Arthur Elsenaar and Remko Scha, "Electric Body Manipulation as Performance Art: A Historical Perspective." *Leonardo Music Journal*, Vol. 12 (2002), pp. 17–28.

26: Mixer, Matrices, and Processing

For more on Y-cord mixing see: Andy Keep. "Audio Y-connectors: My Secret for Instant Guerrilla Oscillators, Raw Synthesis and Dirty Cross Modulation." *Leonardo Music Journal*, Vol. 17 (2007), pp. 30–31.
Jon Satrom: see notes to Art & Music 10 "Visual Music," Chapter 24.
James Murray: www.collateralwork.blogspot.com.
Douglas Ferguson: www.douglasferguson.us/.
Steve Marsh: www.myspace.com/stevemarshtx.

28: Analog to Digital Conversion, Really

Leroy Anderson: see "The Typewriter" on www.leroy-anderson.com/html/hearthemusic.htm.
I-Cube: http://infusionsystems.com/.
MidiTron: http://eroktronix.com/.
CH Products: www.chproducts.com.
SensorLab, JunXion and JunXionboard: www.steim.org/steim/products.html.
Arduino: www.arduino.cc/.
Gluion: www.glui.de/.
Some good resources on hacking the Wii:
http://createdigitalmusic.com/2007/03/20/free-mac-looper-for-wii-controller-wii-midi-hacking-round-up/.
www.hackszine.com/blog/archive/2008/06/wii_guitar_hero_guitar_as_a_re.html.
Mattel Power Glove: http://en.wikipedia.org/wiki/Power_Glove.
As long as you're hacking game controllers you might read up on their history: J. C. Herz, *Joystick Nation*, London: Abacus (Little and Brown), 1997.
For advice on hacking mice to make motion trackers see: www.contrib.andrew.cmu.edu/~ttrutna/16–264/Vision_Project/.

An excellent overview of the general theory of the design of physical interfaces for computers can be found in: Bert Bongers, "Interaction with Our Electronic Environment—an E-cological Approach to Physical Interface Design," Cahier 34, Academie voor Digitale Comunicatie, Hogeschool van Utrecht, Utrecht (NL), 2004.

The annual NIME (New Instruments for Musical Expression) conference gathers dozens of practitioners in the field. Many of the papers can be downloaded through the website: www.nime.org/.

Art & Music 11: The Luthiers

Robert Moog: www.moogmusic.com.

Donald Buchla: www.buchla.com.

Serge Tcherepnin: www.serge-fans.com/history.htm.

Craig Anderton: www.craiganderton.com/.

Craig Anderton, *Electronic Projects for Musicians*. New York: AMSCO Publications, 1975.

Bob Bielecki: www.bard.edu/academics/faculty/faculty.php?action=details&id=120.

Bert Bongers: www.xs4all.nl/~bertbon/.

Bert Bongers. "Electronic Musical Instruments: Expriences of a New Luthier." *Leonardo Music Journal,* Vol. 17 (2007), pp. 9–16.

Laetitia Sonami: see Art & Music 8 "Composing Inside Electronics," Chapter 14.

Jonathan Impett: www.uea.ac.uk/mus/staff /academic/impett.html.

Michel Waisvisz: see Art & Music 7 "The Cracklebox," Chapter 11.

www.crackle.org/.

Sukandar Kartadinata: www.glui.de.

STEIM: www.steim.org/steim/.

Jane Henry: www.janehenry.com.

Machine Collective: www.machinecollective.org/.

Monome: http://monome.org/.

Ben Neill: www.benneill.com/.

Chikashi Miyama: http://homepage.mac.com/chikashimiyama/.

29: Power Supplies

Suggestions for making batteries from fruits, vegetables, spare change, salt, and wet paper towel can be found in: Cy Tymony, *Sneaky Uses for Everyday Things*. Kansas City, KS: Andrews McMeel Publishing, 2003.

Listening to the potato's song was suggested by Sally Shepardson.

Ithai Benjamin: www.ithaibenjamin.com.

Alejandro Abreu: www.alexabreu.com/.

Phil Archer: www.philarcher.net/.

30: The Future Is Now

For more information on Reed Ghazala and Circuit Bending see notes to "Circuit Bending" in Chapter 15.

Kapsar König: www.kasparkoenig.com/.

Charles McGhee Hassrick: http://home.uchicago.edu/~chasmh/.

Chester Udell: http://grove.ufl.edu/~cudell/.

Ian Baxter: www.ianbaxter.net.

Haco: www.japanimprov.com/haco/.

Zach Lewis: www.disposablethumbs.com/.

Jeffrey Byron: www.mae-shi.com/.

Peter Blasser: www.ciat-lonbarde.net/.

Radiothizer: www.ciat-lonbarde.net/rzither/index.html.

Ben Neill: www.benneill.com/.
Lesley Flanigan: www.leseleyflanigan.com/.
Speaker Synth: : www.leseleyflanigan.com/speakersynth.html.
Minoru Sato: www.ms-wrk.com/.
www.youtube.com/watch?v=e30xL1yiqoI.
www.archive.org/details/Transmediale08_ValveMembrance.
Vic Rawlings: www.vicrawlings.com.
Toshimaru Nakamura: www.japanimprov.com/tnakamura/.
Daniel Schorno: www.pocketopera.info.
Fred Lonberg-Holm: www.lonberg-holm.info/.
Ithai Benjamin: www.ithaibenjamin.com.
Alejandro Abreu: www.alexabreu.com/.
Emir Bijukic: www.bijukic.org/.
Lorin Parker: www.electricwestern.com.
Information on the steam-powered synth: www.electricwestern.com/stynth.html.
M. R. Duffey "The Vocal Memnon and Solar Thermal Automata." *Leonardo Music Journal*, 17 (2007), pp. 51–54.
NotTheSameColor: http://ntsc.klingt.org/.
(e)-Bombers: http://e-bombers.jugem.jp.
Oscillatorial Binnage: http://vids.myspace.com/index.cfm?fuseaction=vids.channel&ChannelID=250048313.
Owl Project: www.owlproject.com/.
Grace and Delete: www.utrophia.net/utrophia%20site/site/audio/grace_and_delete.html.
Stewart Collinson and Duncan Chapman:
http://expofestival.org/?page_id=23.
http://elists.resynthesize.com/mulch-discuss/2008/05/1948215/-Bent-Radio-Orchestra-EXPO-2008.html.
Scratch Orchestra: Cornelius Cardew, editor. *Scratch Music*. Cambridge, MA: MIT Press, 1974.
Portsmouth Sinfonia: www.portsmouthsinfonia.com/.
RGBToysband: http://bandua.net/info/040908_alg-a/modules.php?op=modload&name=News&file=article&sid=326&mode=thread&order=0&thold=0.
Swiss Mechatronic Art Society: www.mechatronicart.ch/.
Loud Objects: www.loudobjects.com/.
Nick Collins: www.informatics.sussex.ac.uk/users/nc81/.
Nick Collins vs. Nic Collins: www.youtube.com/watch?v=HXBgRKp8ySA.
Nic Collins the potter: www.nic-collins.co.uk/.
Brett Balogh: www.brettbalogh.com.
Alex Inglizian: http://cliplead.com/.
James Murray: www.collateralwork.blogspot.com/.
Chris Powers: www.myspace.com/cpowersbeats.
Jon Satrom: http://jonsatrom.com/.
J. D. Kramer: www.fauxaudio.net/.
Phil Archer: www.philarcher.net/.
Luke Abbott: http://lukeabbottmusic.blogspot.com/.
Phillip Stearns: www.art-rash.com/pixelform/.
BMBCon: www.bmbcon.demon.nl/con/.
LoVid: www.lovid.org.
History of "ghoti": http://en.wikipedia.org/wiki/Ghoti.
Alvin Lucier's comment on the flatness of circuitry was made during the videotaping of Robert Ashley's series of portraits of composer, "Music With Roots In The Aether" (Lovely Music, 1976). The audio portion of the interview can be heard on Ubuweb: http://ubu.artmob.ca/sound/aether/Music-Aether_03_01_Lucier.mp3.

Martin Riches: Carsten Seiffarth, Markus Steffens, Thorsten Sadowsk. *Martin Riches: Maskinerne/The Machines*. Heidleberg: Kehrer Verlag 2004.

Ferran Fages: www.ferranfages.net.

Chris DeLaurnti: www.delaurenti.net/.

Joy Ridderhof: www.missionfrontiers.org/1999/08/joy.html.

Guillermo Galindo: www.galindog.com/.

Marc Berghaus: www.marcberghaus.com/.

Sebastian Tomczak: http://little-scale.blogspot.com.

Christian Haines: www.music.adelaide.edu.au/staff/musictech/christian_haines.html.

Tuomo Tammenpää: http://misusage.org.

Alejandra Perez Nuñez: www.elpueblodechina.org/.

Douglas Ferguson: www.douglasferguson.us/.

Eric Archer: http://ericarcher.net/.

Steve Marsh: www.myspace.com/stevemarshtx.

Adachi Tomomi: www.adachitomomi.com/.

Florian Kaufmann: www.floka.com/cmos/cmossounds.html.

Osamu Hoshuyama: www5b.biglobe.ne.jp/~houshu/synth/.

Todd Bailey: www.narrat1ve.com/.

Beavis Audio Research: www.beavisaudio.com/bboard/.

hans w. koch: www.hans-w-koch.net/.

Tom Zahuranec was one of the first artists to experiment with the musical application of plant response to human presence. He was active in the mid-1970s, primarily in the San Francisco Bay area. See www.psychobotany.com/projects/Tom%20Zahuranec.htm. He appeared in the Walon Green's 1976 documentary film, "The Secret Life of Plants," which is available on Google Video: http://video.google.com/videosearch?q=%22The+Secret+Life+of+Plants%22&emb=0#.

No mention of the musical use of printers would be complete without The User, a Canadian artists' collective (Thomas McIntosh and Emmanuel Maddan) who send synchronized, specially-programmed print commands to dozens of obsolete printers in order to create their massive "Symphony for Dot Matrix Printers." See: www.theuser.org/dotmatrix/en/intro.html.

Wilson hedgehog video: www.youtube.com/watch?v=RuhFP3JNFFW.

Alex Baker: www.alexbaker.co.uk/.

Douglas Irving Repetto. "Crash and Bloom: A Self-Defeating Regenerative System." *Leonardo Music Journal*, Vol. 14 (2004), pp. 88–94.
http://music.columbia.edu/~douglas/portfolio/index.shtml.

Dorkbot: http://dorkbot.org/.

Patrick McCarthy: www.rothmobot.com/.
www.rubbermonkey.org/circuitbending/.

Nyle Steiner: www.sparkbangbuzz.com/.

Peter Blasser paper circuits: www.ciat-lonbarde.net/paper/index.html.
Peter Blasser. "Pretty Paper Rolls: Experiments in Woven Circuits." *Leonardo Music Journal*, Vol. 17 (2007), pp. 25–27.

Grégoire Lauvin: www.gregorth.net/.

G. Pascal Zachary. "Digital Designers Rediscover Their Hands." *New York Times*. August 17, 2008. Sunday Business section, p. 4. Quotes from Sennett, Burnett, and Tulley are from this article.

Make magazine: http://makezine.com/.

Bug Labs: http://buglabs.net/.

Lawrence Downes. "Mr. Jalopy Wants to Make a Better World." *New York Times*. August 25, 2008. Editorial Observer, p. A20.

Tinkering School: http://tinkeringschool.com/blog/.

Illustration Credits

All photographs and illustrations by Simon Lonergan except:

Figures 1.4, 5.4, 7.7, 8.3, 15.6 right, 15.16, 17.6, 18.11, 19.5, 20.4 (breadboard), 20.5 (breadboard), 20.9, 20.11, 20.12, 20.13, 21.11, 21.12, 21.13 (drawings), 22.5, 22.10, 22.11, 22.12, 23.1 (breadboard), 23.2, 23.3, 23.7, 23.8, 23.9, 23.10, 23.11, 23.12, 24.5, 24.15, 29.5, 29.6, 29.8, 29.9 © Nicolas Collins.
Figure 3.4 © Gert-Jan Prins, used by permission.
Figure 3.5 courtesy Ikon Gallery Birmingham, used by permission of Christina Kubisch.
Figure 3.6 © Sawako Kato, used by permission.
Figure 3.7 © Vasco Alvo, used by permission.
Figure 4.2 © Lorin Parker, used by permission.
Figure 7.13 © Andrzej Kramarz, used by permission of Richard Lerman and Andrzej Kramarz.
Figure 7.14 © Eric Leonardson, used by permission.
Figure 7.15 © Adachi Tomomi, used by permission.
Figure 7.16 © Martin Vlcek, used by permission of Ivan Palacky.
Figure 7.17 © Michael Graeve, used by permission.
Figure 8.2 © David Tudor Trust, used by permission of Jean Rigg and the David Tudor Trust.
Figure 8.9 © Ted Blanco, used by permission.
Figure 8.10 © Ute Wassermann, used by permission.
Figure 8.12 © Gert Jan van Rooij, used by permission of Nicolas Collins.
Figure 8.13 © Bob Seay, used by permission of Jesse Seay.
Figure 8.14 © Chris Black and Christine White, used by permission.
Figure 9.3 © Laurie Anderson, used by permission of Laurie Anderson/Canal Street Communications.
Figure 9.4 © César Eugenio Dávila-Irizarry, used by permission.
Figure 11.1 photos by Nicolas Collins, design by Simon Lonergan.
Figure 11.4 © Seth Cluett, used by permission.
Figures 11.5 and 21.13 (photo) © Robert Poss, used by permission.
Figure 14.4 © Paul De Marinis, used by permission.
Figure 14.5 © Norbert Möslang, used by permission.
Figure 14.6 © Christian Terstegge, used by permission.
Figures 15.11, 15.13, 17.1, 30.6 © Phil Archer, used by permission.
Figure 15.12 © Joker Nies, used by permission.
Figure 15.14 © John Bowers, used by permission.
Figure 15.15 © Sarah Washington, used by permission.
Figure 15.19 © Vic Rawlings, used by permission.
Figure 17.7 © David Behrman, used by permission.
Figure 21.6 © Frederic Rzewski, used by permission.
Figure 24.1 © Billy Roisz, used by permission.

Figure 24.2 © Yasunao Tone, used by permission.
Figure 24.3a © Baron Wollman for the Tape Music Center, used by permission of Lowell Cross.
Figure 24.3b © Lowell Cross, used by permission.
Figure 24.4 © BMBCon, used by permission.
Figure 24.6 © Jon Satrom, used by permission.
Figure 24.7 © LoVid, used by permission.
Figure 24.8 photography by Josephine Jasperse, used by permission of Arthur Elsenaar and Remko Scha.
Figure 24.16 "Walk" object © Laurie Anderson; technical development by the Information Technology Research Institute at the National Institute of Advanced Industrial Science and Technology, Japan; used by permission of Laurie Anderson/Canal Street Communications.
Figure 26.8 lower left © Douglas Ferguson, used by permission.
Figure 26.8 lower right © Steve Marsh, used by permission.
Figure 28.1 © André Hoekzema, used by permission of Laetitia Sonami.
Figure 28.2 © Istvan Szilasi, used by permission of Jane Henry.
Figure 28.3 photo by Nitin Vadukul, used by permission of Ben Neill.
Figure 28.6 © Sukandar Kartadinata, used by permission.
Figure 28.20 © Chikashi Miyama and Shingo Inao, used by permission.
Figure 29.12 © Ithai Benjamin and Alejandro Abreu, used by permission.
Figure 30.1 © Neal Spowage, used by permission.
Figure 30.2 © Chester Udell, used by permission.
Figure 30.3 © Zach Lewis, used by permission.
Figure 30.4 © Lesley Flanigan, used by permission.
Figure 30.5 photo by Susanna Bachmann, © Daniel Schorno, used by permission.
Figure 30.7 © Emir Bijukic, used by permission.
Figure 30.8 photo © Isabelle Sigal, used by permission of Loud Objects.
Figure 30.11 © Aaron Zarzutzki, used by permission.
Figure 30.12 © J. D. Kramer, used by permission.
Figure 30.13 © Leah Crews Castleman, used by permission.
Figure 30.14 © Martin Riches, used by permission.
Figure 30.15 © Marc Berghaus, used by permission.
Figure 30.16 © Tuomo Tammenpää, used by permission.
Figure 30.17 © Adachi Tomomi, used by permission.
Figure 30.18 © Todd Bailey, used by permission.
Figure 30.19 © Beavis Audio Research, used by permission.
Figure 30.20 and 30.21 © Dan Wilson, used by permission.
Figure 30.22 © Alex Baker, used by permission.
Figure 30.23 and front cover photo © Douglas Irving Repetto, used by permission.
Figure 30.24 and 30.25 © Phillip Stearns, used by permission.
Figure 30.26 and 30.27 © Grégoire Lauvin, used by permission.

Index

Page numbers in italics indicate illustrations.